COMMUNICATING PROCESS ARCHITECTURES 2003

Concurrent Systems Engineering Series

Series Editors: M.R. Jane, J. Hulskamp, P.H. Welch, D. Stiles and T.L. Kunii

Volume 61

Previously published in this series:

Volume 60, Communicating Process Architectures 2002 (WoTUG-25), J.S. Pascoe, P.H. Welch, R.J. Loader and V.S. Sunderam
Volume 59, Communicating Process Architectures 2001 (WoTUG-24), A. Chalmers, M. Mirmehdi and H. Muller
Volume 58, Communicating Process Architectures 2000 (WoTUG-23), P.H. Welch and A.W.P. Bakkers
Volume 57, Architectures, Languages and Techniques for Concurrent Systems (WoTUG-22), B.M. Cook
Volumes 54–56, Computational Intelligence for Modelling, Control & Automation, M. Mohammadian
Volume 53, Advances in Computer and Information Sciences '98, U. Güdükbay, T. Dayar, A. Gürsoy and E. Gelenbe
Volume 52, Architectures, Languages and Patterns for Parallel and Distributed Applications (WoTUG-21), P.H. Welch and A.W.P. Bakkers
Volume 51, The Network Designer's Handbook, A.M. Jones, N.J. Davies, M.A. Firth and C.J. Wright
Volume 50, Parallel Programming and JAVA (WoTUG-20), A. Bakkers
Volume 49, Correct Models of Parallel Computing, S. Noguchi and M. Ota
Volume 48, Abstract Machine Models for Parallel and Distributed Computing, M. Kara, J.R. Davy, D. Goodeve and J. Nash
Volume 47, Parallel Processing Developments (WoTUG-19), B. O'Neill
Volume 46, Transputer Applications and Systems '95, B.M. Cook, M.R. Jane, P. Nixon and P.H. Welch

Transputer and OCCAM Engineering Series

Volume 45, Parallel Programming and Applications, P. Fritzson and L. Finmo
Volume 44, Transputer and Occam Developments (WoTUG-18), P. Nixon
Volume 43, Parallel Computing: Technology and Practice (PCAT-94), J.P. Gray and F. Naghdy
Volume 42, Transputer Research and Applications 7 (NATUG-7), H. Arabnia
Volume 41, Transputer Applications and Systems '94, A. de Gloria, M.R. Jane and D. Marini
Volume 40, Transputers '94, M. Becker, L. Litzler and M. Tréhel
Volume 39, Transputer/Occam Japan 6, S. Noguchi, M. Ishizuka and M. Ota
Volume 38, Progress in Transputer and Occam Research (WoTUG-17), R. Miles and A. Chalmers
Volume 37, Parallel Computing and Transputers (ATOUG-6), D. Arnold, R. Christie, J. Day and P. Roe
Volume 36, Transputer Applications and Systems '93, R. Grebe, J. Hektor, S.C. Hilton, M.R. Jane and P.H. Welch
Volume 35, Transputer/Occam Japan 5 (OUGJ-5), S. Noguchi and M. Yamamoto
Volume 34, Transputer Research and Applications 6 (NATUG-6), S. Atkins and A.S. Wagner
Volume 33, Transputer and Occam Research: New Directions (WoTUG-16), J. Kerridge
Volume 32, Networks, Routers and Transputers, M.D. May, P.W. Thompson and P.H. Welch
Volume 31, Transputers and Parallel Applications (ATOUG-5), J. Hulskamp and D. Jones
Volume 30, Transputing in Numerical and Neural Network Applications, G.L. Reijns and J. Luo
Volumes 28 and 29, Parallel Computing and Transputer Applications (parts I, II), (Pacta '92), M. Valero, E. Oñate, M. Jane, J.L. Larriba and B. Suarez
Volume 27, Transputer/Occam Japan 4 (OUGJ-4), S. Noguchi and H. Umeo
Volume 26, Transputers '92 - Advanced Research and Industrial Applications, M. Becker, L. Litzler and M. Tréhel

ISSN 1383-7575

Communicating Process Architectures 2003

WoTUG-26

Edited by

Jan F. Broenink

Department of Electrical Engineering, University of Twente, Enschede, The Netherlands

and

Gerald H. Hilderink

Department of Electrical Engineering, University of Twente, Enschede, The Netherlands

Proceedings of the 26th WoTUG Technical Meeting, 7–10 September 2003, University of Twente, The Netherlands

IOS Press

Ohmsha

Amsterdam • Berlin • Oxford • Tokyo • Washington, DC

ISBN 1 58603 381 6 (IOS Press)
ISBN 4 274 90620 5 C3055 (Ohmsha)
Library of Congress Control Number: 2003110990

Publisher
IOS Press
Nieuwe Hemweg 6B
1013 BG Amsterdam
The Netherlands
fax: +31 20 620 3419
e-mail: order@iospress.nl

Distributor in the UK and Ireland
IOS Press/Lavis Marketing
73 Lime Walk
Headington
Oxford OX3 7AD
England
fax: +44 1865 75 0079

Distributor in the USA and Canada
IOS Press, Inc.
5795-G Burke Centre Parkway
Burke, VA 22015
USA
fax: +1 703 323 3668
e-mail: iosbooks@iospress.com

Distributor in Japan
Ohmsha, Ltd.
3-1 Kanda Nishiki-cho
Chiyoda-ku, Tokyo 101-8460
Japan
fax: +81 3 3233 2426

PRINTED IN THE NETHERLANDS

Contents

Preface
J.F. Broenink and G.H. Hilderink vii

Programme Committee viii

Papers

A Comparison of High Performance, Parallel Computing Java Packages 1
N.C. Schaller, S.W. Marshall and Y.-F. Cho

A Development Method Boosted by Synchronous Active Objects 17
C. Petitpierre

Tutorial: Prioritised Service Architecture using Honeysuckle 33
I.R. East

The Grid Block Device 37
B. Arantsson and B. Vinter

Distributed Shared Memory in Global Area Networks 49
H.H. Happe and B. Vinter

Automatic Conversion of CSP to CTJ, JCSP, and CCSP 63
V. Raju, L. Rong and G.S. Stiles

A Single Chip Solution for Distributed Processing Systems 83
B.C. O'Neill, P.W. Moore and S. Clark

The Denotational Semantics of View-Centric Reasoning 91
M.L. Smith, C.E. Hughes and K.W. Burke

Overtures and Hesitant Offers: Hiding in CSPP 97
A.E. Lawrence

Sampling and Timing: A Task for the Environmental Process 111
G.H. Hilderink and J.F. Broenink

Parallel Processing - the picoChip way! 125
A. Duller, G. Panesar and D. Towner

An Introduction to the Kent C++CSP Library 139
N.C.C. Brown and P.H. Welch

Agents for Concurrent Programming 157
E. Gonzalez, C. Bustacara and J. Avila

occam for Reliable Embedded Systems: Lightweight Runtimes and Model Checking 167
D.A. Nicole, S. Ellis and S. Hancock

The Trebuchet 173
J. Campbell and G.S. Stiles

Formal Contracts: Enabling Component Composition 185
M. Boosten

Flexible, Transparent and Dynamic occam Networking With KRoC.net 199
M. Schweigler, F.R.M. Barnes and P.H. Welch

A CSP-based Processing Architecture for a Flexible MIMO-OFDM Testbed 225
H.S. Cronie, F.W. Hoeksema and C.H. Slump

Real-Time and Fault Tolerance in Distributed Control Software 235
B. Orlic and J.F. Broenink

occwserv: An occam Web-Server 251
F.R.M. Barnes

RMoX: A Raw-Metal occam Experiment 269
F.R.M. Barnes, C. Jacobsen and B. Vinter

Scheduling for ILP in the 'Processor-as-a-Network' 289
D.K. Arvind and S. Sotelo-Salazar

Accurate Calculation of Deme Sizes for a Parallel Genetic Scheduling Algorithm 305
M. Moore

A Multimodal Robotic Control Law Modelled and Implemented with the CSP/CT Framework
G.H. Hilderink, Dusko S. Jovanovic and J.F. Broenink 315

Author Index 335

Preface

Communicating Process Architectures addresses many of the key issues in modern computer science and practice. In broad terms, the conference themes will concern *concurrency* at all levels of software and hardware granularity. The goal of the conference is to stimulate discussion and ideas as to the role concurrency will play in the future generation of *scalable* computer infrastructure and applications - where scaling means the ability to ramp up *functionality* (i.e. stay in control as complexity increases) as well as *physical metrics* (such as performance).

Traditionally, *concurrency* has been taught and considered and experienced as an advanced and difficult topic. The thesis underlying this conference is that this tradition is wrong. The natural world operates through the continuous interaction of massive numbers of autonomous agents at all levels of granularity (astronomic, human, sub-atomic). If modern computer science finds concurrency hard, then it is probably not doing it right! Research and development in the field of communicating process architectures show that concurrency simplifies software and hardware architectures in a natural way how we think about systems, while managing complexity and enhancing performance.

The WoTUG forum aims to continue the successful series of yearly conferences, this one, the CPA 2003, being the fourth under the name of CPA, and the 26th in the series of WoTUG conferences. They have consistently proven to be a valuable meeting place for all those interested in the problems and opportunities thrown up by parallel computing and concurrency. The delegates and speakers, and thus the papers in this proceedings, reflect a wide spectrum of disciplines – theory, software architecture, hardware architecture, hardware / software co-design, embedded systems design, languages for concurrency, tooling design and applications.

We would like to thank the members of the IPC and WoTUG committee for their work in reviewing papers and advice for the organisation. Furthermore, we thank our colleagues of the Local Organisation Committee, Dusko Jovanovic, Bojan Orlic and Peter Visser for their help with the organisation.

Being the third time in Enschede, Netherlands, we hope you enjoy the Communicating Process Architectures 2003 and your stay at our Campus. We do our best to provide an atmosphere where we can meet as relaxed, friendly, exciting and productive as at previous CPA's, including the famous fringe sessions and after hours.

Jan F. Broenink
Gerald H. Hilderink
University of Twente, the Netherlands

Programme Committee

Professor Peter Welch (Programme Chair), University of Kent at Canterbury, UK

Dr. Jan F. Broenink (Conference Chair), University of Twente, the Netherlands
Gerald H. Hilderink (Conference co-chair), University of Twente, the Netherlands

Dr. Alistair Allen, University of Aberdeen, UK
Richard Beton, Roke Manor Research Ltd., UK
Dr. Alan Chalmers, University of Bristol, UK
Dr. Barry Cook, University of Keele, UK
Ruth A. Ivemy-Cook, Cherry Hinton, Cambridge, UK
Christopher Jones, British Aerospace, Warton Aerodrome, UK
Professor Jon Kerridge, Napier University, UK
Dr. Adrian Lawrence, Loughborough University, UK
Stephen Maudsley, Esgem Ltd, UK
Professor David May, FRS, University of Bristol, UK
Dr. Henk Muller, University of Bristol, UK
Professor Patrick Nixon, University of Strathclyde, Glasgow, UK
Dr. Denis Nicole, University of Southampton, UK
Dr. Brian O'Neill, Nottingham Trent University, UK
Dr. James Pascoe, Emory University, Atlanta, USA
Dr. Roger Peel, University of Surrey, UK
Herman Roebbers, Philips TASS, the Netherlands
Professor Nan Schaller, Rochester Institute of Technology, New York, USA
Professor Dike Styles, Utah State University, Utah, USA
Dr. Johan Sunter, Philips TASS, the Netherlands
Øvind Teig, Navia Maritime AS, Division Autronica, Norway
Dr. Stephen J. Turner, Nayang Technological University, Singapore
Peter Visser, University of Twente, the Netherlands
Paul Walker, 4Links Ltd., UK
Professor Jim Woodcock, University of Kent at Canterbury, UK

Communicating Process Architectures 2003
Jan F. Broenink and Gerald H. Hilderink (Eds.)
IOS Press, 2003

A Comparison of High Performance, Parallel Computing Java Packages

Nan C. SCHALLER and Sidney W. MARSHALL
Computer Science Dept., Rochester Institute of Technology
102 Lomb Memorial Drive, Rochester, NY 14623-5608, USA
`ncs@cs.rit.edu, swm@cs.rit.edu`

Yu-Fong CHO
R&D Group, Askey Computer Corp.
10F, NO 119, Chien-Kang Road, Chung-Ho, Taipei 235, Taiwan
`yfcho@askey.com.tw`

Abstract. The high-performance computing community has developed numerous Java packages that support parallel and distributed computing. Most of these packages are designed for the typical parallel message passing and shared memory architectural paradigms. This paper presents the results of a recent study that included a web search for such packages, describes the paradigms implemented in them, and evaluates their performance on a parallel, 4-processor SMP machine using three benchmark programs that represent a mix of typical parallel applications, chosen from *The Java Grande Benchmark Suite.* A brief description of each package and a discussion its ease of installation and use are also provided.

1 Introduction

Today, the high-performance computing (HPC) community is more interested than ever in the possibility of using Java. It is becoming more viable to do so. As Pancake and Lengauer recently reported: "Previous Java implementations focused mainly on the portability and interoperability aspects required for Internet-centric client/server computing. Because Java was originally interpreted, it is commonly perceived as being execution inefficient. Recent developments in compiler technology and instruction-level optimisations have done away with many of the sources of this inefficiency. Java's recent execution efficiency improvement is due to static analysis, just-in-time compilation and optimization of the Java Virtual Machine (JVM). Software scientists are also making efforts to improve Java's performance of remote method invocation (RMI), numeric capabilities, and communication mechanisms. These days, Java is competitive with C and C++ for some applications on some platforms, and it is considerably safer to execute and easier to program." [1]

This paper reports on a study that evaluated several publicly available Java packages for HPC with respect to ease of installation, ease of use, and performance. Some of these packages were sponsored by the *Java Grande Forum* [2], a major consortium representing the interests of the Java high-performance computing community. The Forum's benchmark software was used to measure and compare the performance of these packages.

2 Background

Parallel computing platforms are usually either single computers with multiple internal processors or multiple interconnected computers. Two methods for communicating between such processors use message passing or shared memory. In message passing systems, memory is distributed among the processors, each processor having its own address space. In this case, a processor can only directly access its own memory and an interconnection network is necessary for processors to be able to send messages to other processors. Shared memory multiprocessors use a single address space. Here, each location in memory has a unique address that may be used by each processor to access that location. [3]

The HPC community has developed several Java packages that are suitable for these architectural paradigms. For example, for message passing, there are packages based on the Communicating Sequential Processes (CSP) [4], Message Passing Interface (MPI) [5], and Parallel Virtual Machine (PVM) [6] models and for shared memory, there are packages based on the Open Multi Processing (OpenMP) [7] and Linda [8] models. Brief descriptions of these models are provided below.

- **Message Passing using CSP**
  "Communicating Sequential Processes (CSP) is a mathematical theory for specifying and verifying the complex patterns of behavior arising from interactions between concurrent objects. CSP has a formal, and compositional semantics that … encapsulates fundamental principles of processes, networks and communication." [9] CSP programs are written using *processes*, *networks* of processes and various forms of *synchronization* and *communication* between them.
- **Message Passing using MPI**
  "MPI (Message Passing Interface) is a library specification for message passing, proposed as a standard by a broadly based committee of vendors, implementers, and users. The message-passing application programmer interface (API) is combined with protocol and semantic specifications for how its features must behave in any implementation. MPI includes point-to-point message passing and collective operations." [5]
- **Message Passing using PVM**
  "PVM (Parallel Virtual Machine) is an integrated set of software tools and libraries that permits a heterogeneous collection of computers hooked together by a network to be used as a parallel computer." [6][10][11]
- **Shared Memory using OpenMP**
  "OpenMP (Open Multi Processing) is a specification for a set of compiler directives, library routines, and environment variables that can be used to specify shared memory parallelism. The OpenMP Application Program Interface supports multi-platform shared-memory parallel programming on all architectures..." [7]
- **Shared Memory using Linda**
  Linda is a concurrent programming model whose primary concept is that of a `tuple-space`, an abstraction via which cooperating processes communicate ... Linda … provides a shared-memory abstraction for process communication without requiring the underlying hardware to physically share memory." [8][12]

3 A Survey of Packages Found

The purpose of this study was to locate HPC Java packages on the Web, and evaluate their ease of installation, their ease of use, and their performance. After locating many such packages, the study was narrowed to those packages that

- Supported either the message passing or shared memory programming models.
- Were publicly available, i.e., not commercial software.
- Worked with the standard Java Development Kit (JDK).
- Were up-to-date. (For the purposes of this study, JDK 1.4.0 was considered up-to-date.)

The Web search was performed in April 2002 with the evaluation process continuing through September 2002. Packages found under these narrowed criteria are listed below along with a brief description taken from each web site, and are summarized in Table 1.

3.1 Message Passing using CSP

Two Java packages were found that use the CSP model: Communicating Threads for Java (*CTJ*) [13] and Communication Sequential Processes for Java (*JCSP*)[14]. As both *CTJ* and *JCSP* were implemented using the CSP model, they have many concepts in common[15]. An important difference between them is their process scheduling systems.

CTJ's process scheduling kernel is specially designed for programming small, real-time embedded systems. [16][17] "The prototype *CTJ* package provides the thread/CSP model of processes, channels, and composition constructs for Java… It implements the CSP model for the standard Java monitor/threads operations, and enables any Java threaded system to be analyzed in CSP terms." [16]

JCSP also implements the CSP model, but relies on the Java Virtual Machine (JVM) for its thread scheduling. "*JCSP* is a 100% Java class library providing a base range of CSP primitives. It also includes a package providing CSP process wrappers giving a channel interface to all Java AWT widgets and graphics operations." [14]

3.2 Message Passing using MPI

Two packages were found that use the MPI model: *JavaMPI* [18] and *mpiJava* [19].

"*JavaMPI* is a Java binding for MPI... The *JavaMPI* library contains an MPI library that is dynamically linked to the JVM during program execution." [20]

"*mpiJava* is an object-oriented Java interface to standard MPI. *mpiJava* does not assume any special extensions to the Java language. It can be ported to any platform that provides compatible Java development and native MPI environments. The *mpiJava* package is implemented as a set of Java Native Interface (JNI) wrappers to native MPI packages. Platforms currently supported include Solaris using MPICH [21] or SunHPC-MPI [22], Linux using MPICH [21], and Windows NT using WMPI [23]." [19] The *mpiJava* package was developed as part of the *HPJava* project [24] supported by the Northeast Parallel Architectures Center (NPAC) [25] at Syracuse University.

3.3 Message Passing using PVM

The web search found two Java packages that use the PVM model: *jPVM* (JavaPVM) [26] and *JPVM* [27].

"*jPVM* is an interface written using the Java native methods capability that allows Java applications to use the native Parallel Virtual Machine (PVM) [6] software developed at Oak Ridge National Laboratory. *jPVM* extends the capabilities of PVM to the Java architecture-independent programming language. *jPVM* allows Java applications and existing C, C++, and Fortran applications to communicate with one another using the PVM API. " [26]

"*JPVM* is a PVM-like library of object classes implemented in and for use with the Java programming language. It combines two advantages, ease of programming inherited from Java and high performance through parallelism inherited from PVM." [27]

3.4 Shared Memory using OpenMP

The only Java package found using the OpenMP model was *JOMP* [28].

"*JOMP* implements the OpenMP Application Program Interface (API) in Java. It provides OpenMP-like directives and methods consisting of a compiler and runtime library written entirely in Java. *JOMP* implements most of the OpenMP specification.

JOMP uses the fork-join model of parallel execution. A program written using the *JOMP* API begins execution on a single thread called the master thread. The master thread executes in a serial region until the first parallel construct is encountered, whereupon the master thread creates a team of threads, which includes itself. Each thread then executes the code in the dynamic extent of the parallel region..." [29][30]

3.5 Shared Memory using Linda

Two packages were found that used the Linda model: *Jada* [31] and *JavaSpaces* [32].

"*Jada* adds operations to access Linda-like multiple tuple-spaces. *Jada*'s design goal was simplicity rather than performance. *Jada*, like Linda, is a minimalist coordination language. Other Linda-like implementations usually include a pre-processor, necessary because Linda slightly changes the host language syntax. *Jada* is based on a set of classes that are used to access a tuple-space... This allows users to use their standard Java development tools. *Jada* is implemented as a set of classes that allow either Java threads or Java applications to access an associatively shared tuple space using a small set of Linda-like operations." [33][31]

"*JavaSpaces* technology is a simple unified mechanism for dynamic communication, coordination, and sharing of objects between Java technology-based network resources like clients and servers. In a distributed application, *JavaSpaces* technology acts as a virtual space between providers and requesters of network resources or objects. This allows participants in a distributed solution to exchange tasks, requests and information in the form of Java technology-based objects... The design of *JavaSpaces* was strongly influenced by Linda. *JavaSpaces* systems are similar to Linda systems in that they store collections of information for future computation and are driven by value-based lookup." [34] *JavaSpaces* technology was developed by Sun Microsystems [35] and is included in the Jini package.

3.6 Others

Other Java solutions for high performance computing were found. Some of them were replacements for the Java compiler, or the JVM. Furthermore, some of them replaced the Java language itself, using an extended Java syntax to gain the benefits of Java. These were not included in the study.

3.7 Summary of Results

Table 1 provides a summary of the Java packages found during the web search that met the criteria set forth. It lists the JDK version used for development, the programming paradigm, the version tested, and the date of latest update at the time of the study.

Package Name	JDK Version	Programming Paradigm	Version tested	Latest Update
CTJ	Not specified	Message passing using CSP	0.9, r18	October 2000
JCSP	1.1.5 or later	Message passing using CSP	1.0-rc4	Feb. 2002
JavaMPI	Not specified	Message passing using MPI	0.4	November 1998
mpiJava	1.2 or later	Message passing using MPI	1.2.3	October 2001
jPVM	1.1.5 or later	Message passing using PVM	1.1.4	April 1998
JPVM	Not specified	Message passing using PVM	0.2	Feb. 1999
Jada	Not specified	Shared memory using Linda	3.0 beta	April 2001
JavaSpaces	1.2.2_007 or later	Shared memory using Linda	1.2.1	April 2002
JOMP	1.2 or later	Shared memory using OpenMP	1.0 beta	2000

Table1: Summary of Java Packages

4 Installation and Usage

For the purpose of this study, all packages were installed and tested on a Sun Microsystems Enterprise 450 running the Solaris 8 Operating System. This SMP system has four UltraSPARC II 248 MHz CPUs with 2048 Megabytes system memory, and a system clock frequency of 83 MHz. As some of the packages were designed for an SMP system and could not be used on a cluster or distributed systems, testing was limited to this single four-processor system. All benchmark programs were compiled using the standard JDK compiler (`javac`), and were executed using the standard JDK Java virtual machine (`java`).

In several cases, all that was needed to set up the package was to download it, uncompress a file or two, and add the appropriate directory to the `CLASSPATH` environment variable. This was the case for *CTJ*, *JCSP*, *Jada* and *JOMP*. For others, it was more complicated. And, some packages could not be installed on the system. The following subsections contain notes regarding installation and programming with these packages. Table 2 summarizes this information.

4.1 CTJ

Besides needing to understand CSP to program with *CTJ*, special effort is required to run in parallel. The terms "*Thread*" in Java and "*Process*" in *CTJ* are closely related, but do not act the same way. *CTJ* does not perform time slicing by default, even if a programmer creates a multi-process program. This means that *CTJ* does not context-switch between processes unless those processes engage in an event such as communicating with each other through a *CTJ* channel. *CTJ* does provide a `TimeSlicer` class to enable time slicing among a single processor's subprocesses. *CTJ* has its own process scheduling kernel, and is specially designed for small, embedded, real-time systems. In addition, *CTJ* does not automatically distribute processes across processors. Therefore, to run in parallel, the user must manually start a process on each processor. In addition, special `channel` classes must be provided to implement communication on a distributed memory system.

4.2 JCSP

Programming under *JCSP* is similar to programming using Java threads. Because it is based on CSP, *JCSP* purports to provide a superior mechanism to the Java threads mechanism for handling multi-threaded programming. A program unit is called a *process* rather than a *thread*. Programmers can either simply use the *JCSP* process mechanism to replace the Java threads or use the CSP model to implement parallel programs. While, *JCSP* was designed to be efficient on any parallel architecture, the public domain JCSP version tested only runs in parallel on shared memory systems.. The user must implement classes to provide communication on distributed memory systems or purchase the commercial version. (*JCSP* was commercially released by Quickstone Technologies [36] during the time of our study.)

4.3 mpiJava

A native MPI C interface, such as MPICH [21] or SunHPC MPI [22], is required to use the *mpiJava* package. The default interface is MPICH. Our attempts to modify the *mpiJava* startup script to work with SunHPC 3.1, the MPI version already installed on the test system, were not successful. Therefore, MPICH was used instead. It should be noted that we were successful with SunHPC 4.0 later on a different system.

The installation and usage of *mpiJava* requires precaution, especially if, as in our case, the system uses Secure Shell (SSH) services, i.e., telnet services are not available. Such information must be provided before compiling and installing both MPICH and *mpiJava*. In addition, the SSH agent must be set up to automatically log on without requiring a password before running *mpiJava* programs.

However, programming under the *mpiJava* environment is straightforward. *mpiJava* provides wrapper classes for the MPI package as extended Java classes, so using an MPI function is the same using any Java class. And, it is similar to other MPI programming environments.

4.4 Jada

Programming *Jada* is straightforward. *Jada* implements a Linda-like `tuplespace` called `ObjectSpace`. Linda-like programming models use the term *worker* to represent a thread/process. Communication takes place between workers by depositing `tuples` into and withdrawing `tuples` from `ObjectSpace`.

4.5 JavaSpaces

In order to use *JavaSpaces*, users must install the Jini package from Sun Microsystems, Inc. [35]. Installation is easy, but starting *JavaSpaces* is much more complicated, since Jini must be started first. This requires that the following services be started: an `HTTP Server`, an `RMI Activation Daemon`, an `RMI Registry` or a `Jini Lookup service`, a `Transaction Manager`, and then finally `JavaSpaces`. In addition, a Java security policy must be defined before execution. It is complicated to get everything started correctly.

JavaSpaces uses an `Entry` class that acts like a Linda-like tuple, i.e., to be written into and to be taken out of *JavaSpaces*. Each field of an entry must be a public object type, and an `Entry` cannot store primitive types in its fields. Programmers who are just starting to program using *JavaSpaces* technology could easily miss this restriction.

4.6 JOMP

Programming in the *JOMP* environment is similar to programming in other OpenMP programming environments. *JOMP* provides OpenMP-like syntax and programming style. It also provides a *JOMP* pre-compiler that will translate a Java program into a *JOMP* parallel program, which helps programmers focus on their parallel algorithm and not *JOMP* syntax.

4.7 Packages Not Tested

The *JPVM, jPVM* and *JavaMPI* packages, mentioned in the previous section, were not tested. Research had ended for all three packages and their most recent implementations would not execute under the JDK specified in our refined criteria. Furthermore, MPI seems to have, for the most part, replaced PVM as the de facto message passing standard.

Package Name	Additional Required Software	Ease of Installation	Ease of Use	Remark
CTJ	None	Easy	Need knowledge of CSP	
JCSP	None	Easy	Need knowledge of CSP	
JavaMPI	MPI/LAM	Not compatible with hardware	Out-of-date	Not tested
mpiJava	MPICH or SUN HPC	Need to compile both packages	Need knowledge of MPI	
jPVM	PVM		Out-of-date	Not tested
JPVM	None		Out-of-date	Not tested
Jada	None	Easy	Need knowledge of Linda	
JavaSpaces	Jini	Complicated	Need knowledge of Linda and complicated to set up.	
JOMP	None	Easy	Easy	

Table 2: Ease of Installation and Use

5 Benchmarks

The purpose of benchmark testing is to provide a means to meaningfully measure and compare alternative execution environments. Three benchmarks were chosen for this study out of the five available in *The Java Grande Forum Benchmark Suite[37]*. "These algorithms are designed to use large amounts of processing, I/O, network bandwidth, or memory." [37] Thus, each algorithm chosen represents a particular type of application program. "These codes are simple kernels that reflect the type of computation that can be expected to be found in the most computation intense parts of real numerical applications"[38]. However, they are not the optimal implementations of these algorithms. Brief descriptions of the benchmark algorithms chosen follow.

5.1 The Crypt Benchmark

"The Crypt benchmark performs International Data Encryption Algorithm (IDEA) encryption and decryption of an array of *N* bytes. This algorithm involves two principal loops, whose iterations are independent and may be divided between processors in a block fashion." [38] The Crypt benchmark represents parallel applications that are both computation and communication intensive. The parallel version of this benchmark distributes encryption computation to processors, merges the results of encryption, distributes decryption computation to processors, and then collects the results of decryption. An *N* byte array must be transmitted between processors whenever communication is needed. Synchronization is required at the end of the encryption and decryption steps.

5.2 The Series Benchmark

"The Series benchmark computes the first *N* Fourier coefficients of the function $f(x)=(x+1)^x$ on the interval [0, 2]. The most time consuming component of the benchmark is the loop over the Fourier coefficients. The calculation of each coefficient is independent of every other coefficient and the work may be distributed simply between processors." [38] The Series benchmark represents the purely computation intensive parallel application. The parallel version of this benchmark distributes the computation to the processors and then collects the computed Fourier coefficients. Little communication is required. Synchronization is required at the beginning and the end of the benchmark.

5.3 The SOR Benchmark

"The SOR benchmark performs 100 iterations of Successive Over-Relaxation (SOR) on an *N x N* grid. This benchmark involves an outer loop over iterations and two inner loops, each looping over the grid. In order to update elements of the principal array during each iteration, neighbouring elements of the array are required, including elements previously updated. Hence this benchmark is inherently serial. To allow parallelism to be carried out, the algorithm has been modified to use a "*red-black*" ordering mechanism. This allows the loop over array rows to be parallelised. Hence, the outer loop over elements has been distributed between processors by columns." [38]

The SOR Benchmark represents the most communication intensive application of these three benchmarks. In the parallel version, an *N x N* grid must be distributed equally to the participating processors at the beginning of the program, *2N* rows of the grid must be exchanged between processors for each iteration, and the *N x N* grid must be returned to the host processor at the end of the program. Synchronization occurs at the beginning of the program, at the end of each iteration, and at the end of the benchmark.

6 Performance Results

The benchmarks were executed for a variety of dataset sizes on the test platform. It should be noted that the figures that follow show the results for the largest datasets only for each benchmark. The raw data is provided as well in tabular form in Appendix A. Other figures are available from the authors upon request. The execution of the sequential version of the benchmark was used as the basis for speedup calculations. However, as these sequential versions are not purported to be the "best" algorithms, it might be appropriate to consider the speedup measurements below as "relative" rather than "theoretical". Benchmark results are expected to improve as the dataset size increases. In addition, communication overhead is expected degrade performance. However, bigger data sets requiring more computation can be expected to mitigate some of this degradation.

This study utilized a set of Java timing utility classes, available from The *Java Grande Forum Benchmark Suite [37]*. All of the times measured were actual algorithm computation times in seconds excluding initialization and I/O time. Each benchmark test was executed ten times to evaluate the variation of the results; there was little variance from test to test. Thus, an average execution time was used.

Execution time should ideally decrease as the number of processors used increases.

Speedup is calculated using formula 6.1 and indicates how much faster the parallel version is than the sequential version. Ideally, speedup should increase linearly as the number of processors increases. Computation intensive algorithms are expected to perform better, i.e., exhibit more speedup, than communication intensive algorithms.

$$Speedup = \frac{ExecutionTimeOfSequentialVersion}{ExecutionTimeOfMutiProcessorsVersion} \quad (6.1)$$

6.1 *Results for the Crypt Benchmark*

The Crypt benchmark represents applications that are both computation and communication intensive. Benchmarks were run using datasets of 3 MB, 20 MB, and 50 MB arrays. Figure 1 shows the execution time of all packages for this benchmark for the 50 MB dataset and Figure 2 shows the speedup. These figures show that most packages performed well, with the exceptions of *JavaSpaces*, *mpiJava* and *CTJ*.

JavaSpaces performed worse than any other package for this benchmark. However, *JavaSpaces* did show some benefit from a multiprocessor environment. As mentioned above, Jini services, which consume a lot of system resources, must be started correctly before starting *JavaSpaces*. For example, almost 300 MB of system memory was already consumed before benchmark programs were loaded. Furthermore, *JavaSpaces* threw an `OutOfMemory` exception for the 20 MB or 50 MB datasets. Assigning larger memory to the JVM did not solve this problem.

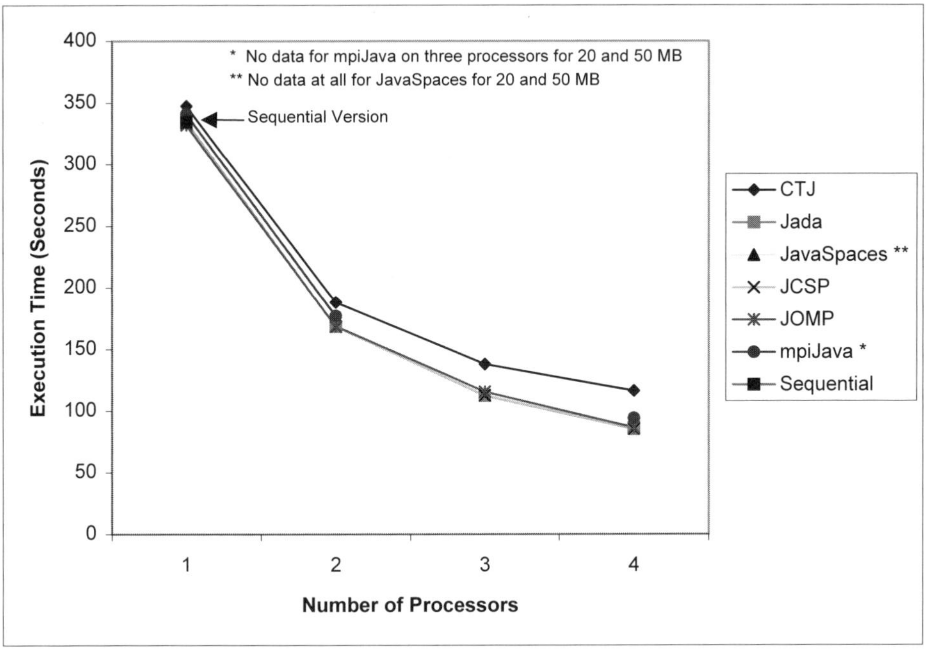

Figure 1. Execution time of Crypt benchmark for 50 MB array dataset

It is interesting to note that the other Linda-like package, *Jada*, performed well for the Crypt benchmark. Although both of *JavaSpaces* and *Jada* are implemented as Linda-like shared memory paradigm, their implementations are different. *Jada*'s implementation takes the advantage of the SMP system, but *JavaSpaces* creates a virtual memory over the network of computers. *JavaSpaces* then uses network-based communication even when shared memory is available. This type of communication is not as fast as but does have better scalability.

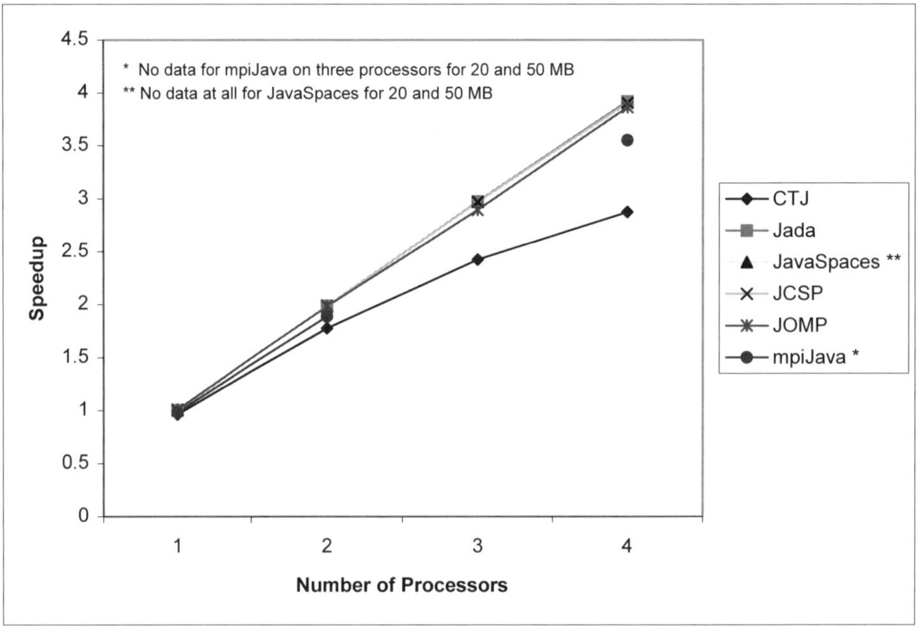

Figure 2. Speedup for Crypt benchmark with 50 Megabyte array dataset

CTJ performed much better than *JavaSpaces* but not as well as the other packages. It was slower, particularly when the number of processors used was increased. This is likely due to the mechanism used to handle communication between individual processors.

We also experienced difficulty with *mpiJava* when using three processors for the 20 MB and 50 MB datasets. We have not been able to identify the cause of the problem, but the MPICH FAQ indicated that it might be due to compiler implementation incompatibility. A new version of MPICH (1.2.4) was released at end of May 2002. More investigation is required to see if this problem is resolved with this newer version, or by utilizing SunHPC's MPI.

To summarize, the Crypt benchmark represented common applications that are both computation and communication intensive. All packages showed reasonably good performance with the exception of *JavaSpaces*, *mpiJava* and *CTJ*. When no other issues arose, as expected, performance was better the larger the dataset.

6.2 Results for the Series Benchmark

The Series benchmark represents a computation intensive application. Dataset sizes were limited to 10K and 100K Fourier coefficients as the time to process a 1000K Fourier coefficient dataset proved prohibitive. Figure 3 shows the execution time of all packages for the Series benchmark calculating 100K Fourier coefficients, while Figure 4 shows the speedup. As is shown in these figures, the performance of all packages was roughly the same. This is because of the benchmark's computation intensive nature, i.e., not much communication is needed. Interestingly, a few of the single processor versions, most notably *JCSP*'s, did execute faster than the sequential version.

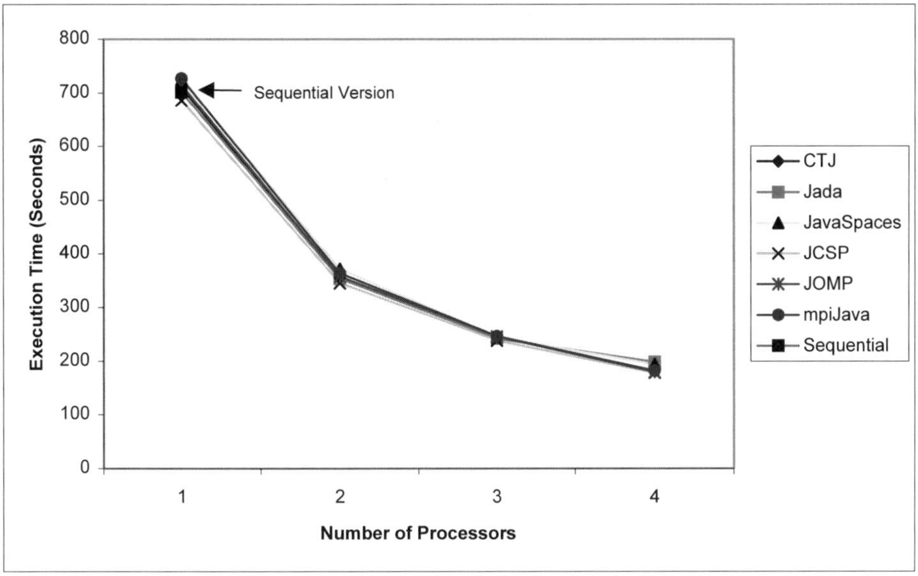

Figure 3. Execution Time of Series benchmark for 100K Fourier coefficients

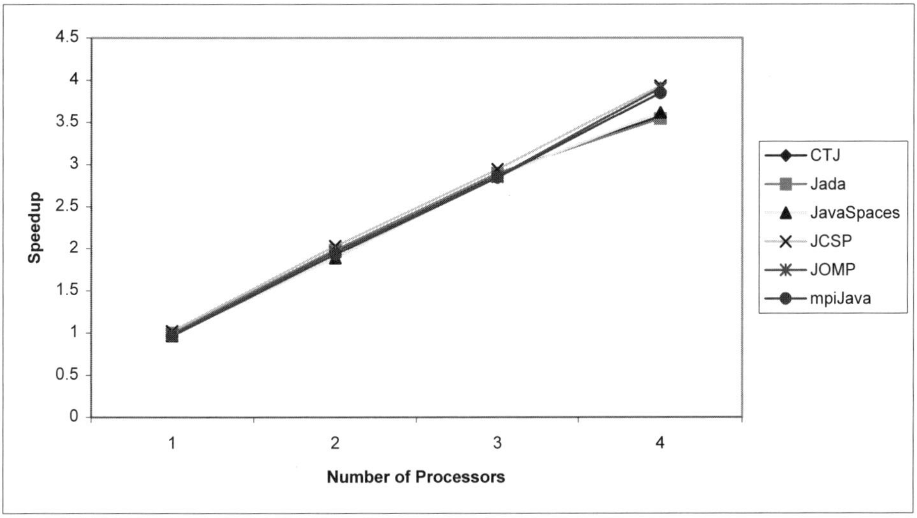

Figure 4. Speedup for Series benchmark for 100K Fourier coefficients

The results varied most when four processors were used. *CTJ*, *JavaSpaces* and *Jada* all deviated more from the ideal speedup than other packages in this situation. All other packages showed near linear speedup. We conjecture that a larger dataset might alleviate this deviation.

In summary, the Series benchmark represented the purely computation intensive application. Packages performed most consistently for this benchmark. As expected, the larger the dataset, the better the performance.

6.3 Results for the SOR Benchmark

The SOR Benchmark was the most communication intensive of the benchmarks used. Datasets of grid sizes *1000 x 1000, 1500 x 1500*, and *2000 x 2000* were used. Figure 5 shows the execution time of all packages for the *2000 x 2000* grid and Figure 6 shows the speedup. Package performance varied more widely for this benchmark than for the other two, demonstrating how communication overhead can undermine the benefit of adding more processors.

The single processor versions were slower than the sequential version for all grid sizes and packages, but most packages demonstrated some speedup when using more than one processor. Performance was better for larger datasets for all packages.

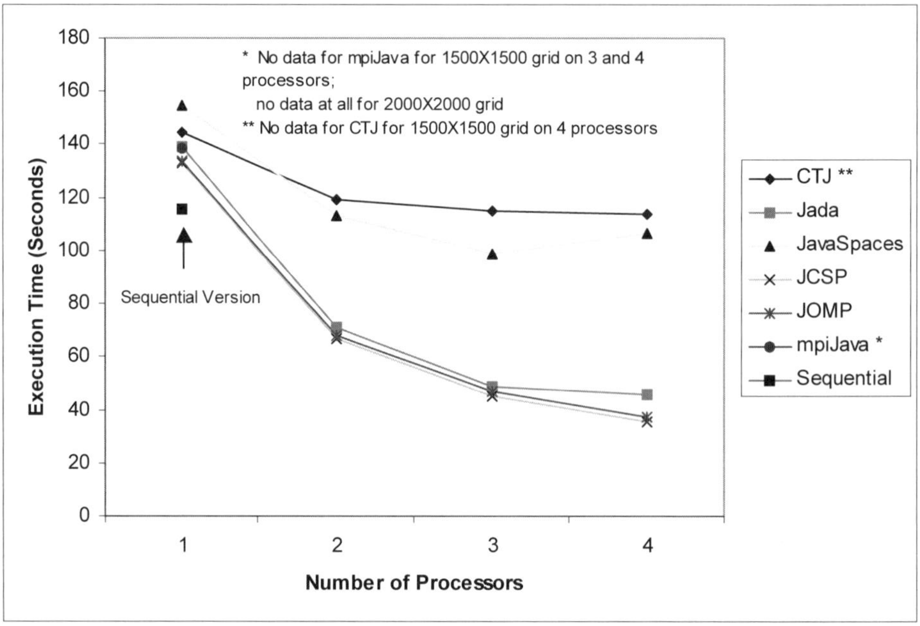

Figure 5. Execution Time of SOR benchmark for a *2000 x 2000* grid

While *CTJ* and *JavaSpaces* did benefit from the multiprocessor environment with the two smaller grids, the more processors used, the worse their performance. In fact, all execution times were slower than that of the sequential version. This is caused by heavy communication requirements along with the way in which communication is implemented for these packages. In addition, we unable to run the *1500 x 1500* grid for *CTJ* with four processors and have not yet determined the cause.

The performance of *JavaSpaces* and *CTJ* improved with the *2000 x 2000* grid. *JavaSpaces*' execution times were faster, in this case, than for the sequential version. This shows that, as expected, communication overhead decreases as the problem size increases.

We were unable to obtain results using *mpiJava* on three and four processors with the *1500 x 1500* grid and obtained no results at all for the *2000 x 2000* grid. We received error messages similar to the ones received while running the Crypt benchmark.

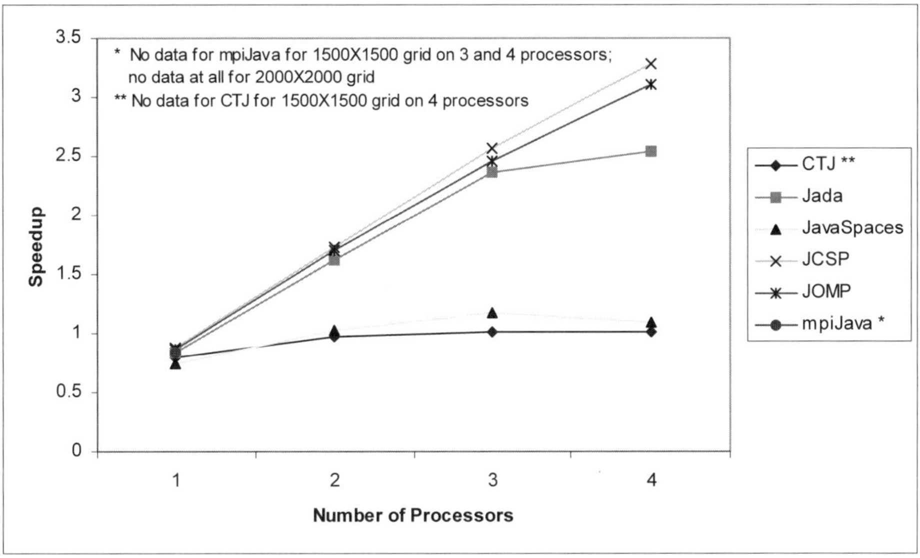

Figure 6. Speedup for SOR benchmark for a *2000 x 2000* grid

Jada performed better when handling larger datasets, but did not show increased speedup when adding a fourth processor for *1000 x 1000* and *1500 x 1500* grids. However, it did run faster on four processors than on one processor or two processors, and it performed much better than the other Linda-like package, *JavaSpaces*.

In summary, the performance of all packages for the SOR benchmark was, as expected, not as good as they were for the Crypt and Series benchmarks. The SOR Benchmark performs more communication operations than either of the other benchmarks, and it performs many bulk synchronization operations. These two factors account for the performance differences. However, most of the packages did perform better with larger dataset sizes.

7 Limitations and Future Work

This study was limited to high-performance Java packages that were up-to-date, non-commercial, freely available, and able to execute using JDK 1.4.0 on the test platform. The packages were evaluated on an SMP system only. While *JOMP* operates only in a shared memory, the other packages could, with some work, be evaluated on distributed memory environments as well. In some cases, the modifications necessary to do this are relatively simple. For example, only the configuration file need be modified for *mpiJava*. In others, it is more difficult. For example, the communication mechanism for *Jada*, *CTJ*, and *JCSP* must be modified or replaced. The commercial version of JCSP from Quickstone[36] does provide this.

Future work will focus on

- Extending the study to include the other benchmarks available in *The Java Grande Forum Benchmark Suite* [37].
- Examining a wider range of platforms, Java environments, and packages. For example, the benchmarks for each packages, when appropriate, could be evaluated on a cluster.
- Evaluating additional packages and newer versions of these packages, i.e., those updated after April 2002.
- Using a variety of different implementations of the auxiliary software. For example, evaluating *mpiJava* with MPICH 1.2.4 and SunHPC MPI 5.0, rather than MPICH 1.2.3.
- Comparing these packages to non-Java packages. For example, running the benchmarks using C and native threads.
- Further investigating the error messages from the *JavaSpaces* and *mpiJava* packages, and determining why there were no results for *CTJ* using four processors for the SOR benchmark.

8 Conclusions

In this study, we evaluated the ease of installation, ease of use, and performance of Java packages found that might support HPC. Three benchmark programs from *The Java Grande Benchmark Suite* [37], representing a mix of typical parallel applications, were used to evaluate this performance. All benchmark tests were run on a common SMP test platform.

Generally, packages that assumed or utilized the shared memory properties of the test system performed better than those that did not. Otherwise, the results from the benchmark tests were as expected: (a) Computation intensive benchmarks showed performance closer to the ideal than those that were communication intensive. (b) Performance results from tests with larger data sets were better than those with smaller data sets.

As a result of this study, we can recommend *JOMP* for ease of installation, ease of use, and performance. It is, however, strictly a shared memory package and thus cannot be used on a cluster or any other distributed memory system. *JCSP* is the next easiest to use. *JCSP* and *JOMP* were the only packages for which we experienced no difficulty during the entire evaluation process. Both had performance results that were close to the ideal.

In contrast, *JavaSpaces* is the most complicated to start and had the worst performance. To be fair, it was not really designed to support HPC. Programming under *JavaSpaces* requires adequate knowledge about its architecture and paradigm, and Jini must be started up before it can be used, not a simple task.

Although easy to install, *CTJ*'s special process scheduling management and real-time kernel are not friendly to the inexperienced programmers. *CTJ* does not automatically take advantage of a multiprocessor environment. Furthermore, the current communication mechanisms do not work well for HPC, but to be fair, that was not its design goal. The underlying communication mechanisms for both *CTJ* and *JavaSpaces* must be improved to be competitive in the HPC arena.

Further study is needed to determine package scalability and to compare performance with that of more traditional languages.

References

[1] C. Pancake and C. Lengauer. High-Performance Java. *Communications of The ACM.* October 2001.
[2] The Java Grande Forum Home Page. http://www.javagrande.org/. 2002.
[3] M. Allen and B. Wilkinson. *Parallel Programming.* 1999.
[4] The CSP Home Page. http://wotug.ukc.ac.uk/csp.shtml. March 2001.
[5] The MPI Home Page. http://www-unix.mcs.anl.gov/mpi/index.html.
[6] PVM: Parallel Virtual Machine. http://www.csm.ornl.gov/pvm/pvm_home.html. April 2002.
[7] The OpenMP Home Page. http://www.openmp.org/. April 2002.
[8] Yale Linda Group. http://www.cs.yale.edu/Linda/linda.html. 2002.
[9] CSP for Java Home Page. http://www.cs.ukc.ac.uk/projects/ofa/jcsp/explain.html.
[10] The PVM system. http://www.netlib.org/pvm3/book/node17.html.
[11] Netlib Repository at UTK and ORNL. http://www.netlib.org.
[12] The Linda System. http://www.netlib.org/pvm3/book/node16.html.
[13] G. Hilderink. Communicating Threads for Java. http://www.rt.el.utwente.nl/javapp. 2000.
[14] P. Welch. The Communication Sequential Processes for Java (JCSP) Home Page. http://www.cs.ukc.ac.uk/projects/ofa/jcsp. 2002.
[15] G. Hilderink, J. Broenink, W. Vervoort, and A. Bakkers. *Communicating Java Threads.* http://www.rt.el.utwente.nl/javapp/cjt/CJT-paper.PDF.
[16] G. Hilderink, J. Broenink, A. Bakkers, and N. Schaller. *Communicating Threads for Java.* 2000.
[17] G. Hilderink, J. Broenink, W. Vervoort, and A. Bakkers. *Communicating Java Threads.* 1997.
[18] The *JavaMPI* homepage. http://perun.hscs.wmin.ac.uk/CSPE/software.html. 2000.
[19] The *mpiJava* Home Page. http://www.npac.syr.edu/projects/pcrc/HPJava/*mpiJava*.html. January 2000.
[20] S. Mintchev. Writing Programs in *JavaMPI.* 1997.
[21] The MPICH Home Page. http://www-unix.mcs.anl.gov/mpi/mpich/. 2002.
[22] Sun Microsystems – HPTC Home Page. http://www.sun.com/solutions/hpc/index.html. 2002.
[23] The WMPI Home Page. http://www.criticalsoftware.com/wmpi/. September 2001.
[24] Northeast Parallel Architectures Center at Syracuse University. http://www.npac.syr.edu/.
[25] The HPJava Project Home Page. http://www.npac.syr.edu/projects/pcrc/HPJava/index.html. January 2000.
[26] The JavaPVM Home Page. http://www.chmsr.gatech.edu/*jPVM*/. 1998.
[27] The *JPVM* Home Page. http://www.cs.virginia.edu/jpvm.html. 1999.
[28] The JOMP Home Page. http://www.epcc.ed.ac.uk/research/jomp/index_1.html. 2001.
[29] J. Bull, M. Westhead, M. Kambites, J. Obdrzalek. *Towards OpenMP for Java.* 2000.
[30] J. Obdrzalek, M. Bull. *JOMP Application Program Interface.* August 2000.
[31] D. Rossi. The Jada Home Page. http://www.cs.unibo.it/~rossi/jada/. 2001.
[32] Sun Microsystems. The JavaSpace Techology Home Page. http://java.sun.com/products/javaspaces. 2001.
[33] D. Rossi and P. Ciancarini. *Jada: A Coordination Toolkit for Java.* 1997.
[34] Sun Microsystems. *JavaSpaces™ Service Specification.* December 2001.
[35] Jini Network Technology Developer Center for Sun Microsystems, Inc. http://developer.java.sun.com/developer/products/jini/. April 2002.
[36] Quickstone Technologies. http://www.quickstone.com/.
[37] The JavaG Benchmarking Home Page. http://www.epcc.ed.ac.uk/computing/research_activities/java_grande/index_1.html. June 2001.
[38] L. Smith, J. Bull, and J. Obdrzalek. *A Parallel Java Grande Benchmark Suite.* November 2001

Appendix A – The Raw Data for Figures 1 – 6

	Execution Time (Sec.)				Speedup			
Package\Proc.	1	2	3	4	1	2	3	4
CTJ	347.46	188.25	137.96	116.34	0.96	1.78	2.43	2.88
Jada	334.58	168.25	112.49	85.37	1.00	1.99	2.97	3.92
JavaSpaces	*	*	*	*	*	*	*	*
JCSP	333.66	168.29	112.68	85.74	1.00	1.99	2.97	3.90
JOMP	332.14	168.80	115.60	86.62	1.01	1.98	2.89	3.86
mpiJava	340.93	177.04	*	94.20	0.98	1.89	*	3.55
Sequential	334.66							

Table 3. Crypt Benchmark Data

	Execution Time (Sec.)				Speedup			
Package\Proc.	1	2	3	4	1	2	3	4
CTJ	709.75	355.43	243.45	196.58	0.99	1.97	2.88	3.57
Jada	701.07	353.48	242.34	198.22	1.00	1.99	2.90	3.54
JavaSpaces	724.96	371.73	245.32	194.26	0.97	1.89	2.86	3.61
JCSP	686.00	345.58	238.72	178.51	1.02	2.03	2.94	3.93
JOMP	714.54	357.47	246.02	179.69	0.98	1.96	2.85	3.91
mpiJava	726.98	363.68	246.70	182.36	0.97	1.93	2.85	3.85
Sequential	701.95							

Table 4. Series Benchmark Data

	Execution Time (Sec.)				Speedup			
Package\Proc.	1	2	3	4	1	2	3	4
CTJ	144.72	119.31	114.71	113.93	0.80	0.97	1.01	1.02
Jada	139.10	71.17	48.87	45.58	0.83	1.63	2.37	2.54
JavaSpaces	154.51	113.01	98.70	106.32	0.75	1.03	1.17	1.09
JCSP	132.76	67.01	45.11	35.32	0.87	1.73	2.57	3.28
JOMP	133.58	67.91	47.07	37.31	0.87	1.71	2.46	3.11
mpiJava	138.50	*	*	*	0.84	*	*	*
Sequential	115.85							

Table 5. SOR Benchmark Data

A Development Method Boosted by Synchronous Active Objects

Claude PETITPIERRE
EPFL, Laboratoire de Téléinformatique, 1015 Lausanne, Switzerland

Abstract. This paper presents a novel development method for interactive and distributed applications. The benefit that this method provides is that the design gives clear guidance towards the implementation. The method is based on three main elements: a concept of synchronous active objects that is closely related to that found in CSP; the Java environment; and a selection of the diagrams defined by UML. This approach alleviates many of the most serious problems that are encountered when using GUI builders, which hide the application structures and so make it difficult to devise sound architectures. The final part of the paper briefly describes an application generator that will help a developer to implement a design which is developed according to the approach advocated here.

1 Introduction

The weak points of most current development methods lie in the coding phases [1]. Although much of the literature describes the implementation process in detail [2] [3], there is a lack of a comprehensive set of approaches which guide the development precisely from the design down through the implementation phase. This paper shows that the concept of synchronous active objects, based on CSP's key concept [4], leads to a simple methodology that works well for the development of interactive and distributed software applications. The approach presented here is tailored to the Java environments and libraries, and based on UML [5]. It is supported by a set of program generators (available on the Web [11]) which make it possible to develop prototypes of interactive and distributed applications very quickly.

After it became clear that the "waterfall" class of development methods was generally unworkable or produced unexpected results, software specialists began to define iterative methods, which cycle over a number of phases such as analysis, design, implementation and test, with more success, but still with many failures too. Taking a cues from the "gurus" who could create wonderful programs by just rolling up their sleeves and working hard until the system works, a further step was taken into what is called eXtreme Programming (XP) [6]. This method stresses the communication between the team members and the customers, the simplicity of the code, the feedback from testing, and the developer's courage and pride. The extreme developer takes pride in an affordable working result obtained in timely fashion, rather than in a dazzling, sophisticated, expensive, and possibly late, solution.

This paper presents concepts that complement rapid-cycle approaches, such as XP, with a set of documents and rules that lead quickly to realistic prototypes. It exploits the benefits of UML by selecting a set of diagrams and concepts from this language and by proposing a precise way of using them. The introduction of synchronous active objects into the method makes it possible to create working prototypes directly from the UML diagrams, which allows a team to discuss the architecture of the application and to involve the stakeholders in the project and get their feedback early. The prototypes can be created, rebuilt, refined, and completed several times, approaching ever-closer to the goal, and they may even persist into the maintenance phase of the product. The proposed approach supports multi-tiered architectures in which the GUIs, the business logic, and the local or remote accesses to the databases are embedded in separate components (or tiers).

Section 2 briefly defines the concept of synchronous active objects. The proposed method, summarized in section 3, iterates over the following steps: design of collaboration diagrams (section 4), design of use cases (section 5), refinement of the collaboration diagrams (section 6), design of state charts for the active components (section 7) and creation of prototypes. The prototypes can be created with the help of a program generator (section 8) and a compiler capable of handling synchronous objects. The generator takes as input a diagram very close to a collaboration diagram. Finally, an example of development that leads to a prototype is presented in section 9.

2 Synchronous Active Objects

The concept of synchronous active objects is described in [7], [8], [9] and [10], but is briefly summarised here for the sake of completeness. A synchronous active object is implemented as a Java active object, namely an object that has a method *run* executed on a thread dedicated to the object. A call to a synchronous active object (either from a passive or an active object) is blocked until the synchronous object explicitly accepts that the called method be executed, by means of the statement written below.

```
accept someLocalMethod;
```

The *accept* statement is also blocked until a call has actually been performed, which realises a rendezvous like the one specified by CSP. The synchronous active objects also define a *select* statement that allows them to specify the set of (blocking) calls and acceptations they are ready to execute. This *select* statement, presented below, exactly implements the CSP alternation shown on the right of the *select* statement.

```
public void myM(int i){
    local = i;
}                                                   CSP equivalent
public void run () {
    . . .
  P:                                    P =
    select {                            (
    case
        activeObject.m(x);                 activeObject_m ! x → Q()
        // statement possible           |
    case
        accept myM;                        thisObject_myM ? x → Q(x)
        // statement possible
    case                                |
        activeObj.met(y);                  activeObj_met ! y → Q()
        // statement possible
    }                                   )
  Q: . . .
}
```

A call to a method is equivalent to sending some parameters into a channel. An *accept* statement, accepting a method that stores the parameters locally, corresponds to the input of the parameters. In the example given above, there is a single local variable, named *local*. *Q(x)* corresponds thus to the object on the left, when its location counter is set after the selection (at *Q:*) and *local* contains *x*. If no value has been passed during the transition from *P* to *Q*, *Q( )* keeps the value previously available in *local*.

3 The Documents Used to Drive a Project

UML is only a language and may be used within various development processes. In the following, we define our own process by selectinga subset of UML documents and their relationship. The approach is supported by the synchronous objects defined above, which make it possible to map the design to a prototype implementation that can be executed. This prototype can be generated with the help of a graphical generator that represents a diagram very close to a collaboration diagram.

- The ***collaboration diagrams*** provide a "geographical" view of the application, a kind of playground showing the components, namely the actors, the sources and the sinks of data as well as the communications established between these elements. The lines that represent the communications can be decorated with the message transmissions or the operations that correspond to the use cases.
- The ***use cases*** describe the scenarios that may be played in the application playground described by the collaboration diagrams. They are written in plain text. There may be one or more use cases per collaboration diagram. Thus, they may be represented by the same set of components on several versions of a collaboration diagram. A typical use case traces the retrieval of some data from a GUI, the forwarding of those data to some central server, and their introduction into the database.
- The ***class diagrams*** define the datasets and their structures. They specify passive classes, with their attributes, methods, and the aggregations they form. We will not be concerned with class diagrams further in this article, because they are well treated in the relevant literature.
- The ***state chart diagrams*** are used to specify the behaviours of the components of the collaboration diagrams, namely the sequences of events the components receive or produce. A state chart diagram is thus defined at the intersection of the use cases and a component. State chart diagrams are particularly well adapted to describing the behaviour of active components.
- If the aforementioned diagrams have been correctly designed, they should be easily amenable to a ***prototype***. A prototype helps to verify that the assumptions made by the developer are correct and that she or he masters the technology on which the project will be based. A prototype is the key "document" to which the members of the project team can refer in order to understand and extend the application. Every (new) member of the development team should be able to view the early prototypes, so as to understand the structure of the application. While the project is progressing, the prototypes may, step by step, become the definitive application.

All the diagrams listed above, as well as their relationships, are detailed in the following sections.

4 Collaboration Diagram

Most development methods suggest starting a project with the design of the use cases. However, use cases involve actors, or components. We therefore propose to start the design by drawing a collaboration diagram, which corresponds, as previously mentioned, to the playground on which the actors and the components will exist. The first collaboration diagram should not represent an abstract situation, but a concrete one, with the central system, some terminals, the databases, the GUIs, special devices, and so on. A concrete example makes it easier for the developer to imagine the operations that must be carried out on the elements. For example, instead of representing all clients by the client class, it is better, at least at the beginning of the project, to represent them by a few hypothetical client components.

A collaboration diagram contains the following components:

- The ***sources*** and the ***sinks*** of data (the keyboard, the display, the printer).
- The main ***data pools***, such as the database servers and the aggregations of data kept in the local memory. A data pool is both a source and a sink of data.
- The ***controllers***, which include the data conveyors, the data transformers and the schedulers.
- The ***actors*** (human, devices) who create events and start sequences of operations.

Figure 1 is an example of a collaboration diagram. The rectangles correspond to the components listed above. The rectangles with thick borders represent units that, unlike passive objects, can react on their own, not only when they are being called (humans, devices, active objects). The rectangles with the thin borders correspond to passive objects. Passive objects should only be used as terminals. Their internal structures must be described in the class diagrams. The behaviour of the standalone components may be described by state chart diagrams, as explained below.

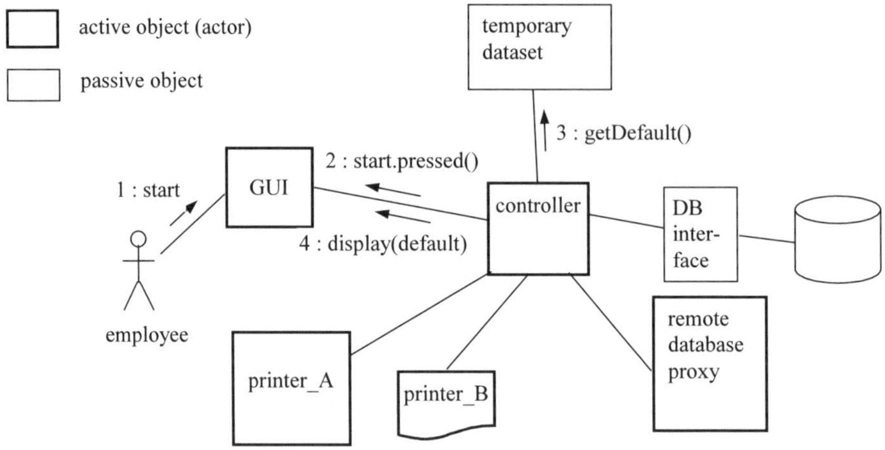

Figure 1. An example of a collaboration diagram.

5 Definition of the Use Cases

A use case is a scenario played on a collaboration diagram. In a system based on synchronous objects, all actions performed by the components can be expressed by calls or acceptances. A use case is defined by a name and by the flow of events and operations that are exchanged between the components of a collaboration diagram. A use case specifies part of the functionality of a system. The flow of events is written in plain language. The actors involved in the events use the names of the components defined in the collaboration diagram. The description of the events should follow the format given below as much as possible, to assure the coherence of the use cases. They are attributed sequential numbers.

Use case X

1. actor verb element acted upon.
2. actor verb element acted upon
3. . . .

For example, with respect to Figure 1:

Entering a new person

1. the employee clicks button start in the GUI
2. the controller executes *start.pressed* (namely, has its call accepted)
3. the controller gets the default person frame
4. the controller displays the default frame on the GUI
5. and so on

The reader may wonder why the use case diagrams defined by UML are not presented. The reason is that these diagrams carry very little information, and besides being shown as illustrations on the cover of the report to the customer, they have very little utility. The collaboration diagrams include the same information, and, unlike the use case diagrams, they can express scenarios that involve a succession of actors and components.

6 Refinement of the Connections Between the Components

The collaboration diagrams contain several kinds of communications that must be specified within a distributed application: the communications between the GUI and the local body of the application, between the body of the application and the remote sites (possibly in both directions), between the server body and the databases, and so on.

After a rough version of the collaboration diagrams has been devised, they must be refined with respect to these communications. This refinement actually amounts to devising the software architecture, namely attributing the application functions to a set of processes and then defining their interconnections to each other and to the devices they use (databases, GUI, sensors). This architecture should be adapted to the performances and security required by the application, taking into account the location of each pair of communicating components: these may be located in the same program (loaded in the same memory partition), in the same computer (but as a program started separately), or on different computers. In this paper, we assume that the system is realised using synchronous active objects.

6.1 Analysis of the Remote Accesses

In the following, we will first discuss the weaknesses of a naive implementation and then show how the introduction of a controller clarifies the program structures and avoids meddling with the *synchronized* keyword, a dangerous concept when it is used at the application level. Note that the sequence diagram used for the explanations is an auxiliary feature, and is not intended to be a part of the core method.

This example, extracted from the application discussed in the second part of this paper, is the following: we assume that an alarm triggered by some sensor (in the event that a temperature reaches some limit) must be sent to a central server, which in turn broadcasts the alarm to several services (for example the police and the fire department).

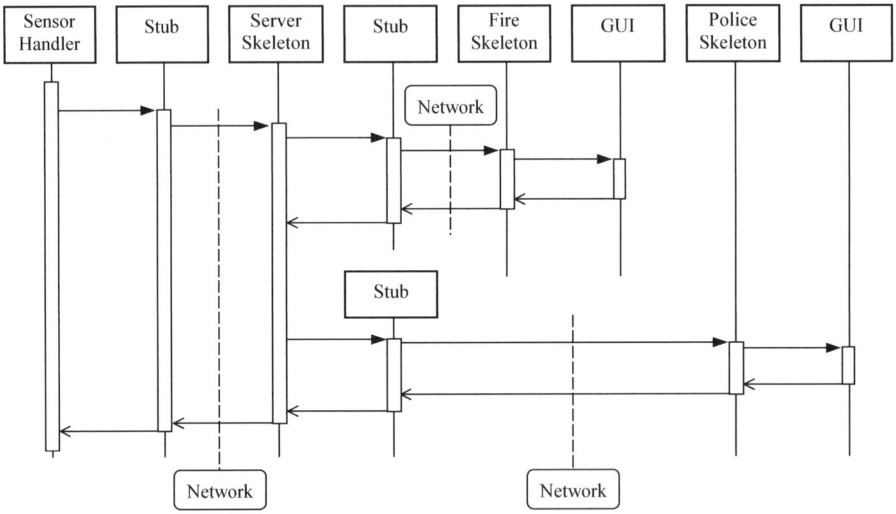

Figure 2. Element involved in the broadcast of an alarm.

The communications illustrated in Figure 2 represent both local and remote calls. The calls that cross a dotted line with the indication *Network* are remote communications, which, we assume, are implemented by remote method invocations (RMI). When the sensor handler triggers an event, its program performs an RMI to reach the server skeleton. The latter calls the stub that leads to the fire department skeleton and then the stub that leads to the police skeleton. As the sequence diagram clearly shows, the sensor handler is blocked until the data have been broadcast to all services concerned with the broadcast alarm. This use case requires several network traversals (three in Figure 2). Thus, if the network is congested, the sensor may be prevented from performing its surveillance task for unacceptably long periods of time.

Moreover, several sensors located at different places may interleave their calls. An RMI skeleton starts a new thread each time a client calls a remote method, and so these methods may access several data structures in the server concurrently, introduce race conditions and corrupt the data structures. These interactions therefore need to be coordinated, which may slow down the operation even further.

The coordination may be done using either of two different philosophies: a defensive philosophy and a constructive one. Under the defensive philosophy, the developer wraps her or his data within monitors defined by the *synchronized* keyword. This approach is simple for simple applications, but it does not scale up well, as it may serialize the requests of all sensor handlers and even lead them to deadlock. The developer must also assure that the stubs are re-entrant, that the accesses to the database do not exceed the allowed number of simultaneous accesses, and so on. This approach is defensive in the sense that it only determines and prevents what is forbidden, but allows the system to run at its own pace otherwise.

Under a constructive philosophy, the developer designs an active distributed structure. In the case of the server defined in the example, one or several active objects may be installed in the server and all alarms may be sent to this or these objects. The objects execute the requested operations, letting the methods of the skeletons return to their callers as soon as they have delivered their messages. Several approaches may again be envisaged to connect a skeleton method to the active objects: creating a new object each time one is required, retrieving one object out of a pool of active objects, or dedicating active objects to the different roles required by the application.

Creating a thread for each new call makes it easy to pass the data to the object that must process them: they can just be stored in the parameters of the constructor. However, the creation of threads has its drawbacks. It is both time and memory consuming, and the number of threads that a program may execute simultaneously is limited by the underlying operating system. Some means must thus be devised to count the number of active threads, to suspend a caller if the upper number of threads is reached, and to resume the activity of that caller once a thread is again available.

Managing a pool of threads is not that much of an improvement, even if it is often described in the relevant literature. First of all, once a thread has been retrieved from the pool, the data to process must be transmitted to the threads, which requires a specific construct. If the threads perform different kinds of activities, it may happen that all threads are used for one kind of activity at the expense of other activities. This problem could be solved by using several pools, but the following solution has the same effect and solves the problem in a simpler manner.

The most effective way to entrust active objects with activities is to use channels or synchronous calls. With this approach, a set of threads is dedicated to each kind of task, which makes it easy to determine the number of threads attached to each one. All the threads belonging to the same group await messages from the same (limited or unlimited) channel, or accept a method dedicated to the reception of messages. An object that wants an active object to perform a given action sends a message to the corresponding channel, or calls the synchronous object that performs that action. If there is a single controller, the latter call may be a simple synchronous call. If the application uses several active objects, the following statements can transmit the operation to the next ready object in the set:

```
select {
    cases (i<N)        // instantiation of N cases, with the values of i : 0 ≤ i < N
        task[i].perform(dataPiece);
}
```

A refined collaboration diagram, corresponding to the situation described in Figure 2, is shown in Figure 3, without the indication of the use case events. The controller disconnects the incoming calls from the outgoing calls.

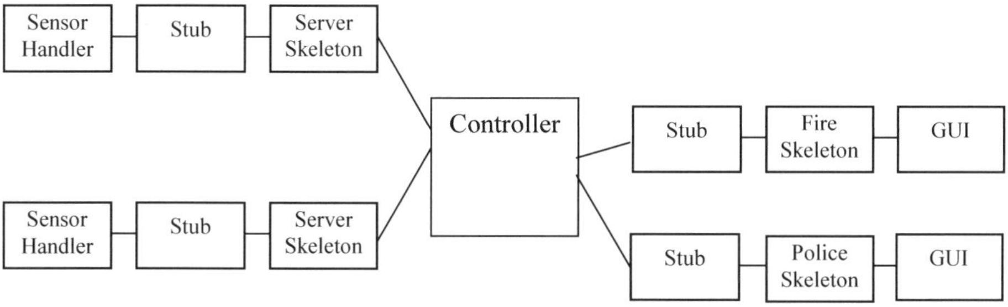

Figure 3. Architecture based on a controller.

6.2 Connecting the GUI to the Program

The use of synchronous objects to manage a GUI has been described in [7]. The source code provided below illustrates the basic principle.

```
select {
case
    buttonA.pressed();
    // perform operation triggered by button A
case
    textField.read();
    // treat the text entered in this text field
case
    buttonB.pressed();
    // perform operation triggered by button B
}
```

This code prepares three parallel calls to the GUI. Then, as soon as one of the GUI elements defined in the *select* statement has been activated by the user, the GUI element (namely its listener) accepts the corresponding method, which in turn releases the case of the *select* that contains that method. This allows the program to "read" the GUI, which is exactly the inverse of what an application based on listeners does.

The use of a controller to retrieve the GUI events has many advantages. It allows the program to perform the requested operations independently from the GUI system thread, and the GUI to handle the subsequent events (window update, data type-ahead and so on) even if the requested operation lasts a long time (such as may occur with remote operations). It also provides a simple means for implementing an automaton that defines the sequences of operations that may be requested by the end-user. The task of specifying this automaton may appear to be an annoyance within the method, since it could be avoided by the use of listeners (because listeners allow a developer to consider each operation independently of the other). However, an automaton allows an exact specification of what the end user can do on her or his terminal. It is a direct implementation and documentation of the use case that describes the behaviour of the end user actor. Finally, it can be represented directly by a state chart diagram (see below).

6.3 Active Stub

In the standard Java environment, the stubs use the threads of the callers to perform the remote call. Thus, if a call to a stub is performed from a listener, the whole GUI is blocked for the duration of the remote call. If a controller is used, as described in the previous paragraph, the GUI is released during the treatment of the events, which is sufficient in simple cases. However, the controller is still blocked, which does not allow it to handle several stubs in parallel or to get a cancellation requested by a user. Therefore, we propose to attribute a thread to each stub. Actually, each time a remote object, namely its implementation on the server side, is called by a client, a new thread is instantiated. There are then as many active threads as calling clients on the server's side, so it is legitimate do the same on the client side.

We have now devised a synchronous active stub, which possesses its own thread. Our stub defines the three following calls:

```
activeStub.remoteMethod(x);

activeStub.post_remoteMethod(x);
activeStub.ready_remoteMethod();
```

The method defined on the first line is blocking, like the method available in the basic RMI library, but it can be decomposed into the two calls defined in the last two lines. The *post* call transmits the parameters to the active stub, which carries out the message transfers in parallel with the caller. Method *ready* is accepted when the data have returned and the call is completed. These two calls can be used in relation with a *select* statement in the following way.

```
activeStub.post_remoteMethod(x);
select {
case
    activeStub.ready_remoteMethod();
case
    cancelled.pressed();
    activeStub.cancel();
    break currentLoop;
}
```

In this code, the remote call is started and its termination is awaited in parallel with the awaiting of a cancel request. If the *ready* event occurs first, the program continues normally. If the cancellation occurs first, the program executes the second case, discontinues the normal execution and takes whatever action is required in this situation. The two kinds of calls only differ by the way they are used, which may be specified in the kind of diagram defined in the following.

7 State Chart Diagrams

The same component may appear in different use cases and thus in several collaboration diagrams. The sequences of events related to this component define a state machine that can be specified by a state chart diagram. Figure 4 represents a set of collaboration diagrams and a representation of the state chart diagram attached to one of its components.

The specification by state charts is mainly useful for the controllers. If these are based on synchronous active objects, all transitions defined by the state charts are either synchronous calls or *accept* statements. The refined collaboration diagrams may contain auxiliary elements, such as the channels, the stubs, the data pools, sinks and sources, but the latter do not need to be specified by state chart diagrams, because they would be trivial.

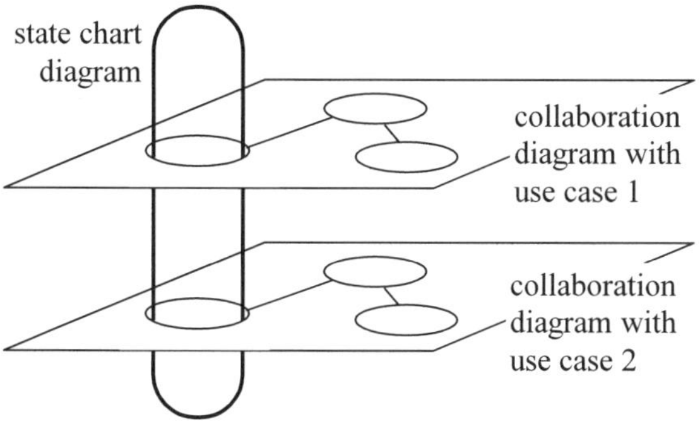

Figure 4. Intersection of the state chart and the collaboration diagrams.

7.1 Implementation of the State Chart Diagrams with Synchronous Objects

The components of a collaboration diagram are implemented either as passive or as active objects, insofar as they do not model unimplemented actors, such as humans and devices. The passive objects contain the data. They are specified by the data types and by the methods used to access and transform the data (class diagrams). The active objects take care of the application behaviour. They may be defined by state chart diagrams.

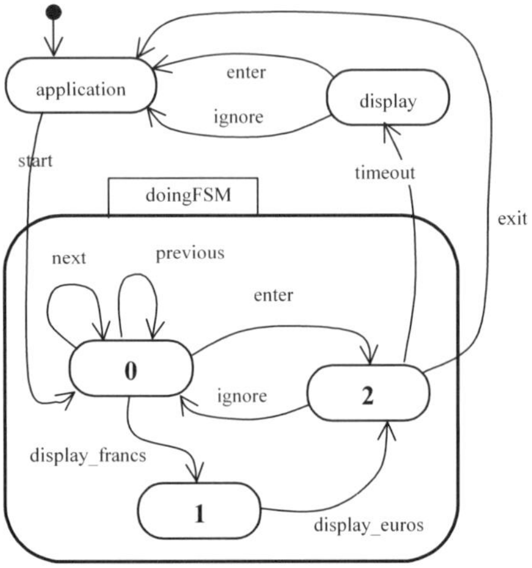

Figure 5. State chart diagrams.

Figure 5 presents an example of state chart implemented in the source code written below. This code is assumed to run in the body of a synchronous object. Method *run* implements the external state machine, while method *doingFSM* implements the composed state. Each finite state machine is placed within a *switch* statement, itself placed within an infinite *for* loop. The states are represented by the values of variable *extState* for the external machine, and by variable *state* for the internal state machine. Of course, in some cases, other statements may also be used, but jumping to another state clearly amounts to performing a *goto* statement, which is naturally not available in Java. Note that a state machine is allowed to perform a goto, since it may jump from any state to any other state. The *break label* statement in Java is the thing closest to the *goto*, but still admitted as a statement compatible with structured programming.

```
void run () {
    for (;;) {
        switch (extState) {
        case application:
            start.pressed ( );
        case doing:
            switch ( doingFSM() ) {
            case exitTrans:
               extState = display;
               break;
            case timeoutTrans:
               extState = application;
               break;
            }
```

```
        case display:
            select {
            case
                ignore.pressed( );
                extState = application;
                break;
            case
                enter.pressed ( );
                extState = application;
                break;
}   }   }   }

int doingFSM ( int state ) {
    for (;;) {
        switch (state);
        case 0:
            select {
            case
                next.pressed ( );
            case
                previous.pressed ( );
            case
                francs = text.readInt ( );
                state = 1;
            case
                exit.pressed ( );
                return exitTrans;
            case
                waituntil (currentTimeMillis(10000);
                return timeoutTrans;
            }
            break;
        case 1:
            text.setText ("€ "+francs / 1.5);
        case 2:
            select {
            case
                enter.pressed ( );
                state = 0;
            case
                ignore.pressed ( );
                state = 0;
            case
                exit.pressed ( );
                return exitTrans;
            case
                waituntil (currentTimeMillis(10000);
                return timeoutTrans;
}   }   }   }
```

One could also resort to the state design pattern [2], but experience shows that it makes the writing of code more difficult than necessary.

The implementation of the state chart diagrams is particularly straightforward if the GUI, the stubs, and all device interfaces are realized with synchronous active objects.

7.2 Validation of the State Charts

The implementation of the state chart diagrams in the various components must naturally be capable of executing every flow of events defined in the use cases. Actually, it may happen that some new event flows, which did not appear in the specification, may be executed. These new event flows should not impede the functioning of the program if the user happens to execute them. When the application is not too large, namely early in the development, automatic validation techniques may help to study and isolate those functionalities.

8 An Application Generator

A program generator, which we have made available on the Web, allows a developer to draw a graph that is very close to the refined collaboration diagram presented in this paper, and to generate a program directly from this graph (Figure 8). The following components may be introduced into the application:

Initialisor	An object that collects the instantiation statements of all the parts of the application.
Controller	A synchronous active object that contains the main body of the application, most often described by a state chart diagram.
Display	A graphical user interface with synchronous accesses. The elements that must appear on this display may be specified by a limited GUI generator.
Implementation/Stub	The ends of an RMI communication. The developer defines the interface of the remote object. A sub-generator then creates an active stub and a standardized remote object. The program generator also calls the *rmic* (Java RMI compiler) to generate the basic stubs and skeletons.
Database	An SQL interface to a database. The developer defines SQL tables and a set of SQL operations and inputs them to a generator. The latter creates classes that can contain the data of the records stored in the tables of the database, and a class that allows a local program respectively to introduce the data of these objects into the database, and retrieve them from the database. The interface executes the SQL commands provided by the developer with the actual data. In the case of retrieval commands, the interface extracts the data from the text returned by the SQL commands. The generator also creates a remote object that can provide a remote access to the database interface described above.
User Object	The developer can enter objects that she or he has defined elsewhere into the graphical editor of the application.

This generator delivers programs based on synchronous active objects. It compiles them and creates the scripts and auxiliary files needed to deploy and run the application.

9 An Example – The Information System of the Town of Verdun

The following presents a development example. We assume that the town of Verdun has asked us to propose an information system for the management of its infrastructures. We want initially to define and validate a basic architecture for this system, then to create a first prototype that shows what kind of functionalities such a system may offer, and finally in this way to have an evolved prototype to present to the officials responsible for the project.

9.1 Requirement elicitation

The definitive system will be used by the full set of town services (public works, police, fire department, environmental and agriculture services, and the planning service responsible for granting construction permits). They will use it to manage the town infrastructures and to coordinate the interventions on these infrastructures, sites and roadwork, to alert the response teams, and to log the alarms provided by the various kinds of infrastructures, for latter analysis.

The infrastructures include the public buildings, the roads, the water networks (drinking water, sewage water, rain water, as well as rivers and channels), gas and electricity distribution, telecommunications, and so on. The proposed system must allow a continuous remote supervision of the installations (for example, level of the water in the channels, traffic overload, fire alarms) and transmit the alarms to a central server and to the responsible services. It must also handle the documents related to these infrastructures, the planning, and the coordination of the works they must undergo.

9.2 First iteration

The first prototype will handle an alarm database (descriptions of the possible alarms, actions to undertake when they occur, alarms that have occurred with time of occurrence), and the broadcast of the alarms to the services that have logged in.

9.3 Services considered

The following services will be considered for the prototype, but at this level of detail there will not be much difference between them.

- the public works
- the police department
- the fire department

9.4 Alarms

- The alarms are generated by electronic surveillance systems and by the persons who have access to the system. The surveillance systems will be simulated by programs that generate alarms, either at random or when a button is clicked.
- The alarms must be transmitted to a central server, which must forward them to the services and to the response teams, and store them in a database.
- The system will have facilities to show which infrastructures have produced which alarms.

9.5 Basic collaboration diagram

According to the method presented at the beginning of this paper, the first step requires the design of a rough collaboration diagram defining the actors and components that will play a role in the application. This first diagram (Figure 6) will be completed in the subsequent phases. Thus, it does not need to be very elaborate. This diagram contains two identical sensors and three services. The definitive system will contain more of them, but we assume that if two or three stations may be handled, the extension to n stations will not be a problem.

9.6 Use cases

The actions and events that compose the use cases refer to the actors and the components by the names available in the collaboration diagrams. They have not been represented in the diagram, but to do so is trivial. Here is a list of possible scenarios for the application analysed in this section:

1. Login of a service

1.1 A service opens a connection to the server
1.2 The service sends its location and an identifier to the server
1.3 The server verifies the identifier
1.4 The server opens a connection to the service for the broadcast of alarms

2. Alarm logging and broadcast

2.1 A sensor sends an alarm to the server
2.2 The server logs the alarm in the database
2.3 The server stores the alarm in memory
2.4 The server broadcasts the alarm to the services that have requested it
2.5 Each service displays the alarms

3. Alarm acknowledgement

3.1 A user selects an alarm in a service
3.2 The service sends the alarm identifier to the server
3.3 The server cancels the alarm in memory

The developer may now choose where to continue: refine the collaboration diagrams (subsequent section) or specify the class diagrams (the latter are out of the scope of this paper).

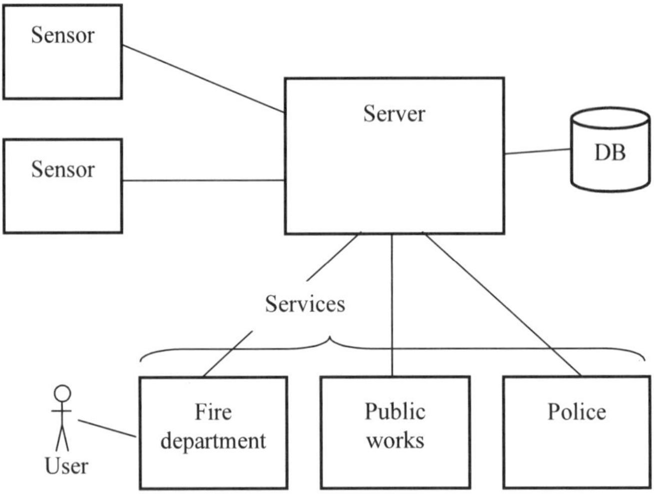

Figure 6. Collaboration diagram of the application.

9.7 Refinement of the remote connections

One of the previous paragraphs explained how to refine the connections generated by the use case scenario 2. The extension of this analysis to the whole application introduces only one additional controller. Figure 7 shows the complete diagram, with controllers in the server and in the clients.

9.8 Controller of the service

We propose that the handling of the events produced by the GUI in the service stations should be performed by the same controller that already handles the alarms. It is obviously safer to have the same controller display the alarms and update the GUI, because it automatically keeps track of the state of the GUI, and thus does not risk trying to display the alarms when the GUI is in an invalid state.

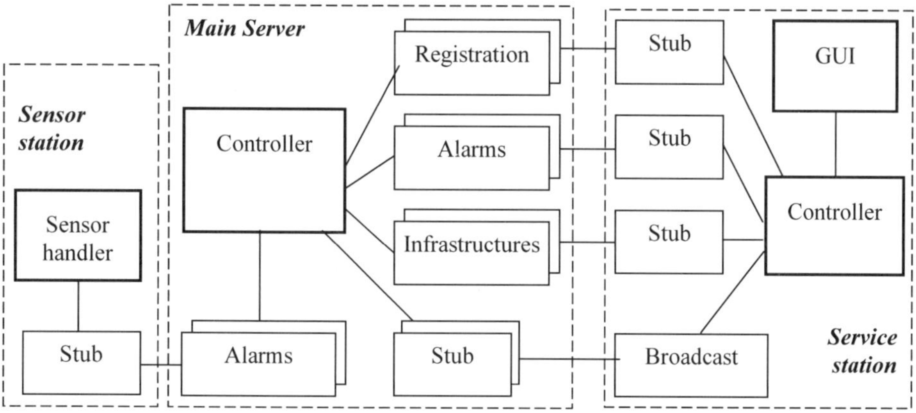

Figure 7. Refinement of the collaboration diagram.

10 Fast Prototyping of the Example

The picture in Figure 8 shows the kind of graphs that appear on the editor of the program generator. The generator creates all the .jar files, scripts and auxiliary files needed for execution.

The approach described in this paper has seen practical use by 7 groups of 5 students within the framework of a one-semester project on software engineering. Using the techniques described here, all these groups were able to present an application having one or several central servers, several services and alarm producers, distributed access to the database, and an on line HTTP access used by the standard browsers. All this was achieved without overstepping the available time. These applications were multi-threaded and handled client-server as well as symmetrical communications.

11 Conclusion

This paper has described a development method that leads directly from design to executable prototypes. It has shown that the synchronous active objects we have defined can greatly help the design of distributed interactive applications. To be fair, other object-oriented approaches derived from CSP could provide similar advantages.

The approach is systematic enough to allow a program generator to produce fast prototypes and the basic architectures on which industrial software applications may be built. In fact, the program generator is itself based on these objects. The program is very flexible and readable. For example, the approach makes it easy to give the ability (to the end-user of the generator) to interleave the handling of the various components. For example, one could begin the creation of a component, create another component needed by the first one, and then terminate the specification of the first one.

This approach has successfully been used by a substantial set of students.

12 Acknowledgements

The program generator was developed in large part by Riadh Kortebi. My thanks also go to Mark Madsen who carefully revised the paper.

References

[1] J.A. Whittacker and S. Atkin, Software Engineering is not Enough, IEEE Software, August 2002, pp 108-115.

[2] E. Gamma et al., *Design Patterns, Elements of Reusable Object-Oriented Software*, Addison-Wesley, 1994.

[3] D. Lea, *Concurrent Programming in Java*, Addison-Wesley, 2000.

[4] C.A.R. Hoare. *Communicating Sequential Processes*, Prentice Hall.

[5] J. Rumbaugh, I. Jscobson, G. Booch, The Unified Modeling Language Reference Manual, Addison-Wesley.

[6] K. Beck, *Extreme Programming Explained: Embrace Change*, Addison Wesley.

[7] C. Petitpierre, Synchronous Active Objects Introduce CSP's Primitive in Java, CPA'02, Reading, September 2002.

[8] C. Petitpierre, A. Eliens, Active Objects Provide Robust Event-Driven Applications, SERP'02, Las Vegas, June 2002.

[9] C. Petitpierre, Synchronous C++, a Language for Interactive Applications, IEEE Computer, September 1998, pp 65-72.

[10] 1999 - C. Petitpierre, Implementing Protocols with Synchronous Objects, in ed. D. Avresky, Dependable Network Computing, Kluwer, November 1999, pp. 109-140.

[11] http://ltiwww.epfl.ch/sJava

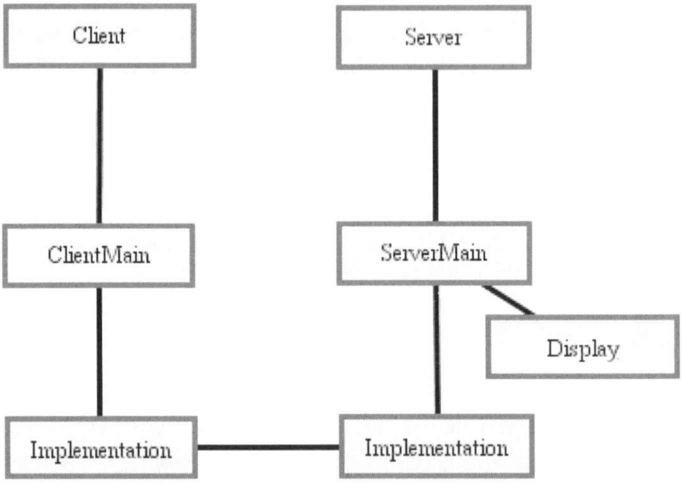

Figure 8. A picture of the kind of graphs handled by the graphical editor.

Communicating Process Architectures 2003
Jan F. Broenink and Gerald H. Hilderink (Eds.)
IOS Press, 2003

Tutorial: Prioritised Service Architecture using Honeysuckle

Ian R. EAST
Department of Computing, School of Technology
Oxford Brookes University, Wheatley Campus, Oxford, England OX33 1HX
ireast@brookes.ac.uk

Abstract. An update will be presented on the progress in establishing the Honeysuckle programming language [1] and its formal foundations. The latter are formally addressed in a paper currently under review for journal publication [2], but will be summarised. They include formal definitions of service protocol plus service network/component (SNC) and the PSA design rule (PSADR), from which a proof of a priori deadlock-freedom emerges directly. Freedom from priority conflict (and thus inversion) is also easily guaranteed. Closure in the definition of system and component guarantees true compositionality under both concurrency and prioritised alternation.

Summary

1. Introduction

The dramatic fall in hardware cost is leading to new software application domains that require the combination of high integrity and low development cost. Traditionally, high software integrity is attained through the use of a posteriori verification of formal specifications using additional development tools. Combined mathematical and programming skills are thus required, which are in short supply. Furthermore, such formal methods do not scale well to large applications. This should not be surprising since they substitute "trial-and-error" for true engineering. Finally, the implied additional cost and risk cannot easily be justified for smaller applications.

An alternative is to employ formal design rules and a priori (static) verification. Such rules can be built into the definition of a programming language and verification built into an appropriate compiler. As a result, no additional tools or skills are required and the method may be expected to scale well.

A second problem addressed is the absence of compositionality at the design level in both process- and object-oriented languages. The specification of any system must remain uncompromised by the substitution of one component by another with the same interface. Achieving compositionality under concurrency while denying the possibility of interference remains a goal in computer science [3]. Here, the primary aim is at least to deny the worst pathological behaviour, namely deadlock, using known effective design rules [4, 5].

Honeysuckle attempts to simultaneously eliminate the possibility of deadlock while programming concurrency and yield true compositionality without complicating abstraction. Human abstraction of the every-day world includes concurrency and prioritized alternation from an early age yet their application to software remains a task for the élite. Honeysuckle is intended to demonstrate that, with the right model for abstraction, it need not be so.

1.1 Service architecture and deadlock-freedom

Proof of *a priori* deadlock-freedom already exists [5], subject to a simple design rule whose verification may be performed at compile-time. However, *ad hoc* conditions are employed. It is shown how appropriate definitions of *service* and *service network* can incorporate all necessary conditions and thus permit a proof of deadlock-freedom free of any additional constraint.

Extending the definition of service network to allow mutual exclusion and dependency of service is shown to achieve closure between service network and service network component (SNC). An adapted proof of deadlock-freedom is then possible, subject to a revised design rule. Recursive parallel (de)composition of systems then becomes possible, retaining deadlock-freedom without need of further proof.

1.2 Prioritised service architecture, compositionality, and abstraction

The same extension is shown also to afford secure (de)composition under prioritized alternation, and a further design rule to ensure freedom from priority conflict. The usual restriction of service (client/server) networks to tree structure need thus not apply.

PSA may prove sufficient alone to express the specification of many systems, negating the need for any meta-level. It is particularly well suited to systems that are distributed and/or embedded. A prioritised service interface (PSI) captures input/output dependency, mutual exclusion between inputs, and their prioritization. It is not limited to listing input and output connections. It can also express how they *relate*.

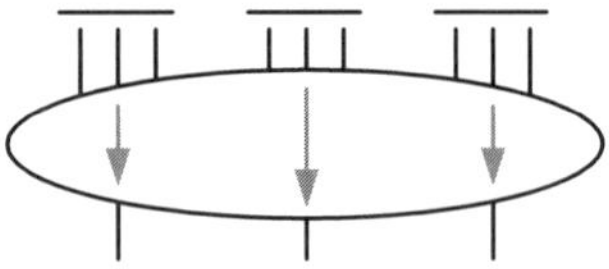

Figure 1 A (black-box) system specified by its prioritised service interface (PSI).

True design-level compositionality requires closure in the definition of system and component, which yields the model for abstracting an interface to either. A PSA design is expressed via an appropriate prioritized service digraph (PSD), which comprises two distinct species of arc which may be drawn in different colours (*e.g.* blue for service, red for interruption).

Process-oriented programming is entirely compatible with object-oriented programming (OOP), though the reverse is an assertion that is less easily supported. There are a number of distinct models for OOP of concurrency, which is usually considered an advanced topic [6, Chapter 30]. (Manfred Broy has recently contributed a general critique of OOP [7].)

In Honeysuckle, the relationship between process and object is simple. Processes *own* objects. Ownership may be transferred between concurrent processes. This mirrors the temporary transfer that takes place between processes in sequence when a parameter is passed by reference to a procedure. (In occam, only values, and not objects, may be passed across channels – a limitation which can give rise to inefficiency through otherwise unnecessary copying.) References to value and object are distinguished implicitly, by operator, avoiding the long-understood problems associated with explicit references (pointers) [8].

1.3 Conclusion

Honeysuckle is intended for the specification, compositional design, and implementation, of concurrent and reactive systems with inherent security against deadlock. In the use of design rules, prioritized service architecture provides for the engineering of systems without additional tools or scarce combination of skills. It should thus be suitable for the affordable development of new computer applications that demand high integrity at low cost (and risk).

A new and consistent formal model for program abstraction exists in the form of prioritized service architecture (PSA). Development tools can easily verify adherence to design rules as the system is incrementally composed *or modified.* (It should be remembered that ~98% of programmer activity is in modifying existing systems.) Thus formal guarantees can be made regarding behaviour *a priori.*

PSA is adequate alone for the specification of many systems and affords true compositionality. Abstraction is natural and powerful, facilitating straightforward capture of the concurrent and reactive behaviour typical of embedded, distributed, and interactive systems.

The complete specification for Honeysuckle will be published openly on the internet. Funding is being sought for the development of a compiler and interactive development environment. Development tools will also be "open source". It is then intended to demonstrate the advantages of the method, perhaps by shadowing a commercial project.

Further work is also planned to investigate the semantics of the language and whether additional guarantees can be made regarding behaviour, for example, freedom from livelock. Another possibility is to allow the introduction *ad hoc* of an additional rule known in advance to guarantee some application-specific safety constraint. The circumstances under which interference can be eliminated, without compromising efficiency of implementation, is also of obvious interest.

2. Acknowledgements

The author acknowledges many valuable, and thoroughly enjoyable discussions with Jeremy Martin, whose D.Phil. thesis addressed many of these issues, and Sabah Jassim, who shared with me the enviable task of supervision.

References

1. East, I.R., The Honeysuckle Programming Language: An Overview. IEE Software, 2003. 150(2), pp. 95-107.
2. East, I. R., Deadlock-Free Programming with Service Protocol, submitted to ACM ToPLAS, 2003.
3. Jones, C.B., Wanted: A Compositional model for Concurrency, in Programming Methodology, McIver and C. Morgan, Editors. 2002, Springer-Verlag. pp. 1-15.
4. Martin, J., I. East, and S. Jassim, Design Rules for Deadlock Freedom. Transputer Communications, 1994. 2(3): pp. 121-133.
5. Martin, J.M.R., The Design and Construction of Deadlock-Free Concurrent Systems. 1996, University of Buckingham: Hunter Street, Buckingham MK18 1EG.
6. Meyer, B., Object-Oriented Software Construction. 1997: Prentice-Hall.
7. Broy, M., Object-Oriented Programming and Software Development – A Critical Assessment, in Programming Methodology, A. McIver and C. Morgan, Editors. 2003, Springer. pp. 211-221.
8. Hoare, C. A. R., Recursive Data Structures. Int. J. Computer and Information Sciences, 1975, 4(2), pp. 105-32. Reprinted as Chapter 14 in [9].
9. Hoare, C. A. R., Essays in Computing Science. 1989: Prentice Hall

Communicating Process Architectures – 2003
Jan F. Broenink and Gerald H. Hilderink (Eds.)
IOS Press, 2003

The Grid Block Device

Bardur ARANTSSON Brian VINTER
Department of Mathematics and Computer Science, University of Southern Denmark

Abstract. In this paper we propose a distributed, replicated block device for The Grid based on the the replication algorithm of Y. Amir[1]. The end goal is to support application-specified semantics, both wrt. replication strategy and the consistency model used for reads and writes.

We list the proposed features of the Grid Block Device and discuss various methods of implementation.

1 Introduction

With the ever-increasing need for computing capacity it has become clear that using clusters of relatively cheap personal computers for some classes of computational problems is much more cost-effective than using mainframe computers.

Similarly, the storage demands of cluster applications, or, more specifically, scientific cluster applications, are increasing at a similar rate. In this paper we describe a proposal for a distributed block device, which, in addition to supporting the very large datasets required for such applications, is also fully replicated, and tolerant of both node failures and network partitions.

Before getting into the specifics, let us first examine traditional cluster computer configurations. The local storage on each individual PC is usually divided between the operating system and a "scratch" area where applications can write intermediate results and temporary data. Currently, the operating system occupies a very small portion of the total available storage – typically on the order of 1% – and the relative amount of operating system data vs. scratch data is only going to grow as storage density on hard drives grows.

In order to access shared data such as user home directories and large sets of experimental data[1], there is typically also a network-accessible drive. This networked drive usually resides on a central server using a large array of disks for redundancy and performance, simply because the administration is much easier that way.

There are a few inherent problems with this approach to cluster computing which we will consider below.

Firstly, the scratch area not explicitly utilized by the particular application that the cluster is running goes to waste. This can amount to quite considerable amounts of wasted storage capacity.

Secondly, and more imporantly, server performance may become a limiting factor for access to the shared data. In addition, if multiple clusters are connected over a Wide Area Network (WAN) then the bandwidth of this link may become a limiting factor for the performance that can be achieved by an application. The incurred extra cost may be unacceptable in many instances, even when the WAN link is a high-speed dedicated link.

Finally, each server is a single point of failure for its part of the file system namespace. As cluster systems grow to several thousands of nodes, such a single point of failure becomes unacceptable.

To solve the problems of using the above approach, the proposed Grid Block Device (GBD) is focused on providing a fault tolerant block device which is used as the sole data storage facility for each node participating in a computation. The proposed block device has many desirable features, among them replication, automatic fail-over, and disconnected operation.

Why a block device instead of a file system? There are a several reasons for this decision, a few of which will be enumerated below.

Firstly, implementing a distributed block device is simpler than implementing a fully-fledged distributed file system, because the semantics of file systems are largely irrelevant to the implementation. We can thus focus on the essentials that are relevant for efficient replication and efficient access to the replicated data. Also, some features, such as global snapshots, are actually much easier to deal with on the block level than the file level.

A distributed block device could also prove useful for distributed database-like applications. Since database applications have much more predictable access patterns than most applications, they usually bypass the file system layer entirely, and often also the caching mechanism of the operating system, avoiding the overhead associated with these. File systems are almost always optimized for much more unpredictable usage patterns which makes them unsuitable for this type of application. This would seem to indicate that it might be more efficient to implement distributed database applications using distributed block-oriented storage, rather than basing the implementation on a more general distributed file system.

Finally, implementing a distributed file system on top of the distributed block device should be relatively simple since the infrastructure for replication and disconnected operation would already exists. To elaborate further on this last point, a very simple implementation of a distributed file system based on the GBD could simply use the old UNIX idea of using trees of i-nodes to represent files and directories. It might seem like a bad idea to have a single root i-node and letting everything else "hang off" that single i-node because each file lookup requires access to the root i-node. However, it has been observed that i-nodes closer to the root have a tendency to be quite static over time[2], whereas i-nodes deeper in the tree have a tendency to change much more frequently. In the context of the GBD this would mean that the root i-node could be replicated to all servers at very little cost because any (expensive) updates to it would occur very rarely. Since all searches for files start at the root i-node, replicating it, and possibly its children, to each node could be significant win. Of course, supporting more high-level features such as truly atomic writes of arbitrary length would require a much more complicated file system structure, but the infrastructure provided by the GBD should make implementing such "extras" much simpler than it would otherwise have been.

By avoiding the complications caused by tying the file system implementation to the replication engine implementation we believe that a much cleaner and more efficient solution to the data distribution problem can be developed.

2 Previous Work

2.1 Spread

The low-level communication layer utilized by the replication algorithm is the group communication system Spread [2], which we give a short overview of here.

Spread provides efficient group communication with reliability and ordering guarantees over WANs. It is based on UDP/IP and implements its own packet ordering and reliability. An overview of how the spread network is structured is given in figure 1.

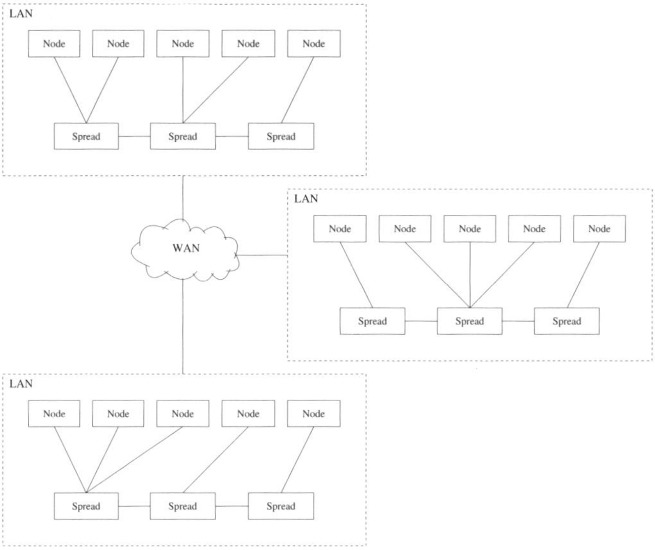

Figure 1: Overview of the Spread network structure. Some (not all!) nodes in a LAN run a Spread daemon (`spreadd`) which communicates with the other Spread daemons in the LAN. Each node initiates a connection to a designated Spread daemon which actually handles all the communication. Spread daemons from multiple LANs communicate via a WAN.

The network is overlaid with a multicast tree network which is based on the network topology and typical latencies in the network, which are, in turn, provided by the user through a static configuration file. The multicast tree is used to implement efficient multicast by simply only sending packets that are absolutely necessary for a multicast message to reach all its intended destinations; see figure 2 for a simple example.

The reason that Spread is based on UDP/IP instead of an ordered, reliable protocol such as TCP/IP is simple: If a reliable point-to-point protocol with built-in packet ordering were used, the latency would suffer in scenarios such as the one shown in figure 3. Looking at the figure, we see that node A sends a sequence of three packets to node C through B. In the example shown, the first packet is lost and must be retransmitted. Observe that all packets transmitted after p1 could easily have been sent on after arriving at B, but when using the ordered reliable protocol they all have to wait for the first packet to arrive at B before being

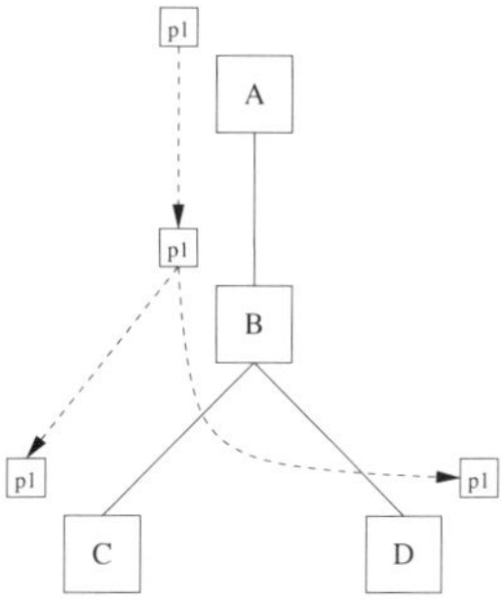

Figure 2: Efficient multicast in Spread: If A wants to multicast a message to C and D, only *one* message is sent from A to B, and B assumes responsibility for transmitting the message to C and D.

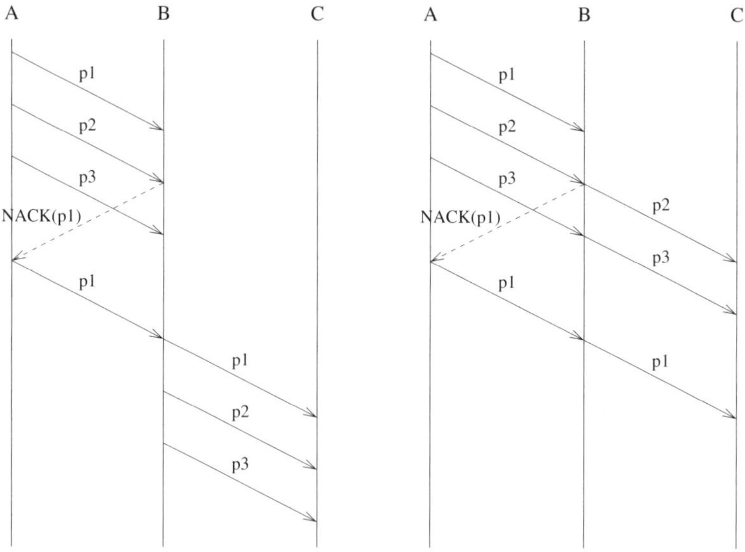

(a) Using a reliable PTP protocol.

(b) Using an unreliable PTP protocol.

Figure 3: Difference in total latency between using a reliable/unreliable point-to-point (PTP) transport protocol for routing packets through intermediate hops along a route.

sent on. As shown, this can lead to completely unnecessary communication delays, even in relatively reliable networks. To avoid such unnecessary delays, Spread assigns responsibility for packet ordering to the final destination of the packet instead of forcing each intermediate hop to perform its own packet sequencing. In the above scenario, this means that all packets following the first packet would be sent to the intermediate node B, and, while p1 was being retransmitted, node B would forward the packets p2 and p3 to C. The packet ordering would happen at the final destination C, just before delivery to the application layer.

Another aspect of the Spread communication layer which is leveraged by the replication algorithm is the fact that the communication layer provides Virtual Synchrony[5] semantics, which effectively means that notification is provided whenever the configuration – i.e. the set of connected servers – changes. There are two types of configuration changes: Transient and regular configuration changes. Transient configuration changes are a signal to the application that the group membership has changed due to a network event, but that the new configuration is not known yet. Regular configuration changes occur when the new network configuration following such an event becomes known. These messages are delivered *interlaced* with the regular data messages and in an order consistent with the "stream" of data messages. This allows for the introduction of a state where the network is "in between" fully connected and partitioned.

2.2 Replication Algorithm

One of the most interesting replication algorithms to emerge in the last few years is the algorithm of Y. Amir [1]. It is based on the Spread group communication system and uses the Extended Virtual Synchrony [7] concept to provide fault tolerant replication whilst considerably reducing the need for – and in the common case avoiding – the most expensive part of most other replication algorithms, namely end-to-end acknowledgements. Avoidance of end-to-end acknowledgements on a per-action basis can yield much higher performance than algorithms such as 2PC and COReL as demonstrated in [3]. Network partitions are also handled correctly.

The replication algorithm is designed mainly for database-like scenarios where the goal is to replicate the complete contents of the database to all participating servers. Instead of taking the traditional approach of replicating *data* to participating servers, it instead replicates all *actions* taken upon the data, and ensures that all these actions are applied across all replicas in a global total order. While ensuring that all participating servers have identical[3] replicas, this also affords a lot of flexibility wrt. allowing different semantics for different actions. For example, an application could choose to use the dirty query semantics over the typical sequential consistency if its needs are more in the realm of interactivity than accuracy. Apart from these two models, the algorithm also supports weak consistency.

The aforementioned actions are small transactional units consisting of a *query* section and an *update* section. Since the replication algorithm cannot give guarantees about the precise timing of the application of updates to the database[4] we must somehow prevent updates from being performed if they are based on stale data. This is achieved using the *query* section of an action as a precondition for performing the updates in the *update* section. As an example of how this works, consider the following: An application which increases values in a database by 10% reads that a particular piece of data that it intends to modify, x, has value 5. To perform the update, the application creates a new action, containing $x = 5$ in the query section, and $x \leftarrow 5.5$ in the update section. The action is submitted to the replication engine which orders it according to the global total order. If, when the action is applied to the database, the value of x has changed, then the action is rejected and the client program is informed of this. Using this technique it can be ensured that the update will only be performed if the value of the data x has not been changed by another action in the time between the read and the update. This can be generalized by associating version numbers with individual blocks and preconditioning operations on these version numbers instead of relying on particular data values.

Originally, the replication algorithm in [1] was restricted to a preconfigured set of replication servers, but it was extended in [3] to allow for dynamic addition/removal of replicas. Our intention is to incorporate this extension into the Grid Block Device implementation.

3 Overview of The Grid Block Device

The overall architecture of the Grid Block Device is rather simple; it simply sits as a layer between the replication engine and the application (as shown in figure 4). Note that the "application" which is shown is not necessarily a user application, but could just as easily be a distributed file system implementation, or even a Linux device driver that would effectively allow using the GBD as a simple means of integrating with Linux filesystems.

In the following sections, the motivation and implementation ideas for each of the proposed features of the GBD will be described in some detail.

4 Features

The main proposed design features of the Grid Block Device (GBD) are as follows:

- Widely Distributed Servers

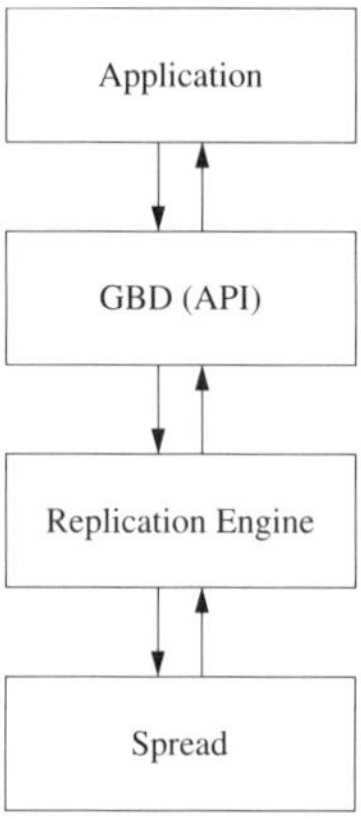

Figure 4: Overview of the Grid Block Device.

- Automatic Replication and Migration
- Disconnected Operation
- Global Snapshots
- Application-specific Semantics
- Legacy Application Support

In the following sections each of these proposed features will be discussed in more detail.

4.1 Widely Distributed Servers

Having widely distributed servers is critical to high availability and low latency for widely distributed clients accessing the system. Using widely distributed servers has traditionally been problematic because the inter-server communication has usually been done using protocols and algorithms (e.g. 2PC) with a lot of overhead, or, which simply could not operate correctly in partitioned networks. Clearly, network partitioning is to be expected in any reasonably-sized WAN, simply because of the distances involved.

Since the algorithm of [1] can operate in partitioned networks and avoids per-action end-to-end acknowledgements, it is relatively cheap to employ widely distributed servers in this context.

While increasing the number of servers usually also increases the performance and availability of the system there is also an increase in the amount of overhead involved in maintaining the replication engine state and in the group communication layer itself. Thus, a balance between the number of servers and the incurred overhead needs to be found through experimentation.

4.2 Automatic Replication and Migration

Automatic replication and migration of data to where it is needed is among the most important features of the proposed Grid Block Device. Much research has been conducted in this area; see [6] for an overview of early work. However, the assumptions underlying many of the proposed algorithms sometimes vary wildly, and may or may not apply practical applications. Obviously, the optimal choice of algorithm may also vary from application to application, and so the best strategy for implementing any replicated/distributed system is probably to have a generally applicable algorithm as the default, but giving the user the option of choosing/implementing a different algorithm if necessary.

Any generally useful algorithm for replication and migration needs to satisfy a few properties: Decision-making must be truly distributed, i.e. with no centralized decision-making, and decisions must be taken quickly. It is also very desirable to have as low overhead as possible, particularly wrt. the space requirements for any bookkeeping data. Finally, the algorithm should be quick to adapt to changes in situation such as network congestion, imbalance in server load, and changes in user-imposed resource utilization constraints.

There are, of course, a number of heuristics that apply to this problem. If, for example, a particular block of data is accessed frequenty by a particular client, then it makes sense to try to move the data closer to that client. Also, if a particular block of data is accessed very infrequently on a particular server, then it may be possible and desirable to simply remove the replicated data. Of course, we must be careful to obey any redundancy constraints imposed by the user, and possibly network partitioning, before removing replicated data.

There are also some randomized algorithms for the replica placement problem that are quite simple to implement which also do well in practice; see [4], for example.

It should be noted that the basic unit of replication in the GBD is the *block*. Each block is therefore – in principle – distributed completely independently of other blocks, which also means, that no single node needs to have sufficient storage for the complete contents of the block device. This makes it possible to support vast amounts of storage on relatively cheap (PC-class) hardware. There may of course be significant benefits to keeping neighbouring blocks together, especially wrt. the amount of metadata required, but there are no absolute constraints on the replication/movement of individual blocks of data.

4.3 Disconnected Operation

We distinguish between two forms of disconnected operation, namely *involuntary* disconnection and *voluntary* disconnection. As the names suggest, the former occurs when the network becomes partitioned or individual network links go down, and the latter occurs when e.g. temporarily moving a few participating servers.

Such events should not have any significant consequences for the performance of the rest of the network. Obviously, there are instances where a performance impact cannot be avoided. As an example, consider a network which becomes partitioned, causing all servers with a given piece of data to become isolated from the rest of the network (see figure 5). In this case it is simply impossible to avoid waiting for the network to become fully connected again before any reads or updates can be performed upon the affected data. Furthermore, aside from replicating all data to all network nodes, there is no possible way of avoiding this scenario, since it is impossible to predict where network partitions may occur. However, it should be possible to give the replication layer some *hints* as to which links are more likely to go down than others. This could allow replication/migration decisions to be based on such external information instead of being based solely upon information which is internal to the system, and, which is thus also limited to what can be gathered through internally observable properties of the network.

Another important consideration wrt. disconnected operation is how to handle updates and reads that occur while the network is partitioned. The replication engine automatically selects a new primary component which is allowed to continue unhindered, but if sequential consistency is to be maintained, actions in other components must be delayed until the network becomes connected again. Clearly, choosing a particular consistency model for all applications to adhere to is either going to limit their efficiency, or cause inconsistencies to arise in the case where the model is too weak for the particular application. Therefore, the

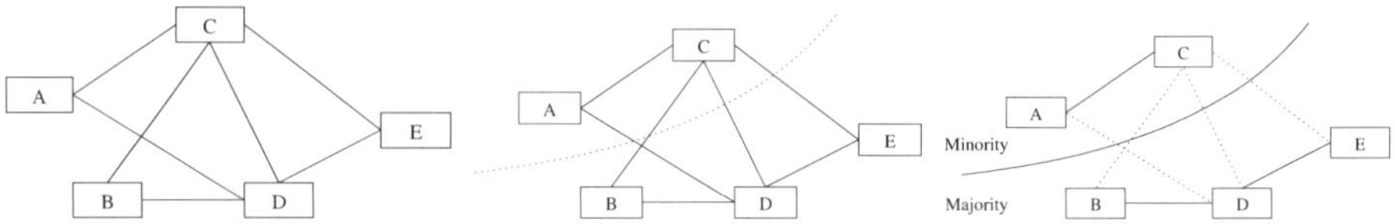

(a) Fully connected network. (b) Partition occurs. (c) Partitioned network.

Figure 5: Impact of network partitions: If node A and node C have some data which is required by node B, D or E, then a network partition may prevent all access to the data until the network becomes whole again. Note, however, that the majority of the nodes may continue issuing and performing updates on data which is *shared* between nodes on either side of the network partition without causing any consistency problems.

choice should be left up to the applications themselves.

There is also the issue of "defensive" replication to consider here: Since network partitions lead to the network separating into two or more connected components there is a likelihood that they each contain a number of replicas of any particular block device. If we are lucky they will contain an equal number of replicas, but we might be unlucky, and one of them might contain only one replica. This could clearly be a dangerous situation for the system to be in since the side containing all the remaining replicas could be completely wiped out from e.g. natural distaster without the other side of the partition knowing anything about such an event before it was too late, i.e. when one of its own servers containing a part of the replica crashed. In order to defend against such failures, we propose a RAID-like redundancy scheme where *transient* replicas are introduced when a network partition occurs. These transient replicas serve as a sort of backup if all (regular) replicas should fail on either side of a network partition.

4.4 Global Snapshots

Supporting global snapshots is not trivial by any means, because snapshots have very precise and strict semantics: A snapshot represents a *consistent* and *unchanging* image of the data, in our case the contents of the block device, at the time the snapshot was taken. Global snapshots are extremely useful because they allow backups in systems which cannot be taken off-line for extended periods of time. Additionally, snapshots can serve as a hint to the system that no further changes to the given data are expected. This could allow the system to optimize for the read-often, write-rarely scenario.

There are two basic techniques for supporting snapshots:

1. Create a read-only duplicate of the data.
2. Avoid overwriting data that is part of a snapshot; instead, write new copies of changed blocks. When the snapshot is released, any subsequent changes can be incorporated back into the "live" block device.

The former of these techniques is *extremely* expensive, since it requires creating a complete copy of a potentially huge data set. This requires a lot of storage, and also requires a lot of time to perform the actual copying operation. Additionally, in a replicated system care must be taken to replicate the snapshot itself to avoid bandwitdth bottlenecks and vulnerability to crashes on the server(s) containing the snapshot. Without such replication a crash of any server containing any portion of such a snapshot could tie up any processes reading from the snapshot. One positive property of using this technique is the fact that actually *accessing* a snapshot is completely trivial and carries no overhead whatsoever, since the snapshot is just a read-only block device containing an exact copy of the original.

If we examine the latter approach, we see that it requires a certain amount of extra bookkeeping, but that it has some significant advantages. Firstly, a global snapshot only requires extra storage on the order of $O(n)$ where n is the number of *modified* blocks since the snapshot was taken. In the worst case this is exactly the same amount of space required using the former technique, excluding the negligable bookkeeping overhead. Under normal circumstances this should result in significant space savings.

A relatively simple way to implement this approach is to apply versioning to individual blocks and use this version information to avoid overwriting/releasing data which is part of a snapshot. Creating a snapshot is then merely a matter of noting the version number at which the snapshot was made, and avoiding making direct changes to the affected blocks while the snapshot is active. Since this requires no copying of data this is also quite cheap time-wise, but it does impose a slight overhead for each request while snapshots are active.

4.5 Application-specifi c Semantics

Application-specific semantics are really a two-part issue. First we consider application-specific replication/migration policies, and then we consider application-specific read and update semantics.

A simple example of an application-specific replication policy would be allowing applications to specifiy a minimal number of replicas which the system must maintain at all times – if at all possible. Note that it may not actually be possible to always guarantee, say, precisely 5 replicas of a given file if the network becomes partitioned. However, upon discovering the partition, the system can start introducing transient replicas of the affected blocks until the replication count reaches 5 again. While this “rebuilding” is taking place, the system runs in a sort of “degraded” mode akin to RAID-like systems. When the network becomes connected again, the transient replicas can be removed.

Supporting application-specific read and update semantics can be an important aid in optimizing for specific access patterns. For example, there may be applications that write sequential blocks a vast majority of the time, while other applications only perform random-access reads of data. Additionally, it may be worthwhile to let applications signal to the system that they only require *dirty read* semantics. This could be useful for an application that continually reads data from a GBD, but which has interactivity constraints. In this case, the read data should converge in some sense, but the interactivity is valued more than showing data which is necessarily guaranteed to be the most up to date. Note also, that network partitions can force normal reads to block until the network becomes connected again. This is obviously a bad idea for such “real-time” data.

4.6 Legacy Application Support

An important parameter for adoption of any new technology is the level of support offered for legacy applications. This support can come in two forms: Either the new techonology supports the legacy applications outright, or it provides an easy-to-implement migration layer/mechanism. Implementing transparent support for legacy applications might be a worthwhile goal for the simple reason that it would promote adoption, but there are several problems associated with it which we will summarize below.

Firstly, the POSIX file system interface[5], which consists mainly of the `open`, `read`, `write`, and `close` operations, does not lend itself well toward a transaction-based view of the file system. These operations all assume in some way that the application is in complete control of the file, i.e. has an exclusive lock on the file. In some distributed file systems this is dealt with by the introduction of shared/exclusive locks which are held by applications while performing read/write operations on any given block of data. But exclusive locking can cause horrendous performance degradation, particularly in the case where applications are accessing completely different regions of the same file. This has been addressed by region-based locking of files through the `fnctl` system call which is part of the POSIX standard, but these locks are quite error-prone and consequently quite hard for programmers to use correctly. Compared to these, the transaction-like semantics of actions seem much more intuitive.

It is unclear exactly how these operations should be mapped to the transaction-like semantics of the GBD. The trivial solution is to simply lock the entire device on `open` and unlock the device on `close`, but this completely eliminates the usefulness of the GBD. The problem is, of course, that since the POSIX file operations were not defined in the context of concurrent reads/updates, their semantics are too weak to have any meaning in a distributed context. Even the *append* file mode in POSIX does not define any sort of mechanism for dealing with concurrent writes to a given file, even though there is the obvoius approach of just appending anything written to the file. This weakness in the POSIX semantics for file operations means that we either have to simply discard the old interface, use locking to achieve POSIX semantics, or make our own assumptions about the intended semantics of the POSIX file operations.

One option is to impose something resembling causal ordering in the following manner: Firstly, version numbers are added to each block. When an application performs a `read`, it causes a certain version of a block of data to become "known" to the application. Any subsequent `write` of that particular block is preconditioned on the block version being the same as that which was read from the device. Subsequent reads of any *other* block are preconditioned on previously read blocks having the same version numbers as when they were read. This stems from the observation that reads are almost always dependent on each other in some way.

Subsequent `writes` of any other block is preconditioned on all "known" blocks having the same version numbers as when they were first read.

Opening or closing a file flushes the "known" blocks for that file. This is based on the assumption that closing and reopening a file means that an application cannot assume *anything* about the file's contents after reopening it. This seems a reasonable assumption since closing a file implicitly removes any locks held on it.

This way of handling the POSIX file operations implements a kind of causal ordering which may give resonable results for applications that do not have strict locking requirements, i.e. most user programs should work properly with these semantics since they usually do not access data which is shared between multiple processes.

Unfortunately, storing version numbers for all these "known" blocks is not something which requires a trivial amount of storage. Actually, if a legacy application – supported using this scheme – were to read a large file sequentially, it would require storage proportional to the file size. It would be prohibitively expensive, probably impossible, to have this much data in memory to handle legacy applications which were streaming in nature. It could be workable for applications which only require a few random accesses to the block device, but in general it would not be feasible nor advisable to use such a scheme.

The best approach is, of course, to rewrite applications to use the native API in which such problems do not occur because dependencies are handled explicitly by using the *query* section as a precondition for the *update* section of actions. This could potentially also result in much higher performance because access patterns can be tailored specifically to the given application.

Another issue is the matter of handling failed updates, i.e. updates which fail because they are performed on stale data, i.e. data which was changed by another process without the application having any way of becoming aware of the change. The return codes for the POSIX file operations simply have no equivalent for this situtation. One would have to make a choice between using one of the existing error codes, and risk problems because of the different semantics of the situations in which the error occurs, or, introducing a new error code and risk problems arising from error handlers not written to cope with this new error code. Either of these solutions is unsatisfactory.

5 Conclusion

Work is currently focused on developing a prototype Grid Block Device based on the various ideas presented in this paper. When finished, it could provide valuable insight as to the merits of the individual ideas/techniques that have been discussed.

When such a prototype is available, the API can be solidified, and a framework for testing performance objectively can be developed to aid in further evaluation. Objective evaluation in real-world scenarios is needed for at least the following aspects of the Grid Block Device: Performance as a function of number of servers and server connectivity, replica placement, and disconnected operation.

In conclusion, a considerable amount of work remains to be done, but we believe the algorithms and ideas presented here to be a sound foundation upon which to base the implementation of the Grid Block Device.

Notes

[1]As an example of what we consider a large data set in this context, the LHC experiments at CERN are expected to result in an estimated 6PB of data annually.

[2]It helps to think of the UNIX file system hierarchy when considering this. For example, the contents of the file system root, `/`, almost never change whereas the contents of a user's home directory, `/home/user/`, change much more frequently.

[3]In the sense that they contain *equivalent*, not necessarily physically identical, data.

[4]Such guarantees are, in fact, impossible to give in partitionable networks since they require the equivalent of a global clock which is provably impossible to maintain correctly in such networks.

[5]Which is also the block device interface, because of the UNIX philosophy of treating everything like a file.

References

[1] Yair Amir. *Replication Using Group Communication Over a Partitioned Network.* PhD thesis, 1995.

[2] Yair Amir, Claudiu Danilov, and Jonathan Stanton. A low latency, loss tolerant architecture and protocol for wide area group communication. In *FTCS 2000*, 2000.

[3] Yair Amir and Ciprian Tutu. From total order to database replication. Technical report, Johns Hopkins University, 2002.

[4] Yair Bartal. *Lecture Notes in Computer Science 1442*, chapter 5. Springer, 1998.

[5] Kenneth P. Birman and Thomas Joseph. Exploiting virtual synchrony in distributed systems. In *Proceedings of the 11th ACM Symposium on Operating Systems Principles*, pages 123–138, November 1987.

[6] D. Dowdy and D. Foster. Comparative models of the file assignment problem.

[7] Louise E. Moser, Yair Amir, P. Michael Melliar-Smith, and Deborah A. Agarwal. Extended virtual synchrony. In *The 14th IEEE International Conference on Distributed Computing Systems (ICDCS)*, pages 56–65, 1994.

Communicating Process Architectures – 2003
Jan F. Broenink and Gerald H. Hilderink (Eds.)
IOS Press, 2003

Distributed Shared Memory in Global Area Networks

H. H. HAPPE and B. VINTER
Department of Mathematics and Computer Science, University of Southern Denmark

Abstract. Distributed Shared Memory (DSM) has many advantages in heterogeneous environments, such as geographically distant clusters or The Grid. These includes: locality utilization and replication transparency. The fact that processes communicate indirectly through memory rather than directly, is giving DSM these advantages.

This paper presents the design of Global PastSet (GPS) which is a DSM system targeted at global area networks. GPS is based on the DSM system PastSet [1] that has been very effective in homogeneous cluster environments. GPS utilizes consistency control migration and replication to scale in heterogeneous environments. This has resulted in a token-based mutual exclusion algorithm that considers locality and an algorithm for locating replicas. GPS has been simulated in multi-cluster environments with up to 2048 nodes with very promising results.

1 Introduction

Message passing systems (e.g. the Message Passing Interface[1] – MPI) have been very successful in homogeneous cluster environments because of good performance and the rather simple programming model. The good performance stems from the fact that the developer explicitly states which processes that should communicate. This means that if the developer has done it right, there will be no unnecessary communication.

When moving to more heterogeneous and dynamic environments message passing applications must be custom made to address the nature of a specific environment. Also, message passing systems rely on node availability and in case of errors the application must handle its recovery. It is not possible to make message passing systems automatically adapt in these dynamic environments because message destinations are explicitly stated and are part of the application's logic.

In DSM systems processes do not communicate directly with each other, but indirectly through memory. This makes it possible for a DSM system to solve several problems transparently.

- **Replication**: The indirect communication makes transparent replication possible because data is addressable.
- **Fault-tolerance**: By using active replication to maintain some level of redundancy transparent fault-tolerance can be achieved.
- **Locality**: Because the reading processes do not have to be known at the time of writing it is possible to prioritize local read requests higher than remote. For example, if many processes were competing to read from a distributed bounded buffer, those processes that are close to the actual location of the next entry could be served first. This makes

[1] http://www-unix.mcs.anl.gov/mpi

the system unfair and introduces the possibility of starvation, but still might increase performance considerably for some applications. Also, it is possible to make a DSM system fair and without starvation, if needed.

- **Focus on functionality**: The developer can focus on the functionality of an application instead of the environment that it has to be deployed in. This is not much of a problem in homogeneous environments where every process can be viewed as equal and communication between processes has the same cost (if each node only runs one process), but in heterogeneous environments such as the Grid this becomes very complex. Also, heterogeneous environments could dynamically change over time, which would be very hard for a developer to address using message passing. DSM systems can adapt in such a dynamic environment because the choice of destination process can be postponed until there is a requesting one. This can be the responsibility of the system so that the developer can concentrate on functionality.

Most research in DSM has been targeted at homogeneous clusters, but some work has gone into making DSM work in a global context. Unify [2] is a *shared virtual memory* system made to scale beyond clusters. Apart from accessing the virtual address space as conventional *random access memory*, it can also be accessed as *sequential access memory* and *associative access memory*. These other types of memory access makes it possible to weaken consistency. Globe [3] is a distributed object system where method invocation can be done as a remote procedure call or by migrating the target object to the calling node. This, together with various consistency models, can be utilized to make Globe scale in a global context. In Global Arrays [4, 5] an extra level of remote memory is introduced to make it scale in multiple distant cluster environments. From a node's viewpoint this gives three levels of memory; local memory, remote memory at a node in the local cluster and remote memory at a node in a remote cluster.

2 PastSet

In PastSet [1], interprocess communication (IPC) is done through what is called an *element*. Basically an element is an infinite virtual array of indexed data-blocks that can vary in size (see Fig. 1). A data-block is denoted a *tuple*, because of PastSet's resemblance to systems based on an abstract tuple space. In the original version of PastSet, tuples had a fixed size for each element, but this is not required.

2.1 The Element

The way one reads and writes an element is like a bounded FIFO-buffer, but it is also possible to read a specific index in the array. There are three variables that control the state of the bounded buffer. *first* points to the index first written but not yet read, and *last* points to the index that should be written next. *delta* is a value that defines the size of the buffer and thus the maximum distance between *first* and *last*. Tuples with an index that is lower than the *first* pointer is not necessarily disposable, it depends on the element's "garbage" policy.

2.2 Operations

PastSet preserves a sequential consistency among all operations on an element. This makes it very user friendly for the developer. There are three main operations that a process can invoke on an element.

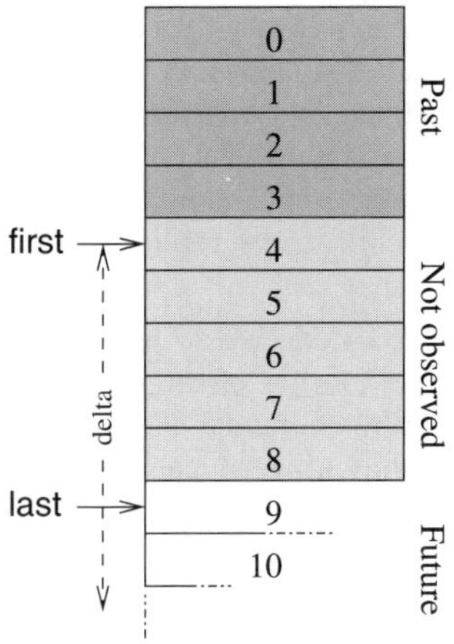

Figure 1: The PastSet 'element'.

- `move(tuple)` writes a tuple to the element at the index referenced by the *last* pointer. The *last* pointer is then incremented by one. If the distance $(last - first)$ between the *first* and *last* pointer is equal to the *delta* value the operation blocks until the distance becomes smaller.

- `observe` reads the tuple referenced by the *first* pointer and the pointer is then incremented by one. If the *first* pointer is equal to the *last* pointer the operation blocks until the *last* pointer becomes greater than the *first* pointer.

- `observe(index)` reads the tuple referenced by the index. If the index has not been written yet, the operation blocks until it becomes available. This operation does not change the *first* and *last* pointer.

Tuples are immutable when first written because no operation makes it possible to update an existing tuple.

3 Global PastSet

This section describes the design of Global PastSet (GPS). To see if DSM makes sense in a global context performance is the objective of this design.

In the original version of PastSet described in [1], an element where located and controlled at one node. Therefore, all communication with an element had to go through this single node. This simple approach worked very well in homogeneous clusters, but the higher latencies imposed by a global environment makes it inefficient.

To avoid the impact of high latencies, GPS distributes elements to the involved nodes. This is done by changing the way an element is controlled and stored:

- **Element control migration**: When a node executes an operation on an element the control of the element is migrated to this node, which then can carry out the operation locally. Also, the node could execute multiple operations while it has the control. This way many operations can be executed at one location (e.g. a cluster) before control is migrated to a remote location.

- **Replication**: When a node receives or writes a tuple it stores a local copy of the tuple. This way the tuples are scattered across all involved nodes and get implicitly replicated when a node reads a tuple. Active replication could also be introduced to address availability and fault-tolerance issues. This excessive form of replication gives no coherence problems because tuples are immutable. The only problem that needs to be addressed, is that of locating the nearest replica.

To gain control of an element a node has to have exclusive access to the state of the element. The state of an element consists of the *first* and *last* pointers together with the *delta* value. A custom made distributed mutual exclusion algorithm has been developed to address this problem. This algorithm utilizes locality by favoring local requests and hereby minimizing the traffic through high latency lines.

3.1 Mutual Exclusion

The algorithm is token-based, meaning that the node holding the token has exclusive access to an element's state. Because an element's state consists of a rather small amount of bytes it is stored in the token. This way an element's state is always available at the node that has exclusive access to it.

The nodes that are using an element are part of a tree topology where the root node holds the token (Fig. 2.a). Each node has a reference to a parent node except the root node. This ensures that there exists a path from each node to the root node and therefore to the token. Also, it makes it very easy to join the group of nodes that are using an element by finding one of the others through some kind of naming service.

When a node executes an operation that needs to update the element's state it most get hold of the token. If it does not already have it, it sends a request to its parent.

When a node receives a token request it forwards the request to its own parent, unless it has a pending request itself. In the latter case the token will eventually arrive at the node and the request is therefore added to a local queue. If the receiving node holds the token the request is added to a queue in the token.

After the root node is done updating the element's state the token must be passed on to the next waiting node, if any. This is done by finding the nearest node in the token request queue and send the token to this new root node. The old root makes the new root its parent. This implies that every node must have a way to decide the cost of communicating with other nodes (e.g. IP addresses, geographical distance, performance monitoring). This way of passing the token shortens the path that the token has to travel to reach all requesting nodes.

When the token arrives at a node, this node's local queue is added to the token queue. Then the pending operations allocates indexes to observes and/or moves, after which the token can be passed on.

Allocating the needed indexes take little time and therefore if there are many pending requests in the system the token will move from node to node rather quickly. The result can be that new requests are chasing the token, which creates a long tail of requests behind the token. Forwarding these requests impose a rather big load on the involved nodes and also on the network. In order to avoid this, a node records the source of the last request it has forwarded. Now when it receives another request it can forward it to this *last seen* node. This node either has a pending request for the token or may already have seen the token in which case the request is just forwarded as described above. If the *last seen* node has not seen the token when the request arrives a distributed queue of requesting nodes will be build in front of the token path. Otherwise, the request will take a shortcut to the token. Though this is very effective, it introduces a locality problem.

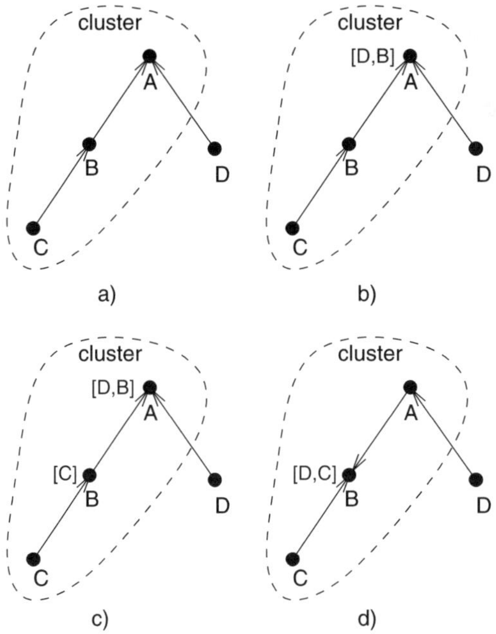

Figure 2: Tree topology. The root holds the token.

In most cases a node will have a local community (LAN, cluster, etc.) and requests from this community should not be forwarded to nodes on the outside, unless it is absolutely necessary. While a node does not know its local community, the cost of communicating with the *last seen* node defines a community in this context. In Fig. 3.a node D (*last seen*) defines node B's local community, by including all those nodes with a lower cost of communication. Now all requests coming from nodes with a higher or equal cost of communication are forwarded to D, because D's request is likely to be served before. Requests from nodes with an equal cost of communication are also forwarded to D because those are likely to be part of D's real local community. In Fig. 3.b C is inside B's community and therefore the request is forwarded to the parent. This is because r_C should not be added to a remote queue (D might be located on the other side of the world). It should rather exploit the locality in the token tree.

Fig. 3.c shows a scenario that could break locality. If C and A are local to each other and remote to the rest, the token might take a detour because r_C is queued at D. This will not always be true, because it is not certain that r_C can catch up with D's request if r_C is just forwarded to B's parent. It would have to be able to do this in order for A to make the right locality decision, when passing on the token. On the other hand, if local communities share the same subtree in the token tree, this would not happen.

3.1.1 Example

Fig. 2 shows an example of how the algorithm works. Fig. 2.a shows the initial configuration of the tree. Nodes A, B and C are part of a cluster and node D is remote to this cluster.

In Fig. 2.b A has received a request first from D and then from B, which both have been added to the token request queue.

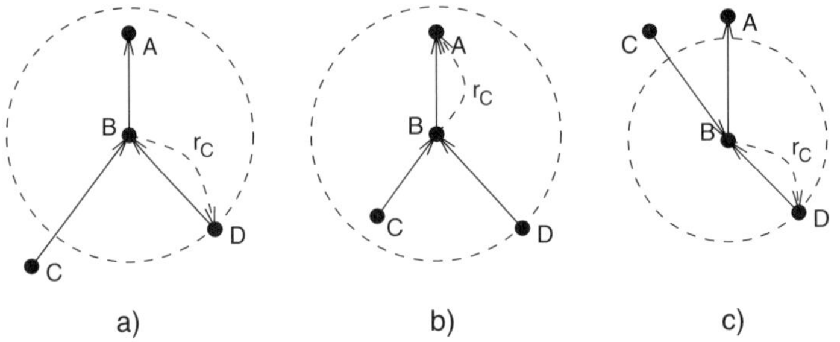

Figure 3: Request forwarding. The dashed circle defines B's local community bounded by the cost of B communicating with D - the *last seen* node indicated by the dashed arrow. a) The request r_C is forwarded to D because C is outside B's local community. b) The request r_C is forwarded to A because C is inside B's local community. c) A scenario that could send the token on a detour, if C and the node that holds the token (A) are local to each other.

In Fig. 2.c C has send a request to its parent B which has added the request to its local queue, because it already has a pending request.

In Fig. 2.d A has passed the token to B even though D's request arrived first at A. This is because B is closer to A than D. At B the local queue has been added to the token queue.

3.1.2 Starvation and Fairness

In most scientific applications starvation and fairness in IPC are not an issue because the main priority is to keep the nodes calculating. If CPU utilization is high all nodes will eventually be served. On the other hand if an application has a high communication overhead, it might as well be carried out by nodes local to each other if possible.

The mutual exclusion algorithm introduced above is not fair and starvation could occur. Fairness could be introduced by removing the utilization of locality and starvation could be avoided by forcing a request to be served if it has been postponed too long (with a counter or some kind of timeout).

3.2 Move and Observe

This section describes how the custom made mutual exclusion algorithm is utilized to implement the `move` and `observe` operations.

The `move` and `observe` operations can only be carried out when the conditions $last - first < delta$ and $first < last$ holds respectively. Therefore, should the token only be passed on to a node that can complete a pending operation (the condition holds).

To distinguish between moves and observes a node explicitly requests a move or an observe, instead of just requesting the token. Nodes also distinguish between pending moves and pending observes when queuing requests locally to avoid dead-locks. A node must also distinguish between move and observe requests when saving a reference to the source of the last seen request.

The fact that the request states which operation to execute also makes it possible for other nodes holding the token to carry out the operation on behalf of the requesting node. Therefore, it is possible to constrain the token to certain nodes for security or dependability reasons.

With these additions the operations can now be specified:

index	references	local
3		true
4	n_2, n_5	false
6	n_4, n_9, n_{12}	true
10	n_4, n_{10}	false

Table 1: Tuple table example.

- `move(tuple)`: The issuing node gets hold of the token. Then it inserts the tuple with *last* as index, given that the move condition holds, and increments *last* by one.
- `observe`: The issuing node gets hold of the token. Then it records *first* as the tuple index to read, given that the observe condition holds, and increments *first* by one.

When the root node is ready to pass on the token, it is passed to the closest node in the token request queue that has a pending request which can be carried out (the condition holds). Also, observes are served before moves if their sources are equally close to the forwarding node. This makes it more probable that the observing node gets a reference to the node with the tuple it needs, because the distance between the *first* and *last* pointer is minimized (see section 3.3).

3.3 *Locating replicas*

To guarantee that a node can locate a tuple with a given index there will have to be a deterministic path from each node to each tuple. This section describes what information a node must maintain to ensure that this path exists and how this information is used to locate a tuple.

3.3.1 *Tuple Table*

A node should maintain a tuple table (see Tab. 1). For each index that the node has information about, there is an entry in the table. An entry holds a list of references to nodes that have the tuple and it tells whether the node has the tuple locally.

In the example there are no references to other nodes at index three, but the node has the tuple locally. For index six the node holds the tuple locally, but it also has references to other nodes that hold the tuple. This could seem redundant but it makes it possible for the node to flush the local copy while still being able to find it again.

To ensure that there exists a deterministic path from each node to each tuple a constraint is imposed on tuple tables. This constraint must ensure that a node that has a copy of a tuple with index i must also know where to find the tuple with the index $i - 1$. For $i = 0$ this is obviously not required.

All that are needed to uphold this constraint is that the token holds a reference to a node holding the last written tuple. Then the next node that moves a tuple can store this reference in its tuple table.

How this constraint together with the token tree is utilized to locate tuples is explained below.

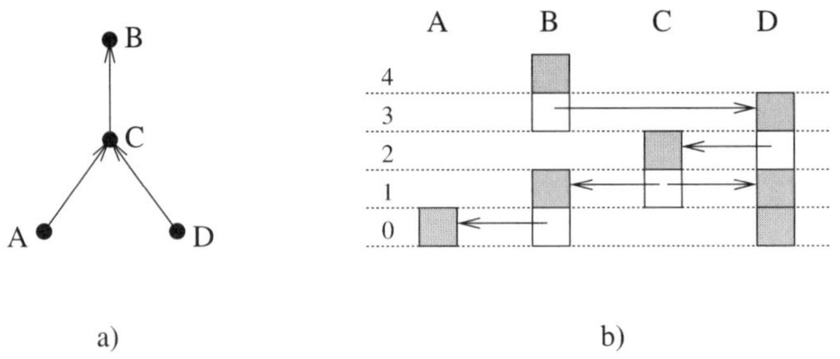

Figure 4: Tuple request paths. a) Logical tree for mutual exclusion for nodes A-D. b) Tuple tables for the four nodes. Dark squares indicate that the node has a copy of the tuple locally and an arrow indicates where it can find it. If there is no square the node does not know anything about the tuple.

3.4 Tuple back-trace

The tuple table constraint described above ensures that for each tuple with index i that a node holds locally there exists a path to the location of each tuple with index $j < i$ (see Fig. 4).

If a node needs a tuple with index i it sends a request to the node that has the tuple with index j, where j is the smallest index so that $j \geq i$. When a node receives a request for a tuple that it does not hold locally it forwards the request in the same manner.

Fig. 4.b illustrates how these paths work. If node C wants the tuple with index $i = 0$, it sends a request to node B because the tuple with index $j = 1$ is the nearest tuple, so that $i \leq j$, that C can find. The request is again forwarded from B to A following the same rule and A has a copy of the requested tuple. If B ceased to exist there would still be a path to the copy at D.

If there are multiple nodes to choose from when forwarding a request (as in the example) the nearest node is chosen.

3.5 Parent forwarding

Fig. 4.b shows that there is not necessarily a path from all nodes to all tuples, but if the logical tree for mutual exclusion is used together with tuple back-tracing, this path exists. This is easy to prove. Following the path to the tree's root a request can find the token, which has a reference to the location of the last written tuple. Now it is possible to back-trace to the required tuple. If for example node C in Fig. 4 wants the tuple with index $i = 3$ it sends a request to its parent B which starts the back-tracing to D.

In many situations the request will be able to start the tuple back-tracing on its way toward the root. This will relieve the root node and make the path shorter.

3.6 Non-existing tuples

In the case of an indexed observe of a tuple that has not been written yet, a process should block until it becomes available. Therefore, all nodes with a pending request for a non-existing tuple should receive a copy of the tuple after it is written, but how is this achieved efficiently?

A tuple request for a non-existing tuple will eventually reach the root node where it discovers that the tuple does not exist. These requests could simply be recorded in the token and be served when the tuple becomes available. When a node writes the tuple it would have to send copies to all waiting nodes listed in the token. For some applications this would result in a huge token payload caused by too many tuple requests, which would slow down token forwarding. Also, the time used for sending copies to waiting nodes would increase proportionally to the number of pending requests.

This problem is solved by parallelizing the copying so that nodes receiving the first copy, forwards it to other waiting nodes and so forth. To accomplish this a tree of waiting nodes is build for each requested non-existing index. In this tree each requesting node knows its child nodes (if any) and the root of the tree is maintained in the token. The token and each node can have an individual fixed number of children and this number defines the max number of copies a node wants to forward. This also helps distribution of existing tuples.

Now how is the tree created? When a node receives a tuple request it checks if it has already got a pending request itself, in which case it adds the requesting node as a child. If it cannot have more children it forwards the request to one of the children it already has. The choice of child to forward to is based on locality, meaning that the closest child will be chosen. If there is a choice of equally close children, requests are forwarded to these in a round-robin fashion. If it does not have a pending request it will forward the request to its parent.

The root of the distribution tree is handled a bit differently. If there is room in the token a request is just added like described above. If not, the existing requests are searched to see if one originates from a node further away from the token's current position, compared to the source of the new request. If such a request exists it is replaced with the new request and sent to the source of the new request. Otherwise the new request is forwarded to one of the nodes in the token. This will build subtrees with nodes that are local to each other, because the token is not forwarded to remote nodes unless it is absolutely necessary. For example when a distribution tree has been build in a cluster where the token exists and the token is transfered to another cluster, local requests in this other cluster will replace the requests in the token. This will have the effect that there will be few edges between the two clusters in the distribution tree.

When a requested tuple is written the writing node has the token and can therefore send copies of the tuple to the children of that index (if any). The tree ceases to exist when the tuple it represents becomes available and it is automatically removed as each node has received a copy of the tuple and forwarded it to its children.

The optimal configuration of a tree would naturally be one that exploited the network's topology to achieve a minimal tuple distribution time, but the information to do this is not available. First of all it is not known which node will write the tuple while the tree is created, secondly the set of nodes requesting the tuple is not predetermined.

3.6.1 Piggybacking

To minimize the number of hops required to locate tuples the tuple tables are filled with extra tuple information. This is accomplished by piggybacking tuple information when sending messages. This redundant information is justified by the fact that bandwidth is increasing every day, while latency improvements are rare and bounded by the speed of light. Therefore, if redundant information can lower the number of hops that requests have to take, performance should be improved considerably.

The information that is piggybacked is small tuples and references to nodes holding larger tuples. Small tuples are sent fully, because they are comparable to references in size. The type of message that piggyback information is attached to dictates in some cases the contents of this information. For example when the token is passed on to a node that needs it for an observe operation, information about tuples with indexes from *first* and up follows the token. For other messages there is no way to connect helpful information in regards to the message

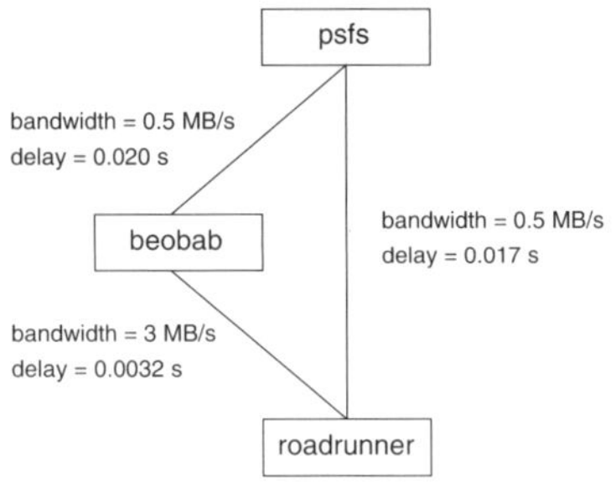

Figure 5: Simulation environment.

itself. Information about the last known tuple is always piggybacked, which helps to find other tuples as described in section 3.4.

It has not been investigated what effect, the number of small tuples and/or tuple references that are piggybacked with each message, has. More piggyback information should lower the number of hops a tuple request has to take. On the other hand, the added payload of more piggyback information would increase the load and take up storage resources.

4 Simulation Results

The performance of the design have been tested by simulating an environment of three clusters from the real world. The simulation has been implemented in the Ptolemy II [6] modeling system. The choice of simulation instead of a real implementation is based on two facts; First of all it has not been possible to get exclusive access to thousands of nodes. Secondly the heterogeneous nature of the environment makes it difficult to compare results.

4.1 The simulated environment

The clusters that the simulation has been based on are *roadrunner* and *beobab* which are located in two different cities in Denmark, together with *psfs* located in Norway.

Fig. 5 shows the three clusters. Also, the bandwidths and one-way null byte latencies (delay) between them are displayed. These numbers have been benchmarked in the real environment.

The real setup of the real clusters have not been simulated. Instead each simulated cluster consists of a number of nodes connected with one 100 Mbit/s switch that has a backplane with infinite bandwidth. The delay from node to node through this switch is 100 μs (benchmarked with a real 100 Mbit/s switch).

4.2 Test Programs and Initial Setups

The performance of GPS has been tested with a variety of communication patterns, instead of actual parallel applications. This is obviously because these applications require too many system resources to be run on a single workstation with Ptolemy. Therefore, a set of communication patterns, that are often used in parallel applications, have been chosen. These are as follows:

- **One-to-one**: This is simply communication between two nodes. This will show how fast a node can communicate with another node, which is not as trivial in DSM systems

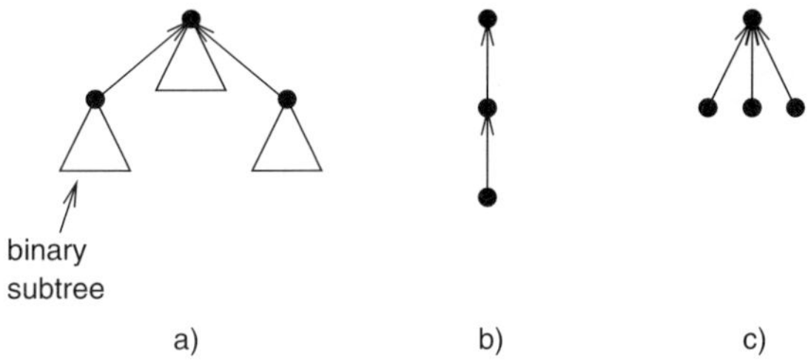

Figure 6: Initial tree configurations.

as it is in message passing systems where nodes explicitly define what node they want to communicate with.

- **One-to-many**: This resembles a broadcast from one node to all nodes.
- **Many-to-one**: All nodes sent data to one node.
- **Many-to-many**: All nodes sent data to all other nodes. This is the same as if all nodes broadcast a message.
- **Global reduction**: This is often used in parallel applications to make a global sum, minimum, maximum, etc. out of local values from each process. When the result is ready it has to be distributed to all involved nodes.

Patterns based on indexed and unindexed observes have been tested. All these patterns have been tested with different numbers of nodes and various configurations defined by these properties:

- Node configuration. Each program has been run with a various number of nodes to show how well GPS scales. One, two (*roadrunner* and *beobab*) and three cluster setups have been tested. The clusters has been given an equal amount of nodes.
- Initial token tree configuration. Fig. 6 shows the initial token tree configurations that have been tested. In the binary configuration each cluster has a binary subtree as in Fig. 6.a. In the linear configuration each node has only one child and there is only one edge between clusters (see Fig. 6.b). In the star configuration all nodes has the root as their parent (see Fig. 6.c).
- Tuple size. The programs has been tested with a tuple size of one byte and 10000 bytes.

4.3 Results

All programs with all the various configurations have been run. This amounts to hundreds of results, which would be to comprehensive to discus in this paper. The general picture is that GPS scales linearly in most runs.

The following will elaborate on some of the outstanding results that illustrates parts of the system that needs improvements.

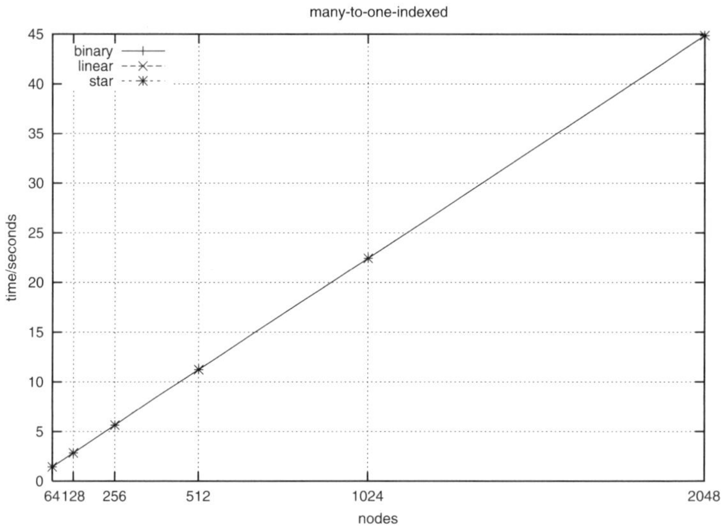

Figure 7: 3 clusters, tuple size=10000.

4.3.1 *Many-to-one Indexed*

In this program all nodes move a tuple, while one node in the *roadrunner* cluster makes n (the number of nodes) indexed observes. Fig. 7 shows the results for three clusters and a tuple size of 10000 bytes.

Though the results scale very well for all three tree configurations the performance is not ideal. The bandwidth between *roadrunner* and *psfs* sets a lower bound of what can be achieved. For $n = 2048$, $2048/3 = 683$ tuples must be transfered from *psfs* to *roadrunner*, which with a bandwidth of 0.5 MB/s takes 13.0 seconds. These poor results are caused by the fact that the observe operation is blocking. The extra time is spend locating tuples before they can be downloaded. This is all done sequentially because only one observe is executed at a time.

A solution to this problem would be to find many tuples in parallel. This could be done by introducing a prefetch operation that requests many tuples at a time.

4.3.2 *Many-to-many Indexed*

In this program all nodes move a tuple after which they make n indexed observes. Fig. 8 shows the results for three clusters and a tuple size of 10000 bytes.

The performance is very bad because many requests for a tuple arrive simultaneously at the node that has moved it. The node will send the tuple to all requesting nodes, instead of forwarding the requests to other nodes that might have or will get the tuple. This is because the design of GPS dictates that a node having a tuple locally must send it directly to a requesting node. Together with blocking observes this results in a very serial distribution of tuples.

Again a prefetch operation would improve considerably on these results, because tuple distribution trees could be build before tuples are moved. Also, a node that has a tuple transfer in progress could forward a consecutive request for the same tuple to other nodes that hold, or will hold the tuple in the future. This could be combined with wormhole routing to improve

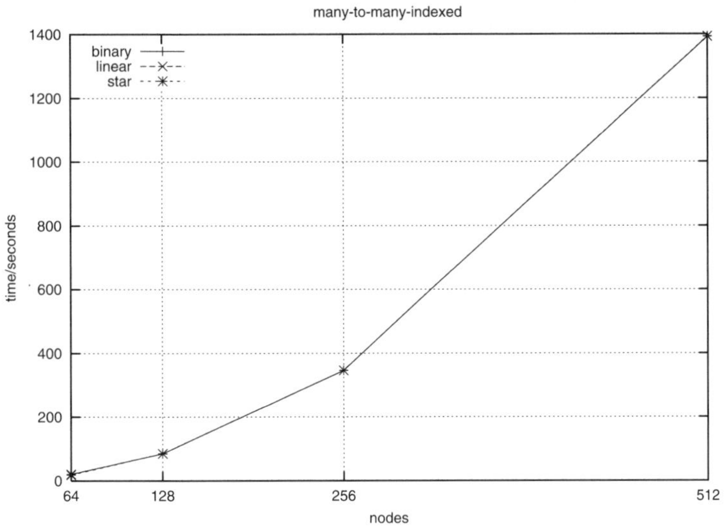

Figure 8: 3 clusters, tuple size=10000.

performance even further.

4.4 Global Reduction

To make a global reduction with GPS one node moves the first result. All other nodes makes an unindexed observe, reduces the received value with the local value and moves the result. All nodes observes the index $n - 1$ to get the result of the global reduction. Fig. 9 shows the results for three clusters and a tuple size of one byte. The results for one and two clusters look almost the same except for slightly better performance, as one would have expected.

Though a global reduction can be made to scale logarithmically, the GPS reduction will at best scale linearly. This is because the token has to visit all nodes. Having this in mind the linear and star configurations are performing ideally. The reduction is carried out locally in one cluster before it is continued in the next. For a tuple size of one byte there is a lower time limit of $2047 \times 100\mu s = 0.2047s$ ($n = 2048$), which is the minimum time for the token to travel to all nodes in a single cluster. This is close to the actual results. Also, all the results show that broadcasting the result does not add much to this time, thus it is done in parallel.

The binary configuration has a problem because the token grows very large as it travels from node to node. Each parent receives one observe request from each child which they queue locally because they have a pending request of their own. When the token arrives at a parent, one observe request is removed from the token, but two are added from the local queue. As a result the token costs more and more to forward until it arrives at the leaf nodes where no new requests are added.

This problem is not easily solved, but a solution could be to set an upper limit on how many requests the token can queue. When the queue is full, remaining requests should simply follow behind the token. The problem with this solution is that the set of nodes to choose from, when picking the next token destination, is smaller.

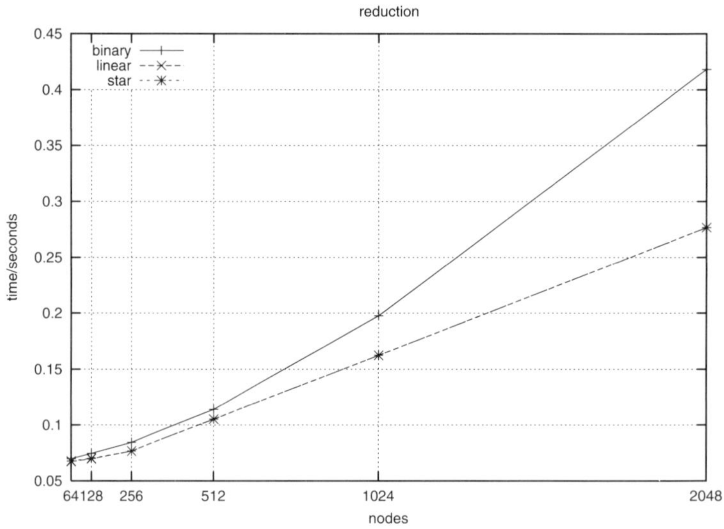

Figure 9: 3 clusters, tuple size=1.

5 Conclusion

This paper has described the design of a DSM system for global environments. The system is based on PastSet [1], a DSM system for cluster environments. The design has been tested in a simulated environment of multiple geographically distant clusters, with up to thousands of nodes.

The design has shown that there are two main problems that need to be addressed, when moving PastSet to this new environment; Some means of upholding consistency and an algorithm for locating near replicas.

The test results have shown that GPS scales linearly, with some exceptions. The problems of these exceptions have been identified and solutions to address them have been suggested.

References

[1] Brian Vinter. *PastSet: A Structured Distributed Shared Memory System.* PhD thesis, Department of Computer Science, Faculty of Science, University of Troms, Norway, 1999.

[2] James Griffioen, Rajendra Yavatkar, and Raphael Finkel. Unify: A scalable, loosely-coupled, distributed shared memory multicomputer.

[3] Philip Homburg. *The Architecture of a Worldwide Distributed System.* PhD thesis, 2001.

[4] Jarek Nieplocha, Robert J. Harrison, and Richard J. Littlefield. Global arrays: a portable programming model for distributed memory computers. In *Supercomputing*, pages 340–349, 1994.

[5] J. Nieplocha and R. J. Harrison. Shared memory NUMA programming on I-WAY. In *Proc. of the Fifth IEEE Int'l Symp. on High Performance Distributed Computing (HPDC-5)*, 1996.

[6] *Ptolemy*. http://ptolemy.eecs.berkeley.edu.

Communicating Process Architectures 2003
Jan F. Broenink and Gerald H. Hilderink (Eds.)
IOS Press, 2003

Automatic Conversion of CSP to CTJ, JCSP, and CCSP

V. RAJU
2861 Tall Oaks Ct. #10, Auburn Hills, MI 48326
varsha_raju@hotmail.com

L. RONG
ASpire Technologies LTD, Hi-Tech Industrial Park, Shenzhen, Guangdong Province, P.R.China
ronglm@21cn.com

G. S. STILES[1]
Computing Laboratory, University of Kent, Canterbury, CT2 7NF, England
dyke.stiles@ece.usu.edu

Abstract. We present tools that automatically convert a subset of machine-readable CSP script to executable Java or C code. CSP is used to design and verify the correctness of large and complex systems of processes that interact only via explicit synchronous messages. These systems can be implemented in Java using CTJ or JCSP, packages that add CSP-like features to Java, or in CCSP, a package that adds similar features to standard C. Implementation of CSP systems can be tedious and error-prone when large numbers of processes and communications are involved, and sorting out errors in channel naming or the ordering of messages can be very time-consuming. The tools we have developed minimize such problems by converting the verified CSP descriptions of communicating processes directly into Java or C code, thus guaranteeing that channels are correctly named and the communications occur in the proper order. This process can significantly cut development time.

1 Introduction

The development of large concurrent applications can be a complex and time-consuming task. The use of formal approaches, such as Hoare's Communicating Sequential Processes (CSP: [1, 2, 3]) simplifies the problem by making the required synchronizations among processes explicitly dependent upon communications. Systems specified in the CSP script can be checked for correctness and freedom from deadlock and livelock with the tools FDR and ProBE [4]. Once verified, the script can then be the basis for developing an actual implementation in languages that directly support the CSP style of concurrency through channels, such as occam [5, 6], and in other languages with the help of packages that add CSP features, such as CTJ [7, 8] and JCSP [9, 10] for Java, CCSP [11] for C, and C++CSP [21] for C++.

The implementation itself can be problematic, however. If there are a large number of processes, connected by many channels, and a reasonably complex pattern of

[1] Permanent address: Utah State University, 4120 Old Main Hill, Logan UT 84322-4120

communications, it may take a substantial amount of time to get the application running correctly. Students compare the complexity to that of wiring by hand a large discrete logic system: there are so many connections that it is nearly impossible to get it right the first time around. The tools developed in this project are designed to avoid these problems by automatically generating the appropriate channel-based Java or C code from the verified CSP script. We have found that this can substantially cut the implementation time.

With this tool we now have a reasonably robust procedure for designing concurrent applications. We begin by specifying the desired behaviour (particularly the communication and synchronization patterns) in CSP. Next we use ProBE and FDR to verify that the implementation satisfies the specification and is free from deadlock and livelock; this typically requires a few iterations, but the turn-around time is faster than when working with actual Java or C code. Once the implementation has been verified we use the conversion tool to generate the appropriate code; this will include the necessary declarations and channel operations for the required communications – correctly named and in the proper order. Depending on the actual problem, additional sequential code may need to be added to the resulting routines. Testing follows and – if the specification was correct – is usually brief.

We are not aware of previous efforts to convert CSP directly into executable code. Many other systems have been implemented, however; e.g., a group at Stony Brook [22] has been developing tools to convert a graphical version of the process algebra CCS into executable Java and Ada'95 code, and at Edinburgh [23] tools are available that convert a stochastic process algebra into Ada.

The following sections discuss first the means of carrying out the conversions from the CSP script to Java or C code and their limitations. This will be followed by several examples of both simple and more complex applications. The paper will conclude with a summary and a discussion of future work.

2 The Conversion

This project originated as a follow-on to a previous exercise [12] that developed routines using Mathematica [13] to automatically convert concurrent CSP structures into completely equivalent sequential versions. The ultimate goal at that time was to build a family of Mathematica-based tools for the manipulation and analysis of CSP scripts. We thus started this project with a Mathematica conversion of CSP to CTJ.

2.1 CSP to CTJ via Mathematica

Mathematica is a very elaborate package, containing a great variety of tools covering the range from numerical and symbolic mathematics through graphics to sophisticated manipulation of strings. The pattern matching and list processing features, when used within Mathematica's procedural programming system, provide a fairly quick path to build a translator. This was the approach used to convert CSP script to CTJ code.

Mathematica is not, however, readily available at all institutions and is fairly expensive at the single-copy level. We thus looked for a more commonly available approach during the second phase of this project.

2.2 CSP to JCSP and CCSP via C++

We wished to implement these translators in C or C++. The bulk of the translation work requires string processing; there is, fortunately, a Standard Template Library (STL) in C++

publicly available [14] that contains some very useful routines in its *String* and *Vector* classes.

In standard C, a string is simply an array of characters that always includes a null terminator. In the C++ String class, string objects associate an array of characters with methods useful for managing and manipulating it. A string object also contains certain housekeeping information about the size and storage location of the data. C++ strings do not include a null terminator. C++ strings greatly reduce the likelihood of common and destructive C programming errors, such as overwriting array bounds and trying to access arrays through uninitialized or incorrectly valued pointers.

The Vector class provides methods for dynamically allocating and reclaiming storage as required. These are useful when working with, e.g., a CSP process, where we must keep track of such details as its name, its channel names and directions, and any internal structures it may have, such as external choice and if-then-else clauses. Different processes will have different amounts of information. Managing the storage required explicitly would be tedious and error-prone. The STL Vector class, however, manages storage automatically.

2.3 Features Converted

The choice of CSP features to convert was based largely on the content of programs that had been developed in our classes over the past few years. These programs tended to model various communication problems. Table I shows the elements that have been translated in the various versions. Semaphores are not explicitly included in CSP, but have been added here since they can easily be modeled in CSP and can be implemented in JCSP and CCSP. The CSP notation in Table I and the rest of the paper is in the machine-readable form used by ProBE and FDR [4].

Table I. CSP Features Translated

CSP	CTJ	JCSP	CCSP
comments: `--`	✓	✓	✓
comments: `{- ... -}`		✓	✓
declarations	✓	✓	✓
prefix: `->`	✓	✓	✓
integer data	✓	✓	✓
`chan ? data, chan ! data`	✓	✓	✓
`chan ? d1 . d2 . ..., chan ! d1 . d2 . ...`	✓	✓	✓
`if ... then ... else ...`	✓	✓	✓
external choice (alternative): `[]`	✓	✓	✓
synchronous (sharing) parallel: `[| {| ... |} |]`	✓	✓	✓
processes	✓	✓	✓
recursive processes	✓	✓	✓
semaphores		✓	✓

This set seems sufficient to put together a good variety of examples. The most glaring shortcoming, from an engineering standpoint, is the lack of floating-point data – but this will soon be remedied. Other features that would be useful are sequential composition, interrupts, timeouts, boolean guards, renaming, and replicated constructs.

2.4 Restrictions

There are two restrictions on the CSP syntax that have been included to simplify the conversion programs; all are compatible with ProBE and FDR. We require first that all CSP statements terminate with the semicolon, and we deal only with integer channels and variables.

3 Examples

We begin by looking at the translation of some of the basic components. This will be followed by increasingly complex examples of full programs.

3.1 Simple Fragments and Programs

We look first at a recursive process `UpHandler` that repeatedly reads an integer into the variable `x` and passes it on. The CSP description is:

```
-- This is a test file for a simple process with recursion;
channel input, output;
UpHandler = input?x -> output!x -> UpHandler;
```

The resulting CTJ code includes the class definition, declarations of variables and channels, the constructor, and the run method:

```
// This is a test file for a simple process with recursion;
import csp.lang.*;
import csp.lang.Integer;
import csp.lang.Process;
import csp.io.*;

public class UpHandler implements Process
{
   private integer x = new Integer();
   private boolean Running;
   private int Identity;
   private Channel_of_Integer input;
   private Channel_of_Integer output;

   public UpHandler ( int Identity,
                      private Channel_of_Integer input,
                      private Channel_of_Integer output)
   {
      this.Identity = Identity;
      this.input = input;
      this.output = output;
   }

   public void run()
   {
      Running = true;
      while(Running)
      {
         input.read(x);
         output.write(x);
      }
   }
}
```

The conversion tool detects the recursion and sets up a loop controlled by a variable `Running` that is set to `true` (and could be modified at runtime by a specific value of the channel input variable `x` if desired). We also automatically add a parameter `Identity` that proves useful when multiple instances of one process are used.

For a process to run an object has to be created and the `run()` method called for this object. This object is created in the main file in the `main()` method of the system class. Calling its `run()` method activates the object. In CTJ the translator generates:

```
...
import csp.lang.System;
public class system
{
   public static void main(String[] args)
   {
      // This is a test input file for a simple process
      private int Identity;
      final Channel_of_Integer input = new Channel_of_Integer();
      final Channel_of_Integer output = new Channel_of_Integer();

      ...

      UpHandler UpHandler =
         new UpHandler(Identity, input, output);
      UpHandler.run();
      java.lang.System.out.flush();
   }
}
```

The next example is the *Stop and Wait* flow-control protocol [3]. The processes *Send* and *Recv* cooperate to manage the flow of data from a channel *in* to a channel *out*. The *Send* process repeatedly accepts an input over the channel *in*, passes it along to the *Recv* process over the channel *mid*, then waits until it receives an *ack* signal from *Recv*. *Recv* does not send the *ack* until it has successfully passed the data to the *out* channel.

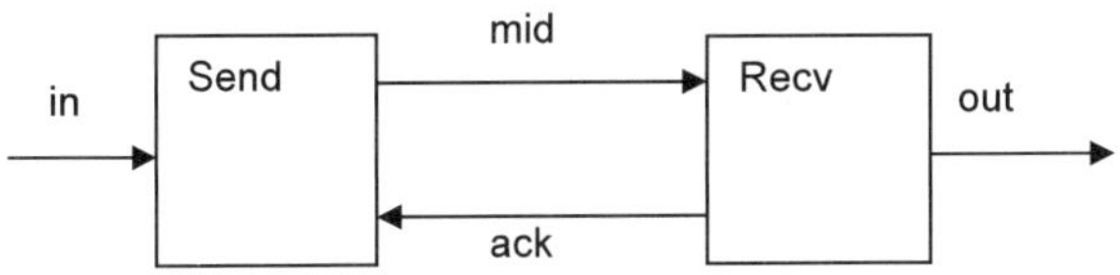

Figure 1. The Stop and Wait protocol.

The CSP description is straightforward:

```
-- Stop-and-Wait Protocol
channel in, mid, ack, out : T ;
Send = in ? x -> mid ! x -> ack ? x -> Send ;
Recv = mid ? y -> out ! y -> ack ! y -> Recv ;
System = Send [| {| mid, ack |} |] Recv ;
```

The complete code, now shown in JCSP:

```
import jcsp.lang.*;
import java.lang.*;
/*-- Stop-and-Wait Protocol */
class Send implements CSProcess
```

```
{
  One2OneChannelInt in;
  One2OneChannelInt ack;
  One2OneChannelInt mid;
  int x;
  boolean Running;

  public Send(One2OneChannelInt in, One2OneChannelInt ack,
              One2OneChannelInt mid)
  {
    this.in = in;
    this.ack = ack;
    this.mid = mid;
  }

  public void run()
  {
    Running = true;
    while (Running)
    {
      x = in.read();
      mid.write(x);
      x = ack.read();
    }
  }
}

class Recv implements CSProcess
{
  One2OneChannelInt mid;
  One2OneChannelInt out;
  One2OneChannelInt ack;
  int y;
  boolean Running;

  public Recv(One2OneChannelInt mid, One2OneChannelInt out,
              One2OneChannelInt ack)
  {
    this.mid = mid;
    this.out = out;
    this.ack = ack;
  }

  public void run()
  {
    Running = true;
    while (Running)
    {
      y = mid.read();
      out.write(y);
      ack.write(y);
    }
  }
}
```

```
public class Target
{
  public static void main (String[] args)
  {
    One2OneChannelInt  in = new One2OneChannelInt();
    One2OneChannelInt  mid = new One2OneChannelInt();
    One2OneChannelInt  ack = new One2OneChannelInt();
    One2OneChannelInt  out = new One2OneChannelInt();
    Send Send_init = new Send(in,ack,mid);
    Recv Recv_init = new Recv(mid,out,ack);
    Parallel System_init =
    new Parallel( new CSProcess[] { Send_init, Recv_init, });
    System_init.run();
    java.lang.System.out.flush();
  }
}
```

The conversion creates the *main* process (which declares the channels) and the *Send* and *Recv* processes, combines the last two processes in a *Parallel* construct, and starts things up. Demonstration of this program does require adding one process to supply data to the *in* channel and one to consume the data produced by the channel *out*.

The variable and channel names are fairly distinct in this example and the JCSP would be fairly easy to type in by hand. Use of the automatic translation from the verified CSP will, however, avoid any typographical errors.

A striking feature of this conversion is the explosion in the size of the code. The CSP source of four lines grows to 46 lines of Java (ignoring comments, blank lines, and lines with braces only). This ratio of about 10 to 1 results from the normal Java housekeeping, the fact that such things as channels are effectively defined in several places, and the simplicity of the individual processes: if there were more internal operations on the data items, e.g., the ratio would decrease.

We look next at the CCSP version of *Stop and Wait*:

```
#include <stdio.h>
#include "/usr/include/ccsp/csplib.h"
/*-- Stop-and-Wait Protocol */
void Send (Process *p , Channel *in, Channel *ack,
           Channel* mid)
{
  int x ;
  bool Running;
  Running = true;
  while (Running)
  {
    x = ChanInInt(in);
    ChanOutInt(mid, x);
    x =ChanInInt(ack);
  }
}
void Recv (Process *p , Channel *mid, Channel *out,
           Channel * ack)
{
  int y ;
  bool Running;
  Running = true;
```

```
    while (Running)
    {
      y =ChanInInt(mid);
      ChanOutInt(out, y);
      ChanOutInt(ack, y);
    }
  }

  void RunMain(Process* p)
  {
    Channel*  in  = ChanAlloc();
    Channel*  mid = ChanAlloc();
    Channel*  ack = ChanAlloc();
    Channel*  out = ChanAlloc();

    Process* Send_p = ProcAlloc(Send, 65536, 3, in, ack, mid);
    Process* Recv_p = ProcAlloc(Recv, 65536, 3, mid, out, ack);
    ProcPar(Send_p, Recv_p,  NULL);
  }
```

The C version is noticeably shorter – 26 rather than 46 lines – due primarily to the fact that the channels are not mentioned so often. Otherwise the C and the Java versions are equally easy to follow.

The next example illustrates the use of the CSP external choice construct to implement a multiplexer (`Mux_up`) in a system that routes packets among a number of processes. The packets contain message type, source, destination, and data components. The CSP:

```
Mux_up =
      upsendout0?dest.data -> uptodown!1.0.dest.data -> Mux_up
    []
      upsendout1?dest.data -> uptodown!1.1.dest.data -> Mux_up
    []
      uprecack0?ack -> uptodown!0.0.0.ack -> Mux_up
    []
      uprecack1?ack -> uptodown!0.0.0.ack -> Mux_up
```

Each message has four components (all integers). We could define a new JCSP channel type to handle this, but here we simply send each component separately. We use the JCSP `Alternative` construct, followed by a `switch`, to implement the external choice.

```
class Mux_up implements CSProcess
{
  One2OneChannelInt upsendout0;
  One2OneChannelInt upsendout1;
  One2OneChannelInt uprecack0;
  One2OneChannelInt uprecack1;
  One2OneChannelInt uptodown;
  int data;
  int ack;
  int dest;
  int value1 = 1;
  int value0 = 0;
  boolean Running;

  public Mux_up(One2OneChannelInt upsendout0,
                One2OneChannelInt upsendout1,
                One2OneChannelInt uprecack0,
                One2OneChannelInt uprecack1,
                One2OneChannelInt uptodown)
```

```
  {
    this.upsendout0 = upsendout0;
    this.upsendout1 = upsendout1;
    this.uprecack0 = uprecack0;
    this.uprecack1 = uprecack1;
    this.uptodown = uptodown;
  }

  public void run()
  {
    int index;
    AltingChannelInputInt[] in = {upsendout0, upsendout1,
                                  uprecack0, uprecack1};
    Alternative alt = new Alternative (in);
    Running = true;
    while (Running)
    {
      index = alt.fairSelect ();
      switch(index)
      {
        case 0:
          dest = upsendout0.read();
          data = upsendout0.read();
          uptodown.write(value1);
          uptodown.write(value0);
          uptodown.write(dest);
          uptodown.write(data);
        break;
        case 1:
          dest = upsendout1.read();
          data = upsendout1.read();
          uptodown.write(value1);
          uptodown.write(value1);
          uptodown.write(dest);
          uptodown.write(data);
        break;
        case 2:
          ack = uprecack0.read();
          uptodown.write(value0);
          uptodown.write(value0);
          uptodown.write(value0);
          uptodown.write(ack);
        break;
        case 3:
          ack = uprecack1.read();
          uptodown.write(value0);
          uptodown.write(value0);
          uptodown.write(value0);
          uptodown.write(ack);
        break;
      }
    }
  }
}
```

Once again we see the size of the code explode – but the reason here is the number of data items passed in each channel communication. Creation of a new channel type would simplify this example considerably.

3.2 Larger Programs

We look next at several full programs. The first, `Commstime`, is a timing benchmark in the JCSP library [10]. The system (Figure 2) consists of four processes: `PrefixInt`, `Delta2Int`, `Consume`, and `SuccessorInt`. `PrefixInt(0)` initially sends out a 0 over channel `a`, then forwards whatever it receives over channel `c`. `Delta2Int` copies its input value to `Consume` and `SuccessorInt`. `SuccessorInt` receives an integer from `Delta2Int`, adds 1 to it, and passes the result to `PrefixInt`. `Consume` collects and reports statistics. Because there is essentially no work in this system other than the message passing, we can use it to measure the message passing and context switch times.

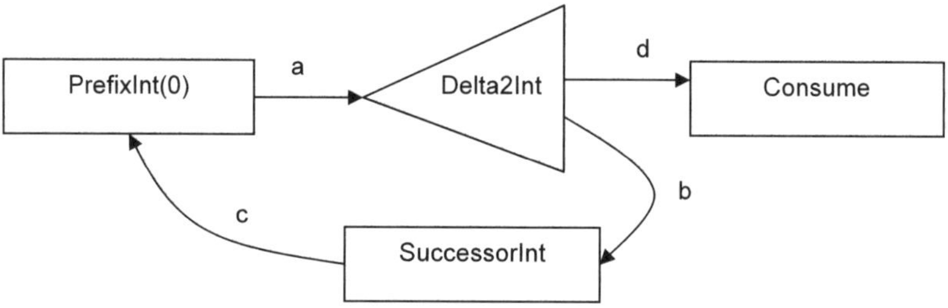

Figure 2. The `Commstime` benchmark.

The CSP description of `Commstime` - with the process names slightly modified:

```
channel a, b, c, d : Integer ;
MyInit = a ! 0 ;
MyPrefixInt = c ? x -> a ! x -> MyPrefixInt ;
MyDelta2Int = a ? x -> d ! x -> b ! x -> MyDelta2Int ;
MyConsume = d ? x -> MyConsume ;
MySuccessorInt = b ? x -> c ! x -> MySuccessorInt ;
System =            (MySuccessorInt
               [| {| |} |]
                        (MyConsume
                    [| {| |} |]
                                (MyDelta2Int
                        [| {| |} |]
                                        (MyPrefixInt
                            [| {| |} |]
                                        MyInit))));
```

The `System` process has been formatted to emphasize its structure. Its component processes are arranged in a horizontal tree, with the topmost operator farthest to the left. The components of this operator – either another simple or composite process – are both shifted right by one level; and so on. Thus the top-level operator is parallel sharing – `[|{||}|]` – and its components are `MySuccessorInt` and another parallel sharing structure. Note that channels are not explicitly required in the sharing operator by the JCSP tool since they can be determined from the definitions of the component processes.

`PrefixInt(0)` is implemented as a combination of `MyInit`, which delivers the intital `0`, and `MyPrefixInt`, which simply forwards its input to its output. The JCSP implementation follows.

```
class MyInit implements CSProcess
{
  One2OneChannelInt a ;
  int value0 = 0;
  boolean Running;

  public MyInit( One2OneChannelInt a)
  {
    this.a = a;
  }

  public void run()
  {
    a.write(value0);
  }
}


class MyPrefixInt implements CSProcess
{
  One2OneChannelInt c ;
  One2OneChannelInt a ;
  int x ;
  boolean Running;

  public MyPrefixInt( One2OneChannelInt c, One2OneChannelInt a)
  {
    this.c = c;
    this.a = a;
  }

  public void run()
  {
    Running = true;
    while (Running)
    {
      x = c.read();
      a.write(x);
    }
  }
}

class MyDelta2Int implements CSProcess
{
  One2OneChannelInt a ;
  One2OneChannelInt d ;
  One2OneChannelInt b ;
  int x ;
  boolean Running;

  public MyDelta2Int( One2OneChannelInt a, One2OneChannelInt d,
                      One2OneChannelInt b)
  {
    this.a = a;
    this.d = d;
    this.b = b;
  }
```

```
  public void run()
  {
    Running = true;
    while (Running)
    {
      x = a.read();
      d.write(x);
      b.write(x);
    }
  }
}

class MyConsume implements CSProcess
{
  One2OneChannelInt d ;
  int x ;
  boolean Running;

  public MyConsume( One2OneChannelInt d)
  {
    this.d = d;
  }
  public void run()
  {
    Running = true;
    while (Running)
    {
      x = d.read();
    }
  }
}

class MySuccessorInt implements CSProcess
{
  One2OneChannelInt b ;
  One2OneChannelInt c ;
  int x ;
  boolean Running;

  public MySuccessorInt( One2OneChannelInt b, One2OneChannelInt c)
  {
    this.b = b;
    this.c = c;
  }

  public void run()
  {
    Running = true;
    while (Running)
    {
      x = b.read();
      c.write(x);
    }
  }
}
```

```
public class Target
{
  public static void main (String[] args)
  {
    /*Channel declaration*/
    One2OneChannelInt  a = new One2OneChannelInt();
    One2OneChannelInt  b = new One2OneChannelInt();
    One2OneChannelInt  c = new One2OneChannelInt();
    One2OneChannelInt  d = new One2OneChannelInt();
    MyInit MyInit_init = new MyInit(a);
    MyPrefixInt MyPrefixInt_init = new MyPrefixInt(c,a);
    MyDelta2Int MyDelta2Int_init = new MyDelta2Int(a,d,b);
    MyConsume MyConsume_init = new MyConsume(d);
    MySuccessorInt MySuccessorInt_init = new MySuccessorInt(b,c);
    Parallel System_init = new Parallel( new CSProcess[]
      {
        MyInit_init,
        MyPrefixInt_init,
        MyDelta2Int_init,
        MyConsume_init,
        MySuccessorInt_init,
      });
    System_init.run();
    java.lang.System.out.flush();
  }
}
```

The JCSP code in this case has about 80 significant lines, about eight times the number in the CSP description.

The last example models a virtual-channel system that connects two processes on an `Upstream` node to two processes on a `Downstream` node. Each process can send messages to any process on the other node; flow control is enforced on the sources by requiring the reception of an acknowledgement to a previous message before a following message is accepted. The top-level schematic is shown in Figure 3. All messages and acknowledgements must be multiplexed over a single channel in each direction. An expanded view of the `Upstream` node is shown in Figure 4.

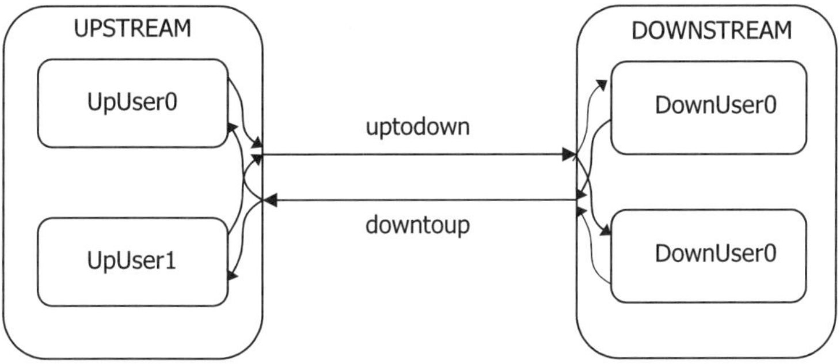

Figure 3. Top-level schematic of the virtual channel system.

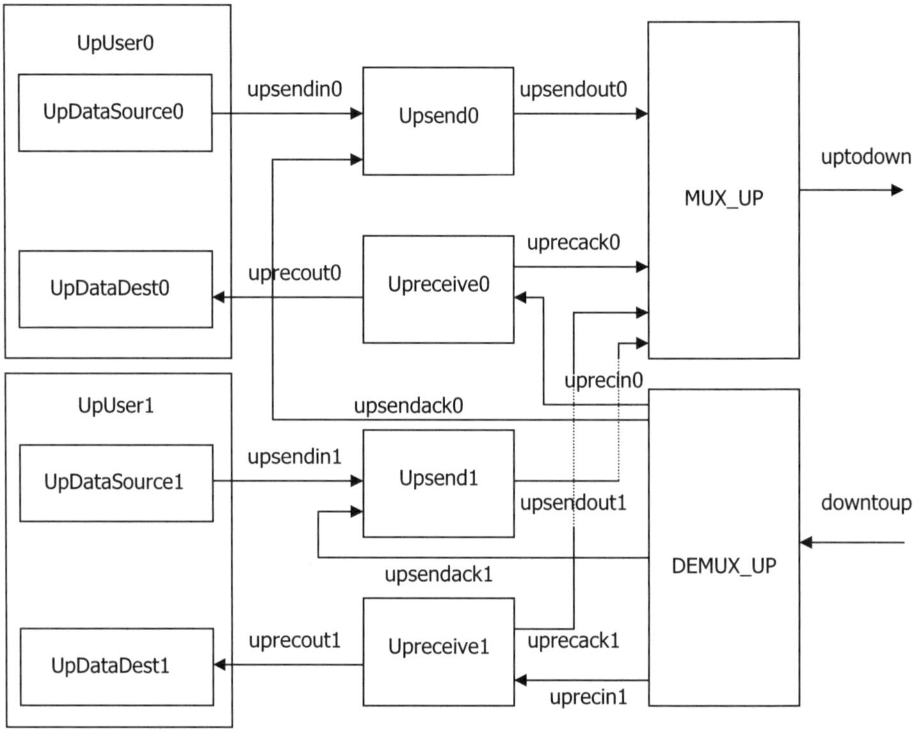

Figure 4. Detail of the Upstream node of Figure 3. Boxes represent processes, and arrows represent channels.

The CSP description of the virtual channel system:

```
-- Virtual channel system ;

channel upsendin0, upsendin1,upsendout0,upsendout1, downsendin0,
        downsendin1, downsendout0, downsendout1 : SourceData ;
channel upsendack0, upsendack1, downsendack0, downsendack1,
        uprecack0, uprecack1, downrecack0, downrecack1 : Ack ;
channel uprecin0,uprecin1,uprecout0,uprecout1,downrecin0,
        downrecin1,downrecout0,downrecout1 : DestData ;
channel downtoup, uptodown : InterData ;

Upsend0 =   upsendin0?dest.data -> upsendout0!dest.data ->
            upsendack0?ack -> upsend0;
Upsend1 =   upsendin1?dest.data -> upsendout1!dest.data ->
            upsendack1?ack -> upsend1;

Upreceive0 = uprecin0?source.data -> uprecout0!source.data ->
             ( if (source ==0)
               then uprecack0!0 -> upreceive0
               else uprecack1!1 -> upreceive0 );
```

```
Upreceive1 = uprecin1?source.data -> uprecout1!source.data ->
            ( if (source ==0)
              then uprecack0!0 -> upreceive1
              else uprecack1!1 -> upreceive1 );
Downsend0 = downsendin0?dest.data -> downsendout0!dest.data ->
            downsendack0?ack -> downsend0;
Downsend1 = downsendin1?dest.data -> downsendout1!dest.data ->
            downsendack1?ack -> downsend1;

Downreceive0 = downrecin0?source.data -> downrecout0!source.data ->
            ( if (source ==0)
              then downrecack0!0 -> downreceive0
              else downrecack1!1 -> downreceive0 );
Downreceive1 = downrecin1?source.data -> downrecout1!source.data ->
            ( if (source ==0)
              then downrecack0!0 -> downreceive1
              else downrecack1!1 -> downreceive1 );
-- if the data is the data we want to send from upstream data
-- source to downstream data destination, we insert a 1 in front of
-- it(also the number indicated data source). If it is ack signal
-- for downstream data source, we insert 0.0.0 in front of it;

Mux_up = upsendout0?dest.data -> uptodown! 1.0.dest.data -> Mux_up
         []
         upsendout1?dest.data -> uptodown! 1.1.dest.data -> Mux_up
         []
         uprecack0?ack -> uptodown! 0.0.0.ack -> Mux_up
         []
         uprecack1?ack -> uptodown! 0.0.0.ack -> Mux_up;
-- if testbit is 0 then the data in testbit.source.dest.data is
-- ack signal from upstream to downstream otherwise it is data send
-- from upstream to downstream;

Demux_down = uptodown ? testbit.source.dest.data ->
            (if(testbit == 0)
             then (if (data == 0)
                   then downsendack0!data -> Demux_down
                   else downsendack1!data -> Demux_down )
             else (if ( dest ==0 )
                   then downrecin0! source.data -> Demux_down
                   else downrecin1! source.data -> Demux_down ) );
-- if the data is the data we want to send from downstream data
-- source to upstream data destination, we insert a 1 in front of
-- it(also the number indicated data source). If it is ack signal
-- for upstream data source, we insert 0.0.0 in front of it;

Mux_down =  downsendout0?dest.data->downtoup! 1.0.dest.data ->
            Mux_down
           []
            downsendout1?dest.data->downtoup! 1.1.dest.data ->
            Mux_down
           []
            downrecack0?ack -> downtoup! 0.0.0.ack -> Mux_down
           []
            downrecack1?ack -> downtoup! 0.0.0.ack -> Mux_down;

-- if testbit is 0 then the data in testbit.source.dest.data is
-- ack signal from upstream to downstream otherwise it is data send
-- from upstream to downstream;
```

```
Demux_up = downtoup? testbit.source.dest.data ->
            ( if (testbit == 0)
             then (if (data == 0)
                   then upsendack0!data -> Demux_up
                   else upsendack1!data Demux_up )
             else (if ( dest ==0 )
                   then uprecin0!source.data Demux_up
                   else uprecin1!source.data Demux_up ) );
UpStream =         (Demux_up
           [| {| |} |]
                         (Mux_up
                 [| {| |} |]
                               (Upreceive1
                       [| {| |} |]
                                     (Upreceive0
                             [| {| |} |]
                                           (Upsend1
                                   [| {| |} |]
                                          Upsend0))))) ;

DownStream =       (Demux_down
           [| {| |} |]
                         (Mux_down
                 [| {| |} |]
                               (Downreceive1
                       [| {| |} |]
                                     (Downreceive0
                             [| {| |} |]
                                           (Downsend1
                                   [| {| |} |]
                                          Downsend0)))));
System =           UpStream
           [| {| |} |]
                   DownStream ;
```

The source files, with source and sink processes added, can be found on the paper web page [24]. Both implementations run correctly. The CCSP code runs about 400 lines, and the JCSP code about 560 lines; the CSP specification is about 60 lines. On a similar problem, with traffic moving in only one direction, a CSP specification of about 30 lines resulted in a CTJ implementation of about 160 lines.

3.3 Discussion

The first impression we have from this exercise is that the conversion from CSP to Java or C results in a 5-fold or so expansion of the code. Given that the CSP description of a reasonable application may run to hundreds or thousands of lines, the utility of automatic conversion is clear.

The last example (the virtual channel system) exhibits some of the features of larger systems that make hand-coding of the target code error-prone: there are large numbers of channels and processes that have nearly identical names. The names are explicitly chosen to be descriptive of the location of the channels or processes within the topology of the system, so we are reluctant to do otherwise. However, when it comes to typing in the code, it is rare that errors do not occur. The worst errors are those where an existing but incorrect channel or process is referred to; the compilers will not catch this mistake.

Over the past few years, final class projects have typically required a week or two to get the CSP script correct, and an equal amount of time to get the Java running correctly. The bulk of the Java effort was aimed at getting the wiring between the processes and the ordering of events correct. With the conversion tool described in this paper, the JCSP version of the virtual channel system above (last year's final project) was up and running within less than an hour of the verification of the CSP. The CCSP version came right behind.

At present the translators do not explicitly handle non-CSP code – such as common numerical or textual operations and assignments. These can be included within CSP comments, which will be converted into appropriate comments in the target code. These comments can then in turn be immediately converted in executable code.

Due to the nature of the conversion process, the structure of the Java or C code will closely follow that of the CSP script. Thus some time invested in a clean design of the CSP pays off directly in a well-laid out version of the much larger target code.

4 Summary and Further Work

4.1 Summary

We have developed simple tools that allow us to create applications automatically from verified descriptions in CSP. At this point the tools convert a small but useful subset of the standard CSP script into executable CTJ or JCSP versions of Java or CCSP versions of C code. The CSP features converted seem sufficient to implement applications based primarily on communication. The primary advantage of the tools is that they correctly implement the large amount of process and channel boiler-plate that is tedious and difficult to get correct by hand. Since applications may be five or more times larger than the CSP specifications, this is a significant gain that can greatly reduce development time. This project was relatively straightforward to carry out, and should be easy to expand. We expect to continue this work.

First, however, we plan to review our approach to the development of the conversion tools. Our initial intention was to extend a set of tools [12] for the manipulation of CSP scripts that had been implemented in Mathematica [13]; this would allow us to take advantage of Mathematica's large libraries of procedures for manipulating sets of strings and symbolic computation – with the hope that we would eventually have a single package to develop the CSP and then convert it directly to the target code once it had been verified. Mathematica is not, however, universally available, and during the second phase of this project (JCSP and CCSP) we developed an additional conversion tool based on C. In the next phase we plan to base the conversion tools on *lex* and *yacc* [e.g., 15] to gain added flexibility.

We have not yet run any performance tests on the translations. The examples completed thus far are composed nearly entirely of communications between processes, and the conversion tools order the communications just as they were ordered in the CSP – by design, since that is the part we really wish to have correct. Optimisation of communications would certainly be a worthwhile feature; this would probably require some interesting effort on extracting the communication patterns.

4.2 Future Work

There are a number of extensions and additional features planned. On the CSP side, we will extend coverage to include replication operators (e.g., on external choice), boolean guards and priorities on external choice, and sequential composition. The conversion of complex channel types (those formed from dotted combinations of more primitive CSP types) will be changed to generate appropriate objects in the target languages and channels to carry them. We will also allow specification in CSP (by annotated comments) of standard target channel types such as `int` and `float`.

A number of changes and additions on the target code side will be made. We should be able to include arbitrary fragments of target code in the CSP script by appropriately annotating standard CSP comments. This will give us the ability to manipulate data read over a channel before passing it along another channel; presently this would have to be done by adding code manually to the target after the conversion. Real-time clocks will be included, which will allow the implementation of time-out operations on channels. (Note that CSP originally included untimed timeouts, so we are not certain how this will be handled in the CSP script. Schneider's Timed CSP [3] does allow timed timeouts – as well as other timed features – so we may wish to extend our CSP coverage to Timed CSP.)

The recent addition of networking capabilities via JCSP.net [18] opens up a number of opportunities for distributed and Internet programming. It should be straightforward to trigger the generation of the JCSP.net code by including directives in CSP comments or by appropriately naming the channels themselves. This capability can also be applied to the CCSP version via CCSP.net [11]. This will allow us to specify in CSP applications that are to be distributed across networks, verify the correctness of the communication patterns, and then generate immediately the application code.

The same approach can be applied to embedded multiprocessor systems, such as might be used on a mobile robot with a processor driving each wheel. Once we know how communications are handled in the target code, the conversion tool can be modified accordingly. Systems could thus be quickly built on JStamp modules [19], which run Java byte code directly in hardware, or the Intel PXA250 ARM embedded processor [20] (which includes hardware support for wireless communication) in Java or C.

We are presently developing a new version of the tool that will convert CSP specifications into Handel-C [16]. This will give us a direct path from formal, verified specifications in CSP to implementation in FPGAs.

The three present versions convert directly into CTJ, JCSP, or CCSP. We hope to combine these into a single tool. This may be done by identifying a single intermediate language that can easily be converted into C and Java; this would also give us a smooth path to adding additional target languages. It may turn out, however, that the three conversions are so similar that an intermediate stage would add unneeded complexity.

We will continue to test these packages on increasingly complex applications. Suggestions are welcome.

Finally we note that this tool should work nicely in combination with Hilderink's [17] package that converts graphical diagrams into CSP. This will give us an automatic path from a graphical description of a CSP system directly to runnable code.

The translator programs and examples can be found on the paper web page [24].

References

[1] C. A. R. Hoare, Communicating sequential processes, *CACM*, 21(8), August 1978, pp. 666-677.

[2] A. W. Roscoe, *The Theory and Practice of Concurrency,* Prentice Hall, London, 1998.

[3] Steve Schneider, Concurrent and Real-Time Systems: The CSP Approach, Wiley, Chichester, 1999.

[4] ______, Formal Systems (Europe) Ltd.: http://www.fsel.com/software.html

[5] J. Galletly, *occam 2 – including occam 2.1*, UCL Press, London, 1996.

[6] ______, Kent Retargetable Occam Compiler (KRoC): http://wotug.ukc.ac.uk/kroc/

[7] Gerald Hilderink, Jan Broenink, Wiek Vervoort, and Andre Bakkers, Communicating java threads, in: *Proceedings of WoTUG 20: Parallel Programming and Java*, ed. A. W. P. Bakkers, IOS Press, Amsterdam, 1997, pp., 48-76.

[8] ______, CSP for Java: http://www.rt.el.utwente.nl/javapp/

[9] P. H. Welch, Process oriented design for Java: concurrency for all, in: *Proceedings of the International Conference on Parallel and Distributed Processing Techniques and Applications (PDPTA'2000)*, ed. H. R. Arabnia, CSREA Press, Athens GA, 2000, pp. 51-57.

[10] ______, Communicating Sequential Processes for Java^{TM} (JCSP): http://www.cs.ukc.ac.uk/projects/ofa/jcsp/

[11] ______, Quickstone Technologies Ltd.: http://www.quickstone.com /xcsp/ccspnetworkedition/

[12] W. Zhao, and G. S. Stiles, The automated serialization of concurrent CSP scripts using Mathematica, *Communicating Process Architectures 2000*, ed. P. H. Welch and A. W. P. Bakkers, IOS Press, Amsterdam, Sept. 2000, pp. 15-32.

[13] Mathematica Book Online: http://documents.wolfram.com.

[14] B. Eckel, *Think in C++ 2nd Edition, Volume2*, Upper Saddle River NJ, Prentice Hall, 2000.

[15] John R. Levine, Tony Mason, and Doug Brown, *lexx and yacc*, O'Reilly & Associates, Inc., Sebastopol CA, 1992.

[16] ______, Handel-C: http://www.celoxica.com/tech/handel-c/default.asp

[17] Gerald Hilderink, A graphical modeling language for specifying concurrency based on CSP, *Communicating Process Architectures 2002*, ed. James Pascoe, Roger Loader, and Vaidy Sunderam, IOS Press, Amsterdam, Sept. 2002, pp. 255-284.

[18] ______, Quickstone Technologies Ltd.: http://www.quickstone/xcsp/jcspnetworkedition/.

[19] ______, Systronix, http://jstamp.systronix.com/about.htm.

[20] ______, Intel, http://www.intel.com/design/pca/prodbref/298620.htm.

[21] N. C. C. Brown, An Introduction to the C++ CSP Library, *Communicating Process Architectures 2003*, ed. Jan Broenink, IOS Press, Amsterdam, Sept. 2003, pp. xx-yy.

[22] ______, http://www.cs.sunysb.edu/~concurr/features.html

[23] ______, http://www.dcs.ed.ac.uk/pepa/tools.html

[24] ______, http://www.ece.usu.edu/research/rtpc/projects/JavaCSP/#CSP_to_Java

Communicating Process Architectures 2003
Jan F. Broenink and Gerald H. Hilderink (Eds.)
IOS Press, 2003

A Single Chip Solution for Distributed Processing Systems

B.C. O'NEILL, P.W. MOORE and S. CLARK

School of Engineering, The Nottingham Trent University,
Burton Street, Nottingham, NG1 4BU, UK
Email : Brian.Oneill@ntu.ac.uk

Abstract. This paper describes a processor and an inter-processor communications interface integrated on a single chip for use in a distributed processing system. The system is based on work of the electronic systems design and parallel processing group at the Nottingham Trent University. The four main elements of the chip design are processor, memory, communication interface and packet routing switch all integrated unto one chip. This design is achieved by the use of the ALTERA ARM based Excalibur system on a programmable chip (SOPC) containing an embedded processor and programmable logic. The paper describes the communication features and implementation carried out by the research group to achieve this single chip processor.

1 Introduction

The first major processor that was designed, primarily for a multi-processing system, was the transputer. From the hardware perspective one major advantage was that systems were relatively easy to build. The inter-connection between processors was via a pair of serial links and this interface was an integral part of the processor chip. The design was very successful and in the initial stages it was used in a wide range of applications from large number crunching tasks to embedded systems. However, with time other competing processors diminished the market for the transputer and by the time the second generation of transputer was introduced its market share was not sufficient to be economically viable. There are many viewpoints on reasons for the transputer demise. One reason is the very high cost of processor development, which in the author's opinion is the most significant. Continuous development can only be sustained from the sale of processors for single use applications. The multi-processor market is not sufficient to recover the cost of development. As the technology of fabrication progresses towards higher densities, the initial cost of development are increasing and the unit costs of production are reducing. The consequence of this trend is that there are less processor types under development today than any period in the previous twenty years.

Currently many multi-processing systems are built using standard PCs linked by standard network cards. There are many exceptions [1,2,3,4] to this but the difficulty for designers of multi-processor systems is the high unit cost and lifetime of systems. This paper describes the design and construction of a multi-processor system using 'System On a Programmable Chip' (SOPC) technology which in part overcomes some of the cost difficulties of developing such systems. The SOPC has an embedded processor and programmable logic, which is used in this project to add specialist communication functions. The user of an SOPC device benefits from the efficiency of ASIC design for core functions, in this case the processor, timers and others, and the flexibility of using PLD technology to provide specialist functions. The same SOPC device can be used for a wide range of applications thus obtaining the economy of scale, which any one of these alone cannot provide. To offer the same functionality (processor and specialist logic) on a PLD device would require the use of a very large and costly device and also access to processor IP.

This paper will outline the design of a distributed parallel processor network, which has at its core a single chip processor node. It will focus on the inter-processor communication features designed by the electronic systems design and parallel processing group at the Nottingham Trent University.

2 Background

The research work started by concentrating on the inefficiency of the serial communication network which has point to point terminations. Figure 1 shows a typical network under consideration, which is configured, from a single host processor node. The processor nodes can have a minimum set of common features that are relatively easy to design and implement. Any additional design work need only focus on the I/O interface requirements for a particular specialist application. In the 17 node network of figure 1 any message which is sent across the network will have to pass into and out of several processor node.

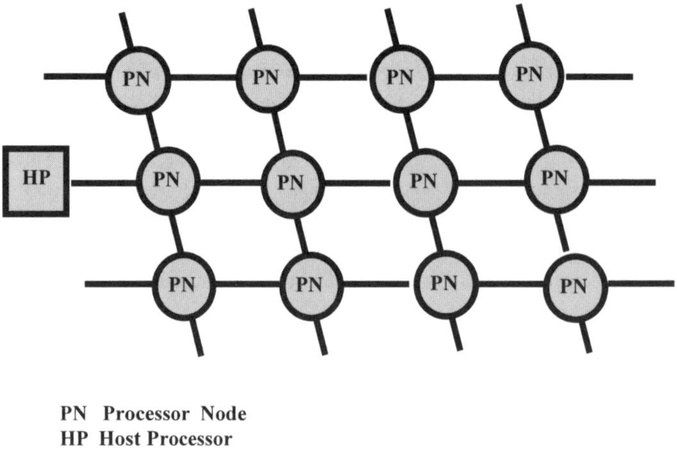

Figure 1- A parallel system with point to point communications.

To address the communication bottleneck the research group developed a packet routing switch. Figure 2 shows a 16 channel packet routing switch (ICR C416) which was designed to interface to the first generation of INMOS transputers [5]. The use of this device, in some real-time control systems, has shown the efficiency of routing as a solution to medium scale, low-cost, inter-processor communications.

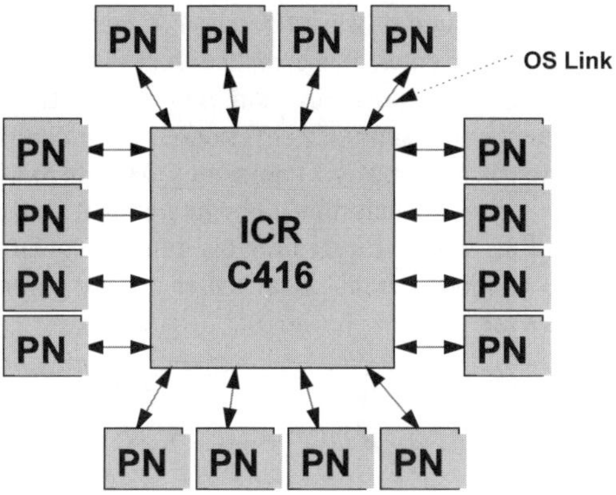

Figure 2- A parallel network with an ICR C416 packet routing switch.

Over a period of several years the group has developed components for this network to build a complete distributed parallel processing systems. The components contain many modular functional block designs captured in VHDL and implemented on FPGA devices. This has allowed for design upgrade and reuse over a period of time and the porting of design to the newer devices. The main features of the current network will now be described. The router network specification is similar to the ESA SpaceWire standard [6].

3 Outline of the communication network

Routers provide direct communication, among processors, similar in function to that of a telephone exchange, supporting the simultaneous transfer of messages from any input to any output of the router. Each packet/message has header bytes, which are used to configure the path of the packet through the network. Control at the core of the device prevents one message from blocking the path of another. If two or more packets request the same output then all except the first are queued in a FIFO. All communications take place over a two wire serial link. The block diagram of figure 3 shows the main functions of the routing device.

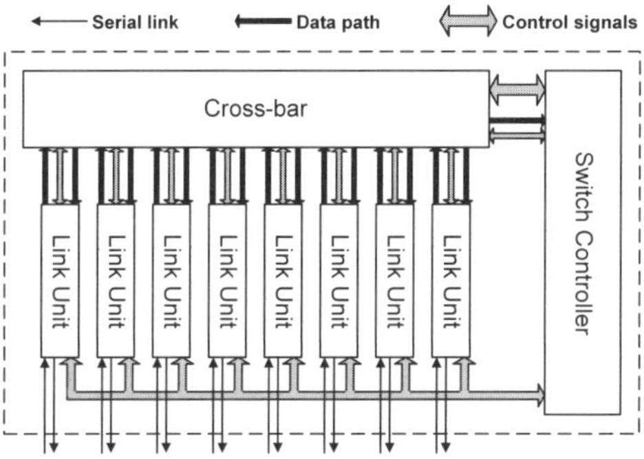

Figure 3 : High level block diagram of the basic router

For large systems routers can be cascaded. A system of up to 14 processors can be linked by two 8 link routers and there is no limit to the size of system that can be built. The system scales without loss of bandwidth for moderate size networks (i.e. networks where the average message passes through 4 or 5 routers). For large networks a strategy of store and forward of packets achieves better performance than the wormhole routing method of this packet switch [7,8]. Figure 4 shows an example of a 12 processor network linked by two 8 link routers and each processor node (PN) can automatically configure the routing switches to transmit a data packet to any other node. Two of the eight links are used to interconnect the router to improve bandwidth. The host processor node (HP) uses the same link connections as all other connections. In most typical applications this would be a standard PC containing boot programmes for all other processing nodes and would handle the external I/O of the network.

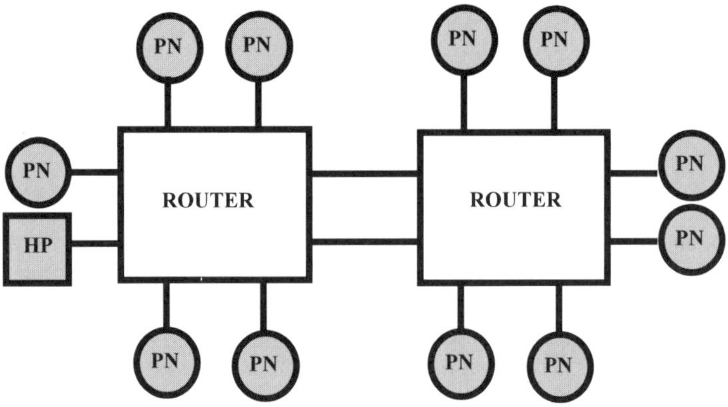

Figure 4: A network configuration of 12 distributed processors

The time taken to connect an input channel to an output channel is approximately 1.5μs. The receiver input link uses an oversampling technique for data recovery of the signal [9]. This has the advantage that a single FPGA device can support many communication links as no phase lock loop circuitry is required for clock recovery. This link is referred to as an OS link and requires only two electrical signals (Tx & Rx) compared to the four required for the DS method of SpaceWire.

3.1. Message types and addressing modes

The router supports both uni and multi cast messages. Uni-cast messages have a single source and destination and multi-cast may start from one of several sources but is transferred to two or more destinations.

Each message contains one or more header bytes of routing information at the start of the serial byte stream. The switch supports two header addressing modes. When physical mode is used the header value directly relates to the hardware value of the output port and this header byte is stripped from the message as the message is transmitted from the device. Logical addressing requires prior configuration of the routing device so that any valid logic value can connect a message to a particular output port. The header value may or may not be stripped at the output. The advantage of logical addressing is to allow several adjacent routers to be configured as one virtual router.

3.2. Packet Protocol/Tokens

The low-level link protocol defines a full-duplex, two-wire system (Tx and Rx), that utilises eleven-bit tokens to transmit information on a serial bit stream. Tokens have eight data bits, a start bit, a stop bit and a control bit, as shown in Figure 5. The start (always logic high) and stop bit (always logic low) are used for framing, the type bit is used for differentiating between control and data tokens (logic low and high respectively). The stop bit can be extended by any amount when there is no further information to transmit.

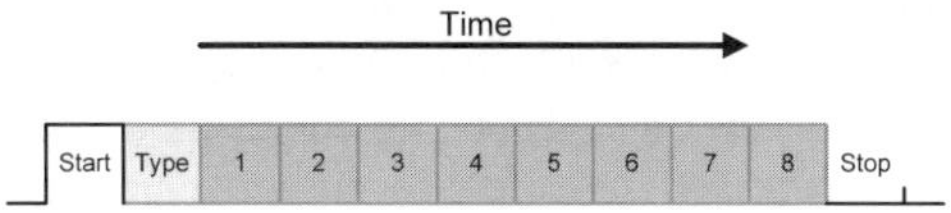

Figure5: Physical link token format

Using this format for the token it is possible to define 256 separate control tokens. However, at present only six tokens are defined. They include an initialisation token, two flow control tokens, and three message terminator tokens. The initialisation token is used to establish a validated connection between the link ends. As the system uses a two-wire asynchronous link, a flow control technique must be employed to prevent buffer overflow that utilises the same resources, but is transparent to the data connection. The protocol uses two flow control tokens to implement a stop/go (X-ON/X-OFF) technique compared to the credit based method of SpaceWire. Thus, if the buffer of a receiving device is approaching maximum, a 'stop' token can be sent on the return line to inhibit further transmission. Once enough space in the buffer has been cleared a 'go' token is sent to re-enable the transmission. The stop/go compared to credit based technique was found by simulation to be more efficient for many bi-directional flow conditions [10]. The flow control tokens are also used in the fault detection mechanism, which will be described later. Finally, the three message terminator tokens are defined for used in the network layer of the protocol, that is, 'end of packet' and 'end of message' and 'bad end of packet' (BEOP). The BEOP token is used in the recovery mechanism of the network after an error has been detected.

3.3. Cascading of router

When routers are connected together it is possible to use several link connections between two routers and group these connections together too logically appear as one link. This technique is referred to as 'Group Adaptive Routing'. Control functions on the router will automatically allocate a message to any free link of the group. This effectively increases the bandwidth of the connection between two routers at the expense of using additional link connections.

3.4. Configuration Port

The routing device contains an internal configuration port that can be accessed from any input port of the device. On power up the default configuration allows physical mode addressing and therefore its is possible to remotely configure the device from any node in the network.

3.5. Fault Detection and response

The router provides a level of fault tolerance for events that were thought to be the predominant causes of failure; namely, disconnection, host failure, device failure, and FIFO overflow. Detection of faults is based on knowing the status of the link at all times. By monitoring activity at the receiver, any operation that did not conform to a basic set of rules could identify faults easily. The two main states are identified as the basis of the detection mechanism namely, 'asleep' and 'awake'. The link enters the 'asleep' state on reset or on the occurrence of a fault. The link remains in the 'asleep' state until the successful completion of an initialisation sequence when the 'awake' state is entered.

A failure is detected in one of four ways. The first method is a synchronisation error. As previously described, the start and stop bits frame each token. If the sampling circuits detect an incorrect frame, a synchronisation error is flagged. The second method of failure detection is a 'receiver buffer overflow' flag. Although the flow control system should prevent such a situation occurring, a buffer overflow signals a breakdown of the flow control mechanism, which indicates a corrupt interconnection. The third relates to the frequency of link activity, where activity refers to the movement of any tokens over the link. Each side of the link monitors the link activity, and if no activity occurs within a predefined time period, the connection is said to be lost and an error is flagged. If there is no data to transfer across the link then idle tokens are inserted to maintain the link active flag. The final detection method triggers if the link receives an initialisation token while in the 'awake' state. As this token is not expected in normal operation, it indicates a failure in the transmission line at the other transceiver.

On detection of a fault a sequence of events occurs on the faulty link. The link is first flagged as unavailable and further input data is ignored. Then no further data is transmitted out and the remainder of the packet is spilt. On the receiver side of the link a BEOP token is added to the receiver FIFO. Finally a reset token is transmitted to signal failure to the other end of the link pair. If the fault is transient the link will recover from this state and resume transmission of further messages. The main use of this fault recovery procedure is to prevent messages sent to non-operating links from blocking the rest of the network.

3.6. Interface to processing node

To build a communication network, in addition to routers, processor interface components are required. Previously, an interface to the PCI bus was developed to connect to a host PC and provided the base for the specification of an interface to the AMBA bus to connect to the ALTERA ARM Excalibur [11] processor. A block diagram of the PCI interface is shown in figure 6.

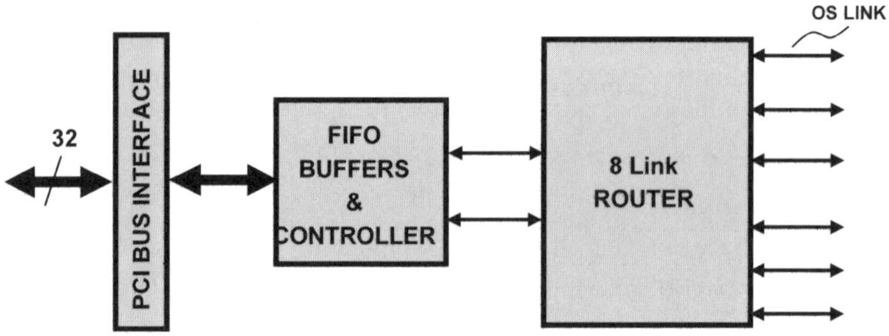

Figure 6: Block diagram of PCI bus interface to router.

The PCI interface supports two link connections between the FIFO buffers and the router. Each link has a separate transmit and receive channel and has a 32bit wide by 64 deep FIFO. Messages are transferred by DMA in data bursts of up to 64 double words, until the message is completed. The rate of transfer is approximately 70Mbytes/sec across the PCI bus interface. The overhead on the processor is very low and, because of the high rate of transfer, bus utilisation is very low. The hardware is optimised for the efficiency of data transfer and the latency of transfer was not part of the design specification.

The data is then transferred to a byte stream FIFO, byte token information and control tokens are added, before the packet is transferred to transmitter module. There is a similar structure for the received data. Currently the baud rate of the serial data link is 22Mbit/s and in addition some successful work has been carried out at link speeds of 44Mbit/s at cable lengths of 100 m [9].

The two link interfaces from the host processor are connected internally to an eight link router which is integrated in the same FPGA device.

At the block diagram level of figure 6 the AMBA bus interface is similar to the PCI interface but at a lower level the AMBA bus contains four separate 32 bit interfaces for data transfer. Nevertheless the conversion of the PCI interface to the AMBA interface of the ALTERA ARM Excalibur was relatively straightforward. Many of the modules were reused with little modification. The number of PLD logic elements (LE) of the Excalibur EPXA1 was half that used for the PCI interface device and therefore contains only one link connection between the FIFO buffers and a five link router to provide four external link connections. Full measurements of the data transfer rate across the AMBA bus have yet to be carried out. It is anticipated that a rate of transfer, similar to that obtained with the PCI bus, will be measured as the same 33 MHz clock frequency and FIFO size is used.

To transmit a packet of data it is necessary to write to three registers of the interface module. First, a 32 bit header containing routing and packet ID information second, a 32 bit start address of the data packet and finally, a 20 bit length value measured in bytes. The start of each packet must be double word aligned but the length can be of any value within the 20 bit field. The transfer is then carried out automatically under DMA by the PLD module allowing the CPU to proceed with other tasks. On completing, an interrupt flag is asserted to flag the end of the transfer. A similar structure is used for the receiver. Any packet entering the receiver is stalled pending writing to the two receive registers (address & length). Once a packet has been received and the interrupt asserted the process must be repeated for any further next packets.

The design of the communication interface is optimised to have a minimum overhead on the processor node to setup and receive messages and also to produce a low overhead on the processor's data bus. The design provides hardware support for different messages types (i.e. virtual channels) in contrast to the single message type of the transputer. However the addition of the FIFO buffers needed to obtain the high bust rates of 70Mbytes/s has added to the latency of transfer.

4 Implementation

The Excalibur EPXA1 device is used as a stand alone processor with the communication interface implemented in the programmable logic elements of the device to provide a single chip solution for the processing node. These elements are configured from a small non volatile memory device (ECP2). Programmes can be booted from the link on power up. Addition SDRAM can be added via a standard connection. The Excalibur EPXA1 without router function requires 933 LEs or 22% of total. With a five link router 2838 LEs are required which is 68% of total.

To obtain access to a host PC a PCI interface is required and this is implemented on an ALTERA APEX 20k 200 device [11]. The two link interface without the router requires 2765 LEs or 33% of total. With an eight link router 5722 LEs are required which is 68% of the total LEs.

5 Conclusions

The network provides a scalable solution for inter-processor communication in a distributed multi-processor network. The benefits of SOPC technology allow the integration of embedded and custom functions in a single chip processing node similar to that of the transputer. The level of integration achieved in the Excalibur SOPC device and development tools have led to cost effective solution for this project.

The porting and reuse of VHDL design modules has been demonstrated to effect.

6 Acknowledgement

The authors would like to thank ALTERA Corporation and IC-Routing LTD for their financial support for this project.

Reference:

[1] N.J. Boden, et al.; "Myrinet – A Gigabit-per-Second Local Area Network"; IEEE Computer, vol. 15, No. 1; 1995; pp. 29-36.

[2] Analog Devices, URL: http://www.analog.com/products/descriptions/2154_0.html, summary: The ADSP-21061 is a member of the powerful SHARC ® family of DSP processors.

[3] TEXAS Instruments, URL: http://dspvillage.ti.com/docs/dspproducthome.jhtml, DSP Platforms both fixed- or floating- point platforms to support many design needs.

[4] CRAY RESEARCH INC.; "The Cray T3D System Architecture Overview"; Cray Research Inc.; Technical document HR-04033; 1993.

[5] J W Ellis, B C O'Neill & S Clark, 'A Router Design for T800 Compatible Transputer Arrays' - Transputer Applications, ISSN 0969-9341, 1993, Vol. 1(2) pp12-18.

[6] European Space Agency, ECSS-Space Engineering SpaceWire – Links, Nodes, Routers and Networks (ECSS-E-50-12) URL: http://www.estec.esa.nl/tech/spacewire.

[7] A.M.JONES, N.J.DAVIES, M.A.FIRTH, C.J.WRIGHT; "PACT The Network Designer"s Handbook"; IOS Press, Ohmsha, Concurrent systems engineering series, Vol. 51; ISSN 1383-7575; 1997; pp 17-18

[8] L.M. Li, P.K. McKinley; "A survey of wormhole routing techniques in direct networks"; IEEE Computer, vol. 26, no. 2; 1993; pp. 62-76

[9] B C O'Neill, S Clark and K L Wong, 'Serial Communication Circuit with Optimized Skew Characteristics', IEEE Comms Letters, IEEE computer Society, Vol.5, No.6, pp 260-262, 2001.

[10] R Hotchkiss, PhD thesis, Nottingham Trent University, Nov 2000.

[11] ALTERA CORPORATION, http//www.altera.com.

Communicating Process Architectures – 2003
Jan F. Broenink and Gerald H. Hilderink (Eds.)
IOS Press, 2003

The Denotational Semantics of View-Centric Reasoning

Marc L. SMITH
Department of Computer Science, Colby College, Waterville, ME 04901-8858, USA

Charles E. HUGHES
School of EE and CS, University of Central Florida, Orlando, FL 32816-2362, USA

Kyle W. BURKE
Department of Computer Science, Colby College, Waterville, ME 04901-8858, USA

Abstract. Both Lawrence's $\mathcal{HCSP}$ [1] and Smith, et al's $\mathcal{VCR}$ [2] (an earlier version appears in [3]) extend CSP [4] with representations of truly concurrent events. Previously, $\mathcal{VCR}$ was described using an operational semantics, while the semantics of $\mathcal{HCSP}$'s Acceptances model, like those of the predominant CSP models described by Roscoe [5] (e.g., Traces, Failures / Divergences), are denotational. We now present a denotational semantics for $\mathcal{VCR}$ and, in so doing, propose an extension to $\mathcal{HCSP}$ (and possibly other existing CSP models) to support View-Centric Reasoning. This work brings $\mathcal{VCR}$ a step closer to being drawn within Hoare and He's Unifying Theories of Programming [6] for further comparisons.

1 Introduction

This paper represents a first step in the evolution of View-Centric Reasoning ($\mathcal{VCR}$) [2] from a model described by an operational semantics to one expressed in the more mainstream denotational semantics of other existing CSP models. In particular, Rosco [5] describes the modern Traces model ($\mathcal{T}$), the Failures/Divergences model ($\mathcal{N}$), the Stable Failures model ($\mathcal{F}$), the Infinite Traces/Divergences model ($\mathcal{I}$), and the Failures/Divergences/Infinite Traces model ($\mathcal{U}$) — all expressed in denotational semantics, and consistent with Hoare's classic CSP [4]. In addition, Lawrence recently proposed a new Acceptances model in two papers on $\mathcal{CSPP}$ [7] and $\mathcal{HCSP}$ [1], again, using denotational semantics.

The ultimate goal of transforming $\mathcal{VCR}$ into a denotational semantics is to provide a mechanism for drawing $\mathcal{VCR}$ into Hoare and He's Unifying Theories of Programming [6]. Accomplishing this task will provide a formal means of comparison between elements of $\mathcal{VCR}$ and those of the other CSP models. It so happens that $\mathcal{VCR}$ and $\mathcal{HCSP}$ share an important common abstraction: both models provide a means to directly represent simultaneously occurring events in their traces. In $\mathcal{HCSP}$, these are called *merged* events; in $\mathcal{VCR}$, they are called *parallel events*. For this reason, it was natural for us to propose $\mathcal{VCR}$ as an extension to $\mathcal{HCSP}$.

The remainder of this paper is organized as follows. Section 2 discusses the nature of environment in CSP (and other) models of concurrency, and the consequences for what it means to observe interactions between processes in such an environment. Section 3 presents a denotational representation of the $\mathcal{VCR}$ semantics we are proposing as an extension to $\mathcal{HCSP}$. We conclude and contemplate future work in Sections 4 and 5, respectively.

Attribute	Classic CSP	$\mathcal{HCSP}$	$\mathcal{VCR}$
Perfect observer?	Yes	Yes	Yes
Sequentially interleaved trace?	Yes		Yes
Trace with true simultaneous events?		Yes	Yes
Multiple (possibly) imperfect observers?			Yes
Multiple views of history?			Yes
Distinction between history and views?			Yes

Table 1: Taxonomy of attributes for three CSP models

2 Environmental and Observational Perspectives

Hoare and He present theories of reactive processes in their Unifying Theories of Programming [6]. The notion of environment is elucidated early in this presentation, as environment is essential to theories of reactive processes, examples of which include CSP and its derivative models. Essentially, the environment is the medium within which processes compute. Equivalently, the environment is the medium within which processes may be observed. The behavior of a sequential process may be sufficiently described by making observations only of its input/output behavior. In contrast, the behavior of a reactive process may require additional intermediate observations.

Regarding these observations, Hoare and He borrow insight from modern quantum physics. Namely, they view the act of observation to be an interaction between a process and one or more observers in the environment. Furthermore, the roles of observers in the environment may be (and often are) played by the processes themselves! As one would expect, an interaction between such processes often affects the behavior of the processes involved.

A process, in its role as observer, may sequentially record the interactions in which it participates. Recall participation includes the act of observation. Naturally, in an environment of multiple reactive processes, simultaneous interactions may be observed. Prior to $\mathcal{VCR}$ and $\mathcal{HCSP}$, recording conventions required simultaneous events to be recorded in some sequence, including random. Hoare and He thus define a *trace* as "the sequence of interactions recorded up to some given moment in time."

In CSP, interactions take the form of communications between processes across channels. Table 1 gives a taxonomy of attributes across CSP, $\mathcal{HCSP}$, and $\mathcal{VCR}$. The table depicts the gradual departure from perfect observation and a sequentialized interleaving expression of concurrency, toward imperfect observation and a contextualized interleaving that more closely preserves true concurrency. In some sense, the attributes depicted in Table 1 provide a roadmap for the denotational semantics presented in Section 3.

Briefly, classic CSP has one perfect observer who records one trace of atomic events, possibly interleaved in cases of unsynchronized simultaneity. $\mathcal{HCSP}$ has one perfect observer who records one trace of merged events; obviating the need for, and eliminating the possibility of, sequential interleaving in the presence of an environment offering simultaneously occurring events. $\mathcal{VCR}$ has multiple, possibly imperfect observers. (It has been pointed out that with hiding and internal choice, classic CSP and $\mathcal{HCSP}$ can model imperfect observers, but this is true only in a restricted sense.) In $\mathcal{VCR}$, two kinds of traces are distinguished: a history and its corresponding views. The history type of trace is a sequence of unordered parallel events (event multisets). The view type of trace is a sequence of ordered parallel events (ROPEs) derived from a given computation's history.

3 $\mathcal{VCR}$ Semantics

Our goal is to describe the views of a trace, but we do so gradually. Along the way we will address imperfect observation in Section 3.1, and differences in perspective for observing a simultaneously occurring event in Section 3.2. We shall see, in Section 3.3, that representing views of a trace requires further modification to $\mathcal{HCSP}$'s already extended notion of a trace. Finally, in Section 3.4, we relate traces and views to one another for further study.

3.1 Imperfect Observation

To account for imperfect observation, there needs to be a way to represent an observer missing (i.e., not observing) one or more atomic events that occur in parallel with other atomic events. Since both $\mathcal{VCR}$ and $\mathcal{HCSP}$ represent parallel events using multisets, we proceed from this construction.

Let B be a bag, that is, an event multiset. Thus, as a multiset,

$$B = \{\!|\, b_1, b_2, \ldots b_n \,|\!\}.$$

Borrowing the *merge* operator ($\diamond$) from $\mathcal{HCSP}$, we could equivalently represent b as a bag, thus

$$B = b_1 \diamond b_2 \diamond \cdots b_n.$$

We introduce the notion of the $pieces()$ of B as the powerset of B. In this case, the powerset would be the set of all *multiset* subsets of B. Thus, let

$$P = pieces(B) = \{p \mid p \subseteq B\}.$$

The elements of P represent the possibilities of imperfect observation. For all elements $p \in P$, we could enumerate the elements of p, in a fashion similar to the original B, thus

$$p = \{\!|\, p_1, p_2, \ldots p_k \,|\!\},$$

where $0 \leq k \leq n$. It is tempting to wish to represent the elements of P using the merge operator, but this possibility would seem to break down for those cases where p is either empty or a singleton multiset. Certainly, one can represent empty and singleton bags using multiset notation; the challenge is to find a meaningful representation for these two special cases of bags using the merge operator. One way to overcome this problem is to borrow from CSP's hiding operator and employ the invisible event, τ, as described by Roscoe [5]. Using τ in conjunction with $\diamond$, we can represent both empty and singleton bags. Thus, we can represent p as follows:

$$p = \begin{cases} \tau \diamond \tau & \text{if } p = \emptyset, \\ \tau \diamond p_1 & \text{if } |p| = 1, \text{ where } p_1 \in p, \\ p_1 \diamond p_2 \diamond \cdots p_k & \text{otherwise, where } \forall\, 1 \leq i \leq k \leq n . p_i \in p. \end{cases}$$

This use of τ is not entirely within the spirit of hiding internal events as defined by CSP, but the end result is it permits us to continue using the $\diamond$ operator as we continue to define the semantics of views. It should be noted for completeness that τ is an identity for $\diamond$, thus

$$\tau \diamond \tau \equiv \{\!|\ |\!\} \equiv \emptyset,$$

and

$$\tau \diamond p_i \equiv \{\!|\, p_i \,|\!\}.$$

3.2 Different Perspectives

In the previous section, neither representation of p, an element of the set of pieces of a bag or multiset, conveys any information about order. The $pieces()$ function merely accounts for the possibility that an observer need not be perfect. As discussed in Section 2, one of the tenets of $\mathcal{VCR}$ is that the environment of a concurrent computation consists of multiple observers. At a minimum, each communicating sequential process represents an observer of the computation in which that process participates.

In $\mathcal{VCR}$ and $\mathcal{HCSP}$, CSP's perfect observer is free to record the trace of a computation unencumbered by the burden, and consequences, of sequentially interleaving simultaneously occurring events. But the semantics of $\mathcal{VCR}$ more closely model the environment by using these parallel events to generate the many possible perspectives of observers within the environment. $\mathcal{VCR}$ thus has the notion of a ROPE, a randomly ordered parallel event.

Just as a bag B has many possible multisets of pieces, for a given set of pieces $p \in pieces(B)$, there are many possible orderings of the elements of p. The set of possible orderings (perspectives) for an observer of B can be defined using our definition of $pieces()$. Thus,

$$ropes(B) = \{\langle r_1, r_2, \ldots, r_k\rangle \mid r = \{\!| \; r_1, r_2, \ldots r_k \; |\!\} \in pieces(B) \wedge k = |r|\}.$$

The careful reader might be bothered by several points in this definition of $ropes()$. First, the set of orderings is a set of traces! But this is only natural, since an ordering implies some sort of list, and it so happens that a trace is nothing more than a list of observable events. This foreshadows the recursive nature of one possible definition of a view presented in Section 3.3.

The next points of concern with this definition of $ropes()$ have to do with whether the given definition really includes all possible orderings of all possible subsets of bag B? There are two levels of event generation — one explicit, the other implicit — defined by the $pieces()$ function. The explicit level determines the size of the subset, k, and specifically which k events are chosen from B. The implicit level concerns the unordered nature of multisets. For example, if $\{\!| \; a, b \; |\!\} \in pieces(B)$, then it is also true that $\{\!| \; b, a \; |\!\} \in pieces(B)$. Thus, every permutation of any trace found within $ropes(B)$ will also be an element of $ropes(B)$.

In keeping with the spirit of $\mathcal{HCSP}$'s $\diamond$ operator, we introduce an appropriately decorated *ordered merge* operator, $\vec{\diamond}$, which we use to define an alternative expression for $\mathcal{VCR}$'s ROPEs. Thus,

$$\langle r_1, r_2, \ldots, r_k\rangle \equiv r_1 \vec{\diamond} \, r_2 \vec{\diamond} \cdots r_k.$$

Now that we have given the definition for a ROPE, which is nothing more than a partially ordered bag or multiset, we may proceed to Section 3.3, where we construct a new kind of trace out of ROPEs.

3.3 Views of a Trace

CSP denotes a trace as "a sequence of symbols, separated by commas and enclosed in angular brackets." [4]. The meaning of this representation of a trace is that of a sequentialized, recorded history of the observable events of a computation. By introducing the $\diamond$ operator, $\mathcal{HCSP}$ extends the definition of a trace with the capability of recording merged events (bags), though it is common practice to omit the bag (multiset) notation for singletons and record the multiset as if it were an individual event. Similarly, $\mathcal{VCR}$ defines a trace as a list of parallel events. In both cases, this extended notion of a trace is that of a list of event multisets. From this common notion of a trace we can proceed to define a trace's views.

Suppose for some composition of processes, P, that trace $tr \in traces(P)$. Further suppose that tr is a sequence of bags, thus

$$tr = \langle b_1, b_2, \ldots b_n \rangle.$$

Then we could define $views(tr)$ as follows:

$$views(tr) = views(\langle b_1, b_2, \ldots b_n \rangle) = \{\langle r_1, r_2, \ldots r_n \rangle \mid r_i \in ropes(b_i) \,\forall\, 1 \leq i \leq n\}.$$

The definition of $views()$ falls out rather nicely at this point. This just specifies that the set of all views of a trace *tr* consists of views that are formed using a ROPE of each bag in *tr*.

If we dig a little deeper into the appearance of these views, there are two possible representations: a list of lists and a list of ordered bags. Both representations have benefits. The list of lists representation could be flattened, resulting in traces the original CSP observer could have recorded. Furthermore, the recursive nature of the list of lists form is elegant and appealing, obviating the need for a merge operator, and supporting a hierarchical, rather than flat, environment of observers. These authors believe this last point has implications for reasoning about composition. The list of ordered bags representation provides a convenient mapping to $\mathcal{HCSP}$, and warrants further study in the context of the unifying theories.

3.4 Views of All Traces

One of the tenets of $\mathcal{VCR}$ is the ability to distinguish a computation's history (trace) from its views, while relating instances of both notions to each other. To relate an instance of a computation's history to all its views, we introduce the set of TraceViews() of some concurrent process, P. Thus,

$$TraceViews(P) = \{\langle tr, vw \rangle \mid tr \in traces(P) \wedge vw \in views(tr)\}.$$

This says that $TraceViews()$ is a set of trace/view pairs. In particular, it is the set of all possible traces, and corresponding views of each possible trace, of some process P.

This is a very large set. It represents the cross product between every possible computation of process P, and every possible (including imperfect) view of every possible computation of P. From this set one could project just those elements that are the trace/view pairs of a single computation. Recall that knowing all possible views of a computation is not sufficient to unambiguously determine a computation's true history. For further discussion on this topic, see Smith et al. [2].

4 Conclusion

We presented elements of a denotational semantics for View-Centric Reasoning within the framework of $\mathcal{HCSP}$, while preserving a dual representation that links $\mathcal{VCR}$ to its original operational semantics. Both representations of views – as a list of ordered bags using the new $\vec{\diamond}$ operator, and as a list of lists (ROPEs) – should prove useful in different ways. The $TraceViews()$ set encapsulates the entire denotational semantics of $\mathcal{VCR}$, and is the starting point for attempting to draw $\mathcal{VCR}$ within the Unifying Theories of Programming. This will provide the necessary framework to compare $\mathcal{VCR}$ to the other CSP models.

5 Future Work

As a consequence of the analysis presented here, we are investigating whether the $TraceViews()$ set could be incorporated into the other CSP models. For example, the CSP traces model, $\mathcal{T}$, consists of a set of traces for a process, in addition to a set of axioms, closure properties, etc. Replacing the set of $traces()$ with $TraceViews()$ would require further revision, verification, and proofs. Similarly for the failures / divergences model, $\mathcal{N}$, the stable failures model, $\mathcal{F}$, the infinite traces / divergences model, $\mathcal{I}$, the failures / divergences / infinite traces model, $\mathcal{U}$, and the Acceptances model in $\mathcal{HCSP}$. The obvious model to attempt first would be $\mathcal{HCSP}$, since it already supports the notion of true parallel events.

6 Acknowledgements

Rebecca J. Parsons, of ThoughtWorks, Inc., Chicago, was our collaborator during the development of the $\mathcal{VCR}$ model. Her guidance and patience helped lead to $\mathcal{VCR}$'s original expression as a parameterized operational semantics. Professor Adrian Lawrence, of Loughborough University, helped with terminology and clarification of some of the finer points of merged events in $\mathcal{HCSP}$. Professor Jim Woodcock, of the University of Kent, provided valuable feedback for our paper last year [3], as well the roadmap we are now following for $\mathcal{VCR}$. Finally, the authors wish to thank the anonymous referees of this paper for their careful reading and insightful remarks.

References

[1] Adrian E. Lawrence. Hcsp: Imperative state and true concurrency. In J. S. Pascoe, P. H. Welch, R. J. Loader, and V. S. Sunderam, editors, *Communicating Process Architectures – 2002*, Concurrent Systems Engineering, pages 39–55, Amsterdam, 2002. IOS Press.

[2] Marc L. Smith, Rebecca J. Parsons, and Charles E. Hughes. View-centric reasoning for linda and tuple space computation. *IEE Proceedings–Software*, 150(2):71–84, apr 2003.

[3] Marc L. Smith, Rebecca J. Parsons, and Charles E. Hughes. View-centric reasoning for linda and tuple space computation. In J. S. Pascoe, P. H. Welch, R. J. Loader, and V. S. Sunderam, editors, *Communicating Process Architectures 2002*, volume 60 of *Concurrent Systems Engineering Series*, pages 223–254, Amsterdam, 2002. IOS Press.

[4] C.A.R. Hoare. *Communicating Sequential Processes*. Prentice Hall International Series in Computer Science. Prentice-Hall International, UK, Ltd., UK, 1985.

[5] A. W. Roscoe. *The Theory and Practice of Concurrency*. Prentice Hall International Series in Computer Science. Prentice Hall Europe, 1998.

[6] C.A.R. Hoare and Jifeng He. *Unifying Theories of Programming*. Prentice Hall Series in Computer Science. Prentice Hall Europe, 1998.

[7] Adrian E. Lawrence. Acceptances, behaviours, and infinite activity in cspp. In J. S. Pascoe, P. H. Welch, R. J. Loader, and V. S. Sunderam, editors, *Communicating Process Architectures – 2002*, Concurrent Systems Engineering, pages 17–38, Amsterdam, 2002. IOS Press.

Communicating Process Architectures – 2003
Jan F. Broenink and Gerald H. Hilderink (Eds.)
IOS Press, 2003

Overtures and hesitant offers: hiding in $\mathcal{CSPP}$

A.E. LAWRENCE
Department of Computer Science, Loughborough University, Leicestershire, LE11 3TU UK
A.E.Lawrence@lboro.ac.uk

Abstract. Hiding is an important and characteristic part of CSP. Defining it in the presence of priority for $\mathcal{CSPP}$ needs care. The ideas of overtures and hesitant offers introduced here arise naturally in the context of Acceptances. They provide clear insight into the behaviour of hidden processes. And particularly illuminate the origin of the nondeterminism which frequently arises from hiding.

Keywords: CSP; CSPP; Denotational semantics; formal methods; concurrency; parallel systems; occam; hardware compilation; priority; hiding.

1 Introduction

Hiding is the basic abstraction mechanism in CSP[1, 2]. A child buys an ice cream:

$$Buy = pay \rightarrow record \rightarrow ice\_cream \rightarrow Stop$$

Buy is the process in which the money is first handed over, then the transaction is recorded, and finally the ice cream is handed over before the shop shuts. The child is unlikely to be very interested in the event *record*: the fact that the shop needs to keep track of the stock and finance is of no immediate concern. This is irrelevant detail: so the child is likely to dismiss the *record* event as irrelevant. The child's view will be

$$Buy \setminus \{record\} = pay \rightarrow ice\_cream \rightarrow Stop \quad .$$

The notation is $P \setminus H$ where P is a process and H is the set of events to be hidden. The explicit set notation is sometimes omitted when there is no ambiguity, so $Buy \setminus \{record\}$ might be written as $Buy \setminus record$.

In **occam**, we might have something like

```
CHAN OF money payment:
CHAN OF goods    hand:

--  data events hidden here
CHAN OF transaction data:
PAR
  database(data)
  ...
  SEQ
    payment ? pay
    ...
    data ! details
    ...
    hand ! ice.cream
```

So hiding can be used to model the effect of scope in **occam**. In π-calculus[3, 4], there is a notion of *restriction* which uses the keyword **new**: this is the nearest that π-calculus comes to hiding. And it matches the conventional idea of scope rather closely. In that respect it is a little like **occam** in that hiding does not appear explicitly, but is implicit in the idea of scope. But when we need to prove properties using CSP, hiding is very important: the proof of the correcteness of the implementation of JCSP[5, 6] and CTJ channels in [7] uses hiding repeatedly. Indeed it is absolutely fundamental in showing the equivalence between real channels and the JCSP and CTJ implementations with the internal details hidden. A particularly simple example, very like the child buying an ice cream, appears on page 291 of [7]:

$$(write.a!x \rightarrow a!x \rightarrow ack.a \rightarrow PROCESS) \setminus \{write.a, ack.a\} = a!x \rightarrow PROCESS$$

CSP seems to be the only process algebra with such an explicit and comprehensive form of hiding. That in turn guarantees that CSP must also include explicit nondeterminism. A simple example is

$$((a \rightarrow c \rightarrow Stop) \,\Box\, (b \rightarrow d \rightarrow Stop)) \setminus \{a, b\}$$

$(a \rightarrow c \rightarrow Stop) \,\Box\, (b \rightarrow d \rightarrow Stop)$ is a process which is driven by its environment – its parallel partners – into performing *either* the event a *or* the event b initially. It depends on which is offered first: for the moment we only consider offers of single events. If a and b are hidden, they become 'internal': that is unconstrained by the environment. If we are concerned with software, it may depend on the details of a particular implementation which event becomes available first. CSP is intended to cover all such possibilities: it leaves open exactly what happens internally in such cases. Thus

$$((a \rightarrow c \rightarrow Stop) \,\Box\, (b \rightarrow d \rightarrow Stop)) \setminus \{a, b\} = (c \rightarrow Stop) \sqcap (d \rightarrow Stop)$$

where $\sqcap$ is *internal choice*, the primary operator capturing the fullest form of nondeterminism. The point of this example is that the presence of hiding in the language inevitably requires a matching operator $\sqcap$, even if we did not need it for anything else. That in turn ensures that refinement is an inherent part of CSP: we will return to that below.

A pseudo-occam model of $((a \rightarrow c \rightarrow Stop) \,\Box\, (b \rightarrow d \rightarrow Stop)) \setminus \{a, b\}$ using booleans is

```
CHAN OF BOOL a,b :
PAR
  PAR
    a ! TRUE
    b ! TRUE
  BOOL x,y :
  ALT
    a ? x
      c ! x
    b ? y
      d ! y
```

Here we model the internal environment which makes the events a and b available in an unconstrained way by

```
PAR
  a ! TRUE
  b ! TRUE
```

which can be implemented in either order on a sequential machine. Without knowledge of the implementation, and perhaps even with such knowledge, we cannot predict whether the occam fragment will communicate on the `c` or on the `d` channel. Of course, this `PAR` will not terminate in the fragment above: only one of the communications will happen, but this does not matter for the purposes of illustration. Notice that even if `ALT` was replaced by `PRI ALT` it still would not determine whether `c` or `d` was used. But things would be different if in addition `PAR` was changed to `PRIPAR`: examining the effects of priority in hiding is one of the aims of this paper.

A conceptual picture of $P \setminus H$ is of the process P placed inside a box with walls opaque to the events of H. And inside the box, the events of H are entirely free to occur: the only constraints are those imposed by P itself. External events like c and d above can 'pass through' the walls and involve interaction with the outside world: they are not directly influenced by the box. Any external events from H which might be offered are simply ignored: they cannot penetrate the box, and their internal counterparts cannot be seen outside.

The 'bubble diagrams' that most **occam** programmers draw when designing their systems use hiding. It is fundamental to compositional design in which only external behaviour need be considered when putting processes together. Figure 1 is a bubble diagram for an occam

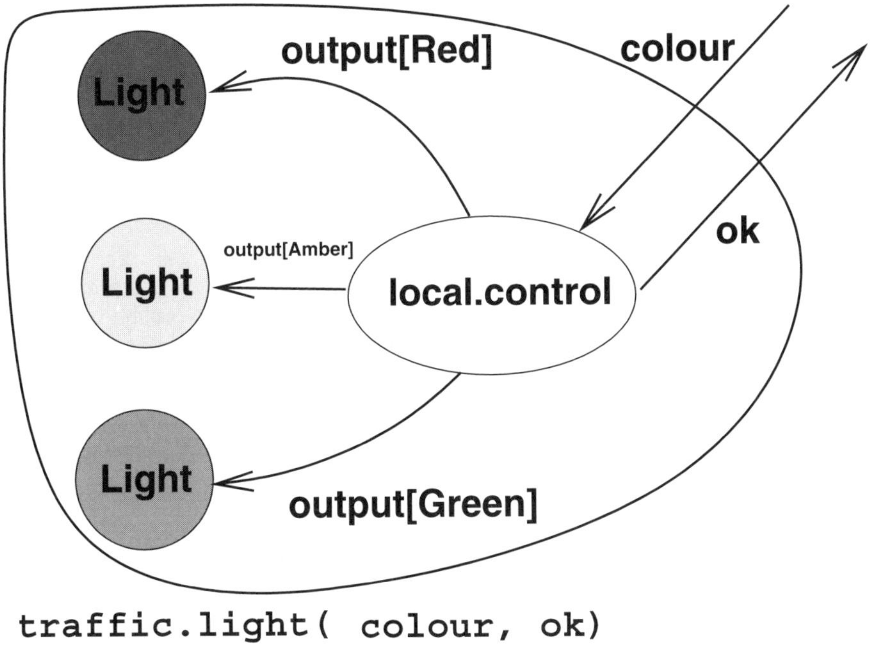

Figure 1: A 'bubble' diagram

model for a single traffic light. Here some of the internal details are shown. But it is understood that for external purposes only the two channels `colour` and `ok` need be considered. In fact the outer boundary matches the idea of the 'box' in the conceptual model above. It is understood that everything wholly contained in the outer bubble is irrelevant and may be substituted by a different implementation provided the same external behaviour is maintained. Indeed the 4 inner bubbles show no internal details: they are already hidden.

So hiding provides

- abstraction, fundamental to compositional design and comprehension; and
- isolation. The external environment cannot interfere with the internals of a hidden process.

Some cases of hiding are less than obvious: in standard CSP

$$((a \to P) \Box (b \to Q)) \setminus \{b\} = Q' \sqcap ((a \to P') \Box Q') = ((a \to P') \sqcap Stop) \Box Q' \quad (1)$$

where $P' = P \setminus \{b\}$ and $Q' = Q \setminus \{b\}$. We will discover that this result is slightly modified in the presence of priority because we have more expressive power. And we must address cases like $\Big((a \to Stop) \overleftarrow{\Box} (b \to Stop)\Big) \setminus \{b\}$.

Hiding also reveals that any complete theory must handle livelock. What happens if we hide a in $\mu p \bullet a \to P$, the process $P = a \to P$ which engages in a continuous stream of a's? It is the essence of divergence, and clearly we must have

$$(\mu P \bullet a \to P) \setminus \{a\} = \mathsf{div}$$

In **occam** we might have something like

```
PROC diverge()
  CHAN OF BOOL c:
  BOOL any:
  WHILE TRUE
    PAR
      c ! TRUE
      c ? any
:
```

The purpose of this paper is to investigate how hiding is modified when priority is included in $\mathcal{CSPP}$[8, 9] and therefore also in $\mathcal{HCSP}$[10, 11]. The next section gives a quick summary of $\mathcal{CSPP}$. Then the Acceptance denotational semantics is surveyed. The ideas of overtures and hesitant offers are then introduced as precise ways to understand hiding with Acceptances and priority. The culmination is the formal definition of hiding in $\mathcal{CSPP}$. Finally we illustrate how this works on some illuminating examples.

2 $\mathcal{CSPP}$

$\mathcal{CSPP}$ is CSP extended to include priority as it occurs in standard **occam**. The `ALT` constructor of **occam** matches external choice, $\Box$ of CSP, at least for guarded processes of the form $c?x \to P(x)$ and for *Skip*. **occam** also includes `PRI ALT` in which textually earlier guards have priority over later guards. $\mathcal{CSPP}$ introduces $\overleftarrow{\Box}$ to capture this: it is a general variant of $\Box$. $P \overleftarrow{\Box} Q$ can also be written as $Q \overrightarrow{\Box} P$.

In **occam**, `PRI ALT` is a refinement of `ALT`: that is, it is a possible implementation. Indeed most **occam** compilers have only the one primitive `PRI ALT` which is also used to implement `ALT`. So in this context, it would be natural to write

$$P \Box Q = P \overleftarrow{\Box} Q \;\sqcap\; P \overrightarrow{\Box} Q$$

and this is what an early version of $\mathcal{CSPP}$ did. This means that the only possible implementations of $\Box$ were biased. As first pointed out by Bill Roscoe, this is at variance with the intuition of the majority of those who do not use **occam** regularly: they normally think of $\Box$

as a neutral choice operator. So $\mathcal{CSPP}$ also includes a *compliant* version of external choice written naturally enough as $\overleftrightarrow{\Box}$. This is another refinement of $\Box$ in the sense that

$$P \Box Q = P \overleftarrow{\Box} Q \;\sqcap\; P \overleftrightarrow{\Box} Q \;\sqcap\; P \overrightarrow{\Box} Q$$

The idea of *compliance* now has an important role in understanding $\mathcal{CSPP}$: see [8, 9].

From the **occam** perspective, both $\overleftarrow{\Box}$ and $\overrightarrow{\Box}$ are refinements of $\Box$ in the sense above. This gives our usual embedding of standard CSP in $\mathcal{CSPP}$. However, there is another simpler embedding in which we *identify* $\Box$ with $\overleftrightarrow{\Box}$. This matches standard CSP and the intuition of many practitioners who use CSP primarily for specification and model checking. But that identification implies that **occam** need not be an implementation of (part of) CSP! Needless to say, we normally work with the first embedding, but the second is simpler and throws a good deal of light on standard CSP. It is useful when priority is not needed.

occam also includes `PRI PAR`, a prioritized version of the concurrency constructor. For completeness $\mathcal{CSPP}$ includes and extends this idea with $\overleftarrow{|||}$, $\overleftrightarrow{|||}$ and $\overrightarrow{|||}$ which are like $\overleftarrow{\Box}$, $\overleftrightarrow{\Box}$ and $\overrightarrow{\Box}$, but operate on every event rather than on just the initial events of a process.

There are really only these two ideas of priority and compliance in the extension from CSP to $\mathcal{CSPP}$, but various derived operators also get biased or compliant variants. So $x : \overleftrightarrow{\{a,b,c\}} \to Stop$ is a process which is neutral about whether it starts with a or b or c. Whereas $x : \{a,b,c\} \to Stop$ might have an implementation which gave priority to b.

3 Acceptances

In order to define $\mathcal{CSPP}$ properly, a new *Acceptance* denotational semantics was invented. This turned out to have unexpected merits and enabled some longstanding problems in describing certain sorts of infinite behaviour to be solved for standard CSP as well as for $\mathcal{CSPP}$. It is described in [8] and [9], so we only highlight some key points here.

The basic idea is to *offer* a process P a set of mutually exclusive events, and record which are *accepted*. So if $a \to Stop$ is initially offered the set of events $\{b,c,d\}$, nothing is accepted. The response, a refusal in this case, is the empty set. We write that as $\langle\rangle : \{b,c,d\} \leadsto \emptyset$. $\langle\rangle$ is the empty trace and indicates that we are at the initial stage of evolution before the process has performed any events. If we offer $\{a,b,c\}$ then a is accepted: $\langle\rangle : \{a,b,c\} \leadsto \{a\}$. In these circumstances, the process can perform a which is recorded in the trace as $\langle a\rangle$. After that, the process behaves like $Stop$ and refuses to do anything else: $\langle a\rangle : X \leadsto \emptyset$ for any offer X.

The idea was to capture priority as in $(a \to Stop) \overleftarrow{\Box} (b \to Stop)$ which gives $\langle\rangle : \{a,b\} \leadsto \{a\}$ in contrast to $(a \to Stop) \overleftrightarrow{\Box} (b \to Stop)$ for which $\langle\rangle : \{a,b\} \leadsto \{a,b\}$. Taking this last compliant process as an example, it is represented as a single *behaviour* p. p is a function: its domain is the set of traces that the process can perform. In this case $\operatorname{dom} p = traces(p) = \{\langle\rangle, \langle a\rangle, \langle b\rangle\}$: in any execution at most one of $\langle a\rangle$ and $\langle b\rangle$ will be involved. p maps a trace to another function: that which records the acceptances. Thus $p(\langle a\rangle)$ is just the function with typical maplet $X \mapsto \emptyset$ which we wrote above as $X \leadsto \emptyset$. So p has type $\Sigma^* \nrightarrow (\mathbb{P}\,\Sigma \to \mathbb{P}\,\Sigma^{\checkmark\text{✗}})$, where Σ is the set of all events. Σ^* is the set of all possible traces: notice that we are now regarding p as a partial function since it is only defined on a subset of 3 traces. $\mathbb{P}\,\Sigma$ is the set of all possible offers which can be made to p. And the response is drawn from $\mathbb{P}\,\Sigma^{\checkmark\text{✗}}$ which is similar except that there is also the possibility of either $\checkmark$ representing termination or ✗

p.

In general, in acceptance semantics, a process 'consists of' multiple behaviours. A very simple case is *Stop* $\sqcap$ $(a \rightarrow Stop)$. This has two behaviours. One matching *Stop* for which $\langle\rangle : X \rightsquigarrow \emptyset$ and the other matching $a \rightarrow Stop$ for which $\langle\rangle : X \rightsquigarrow X \cap \{a\}$ and $\langle a\rangle : X \rightsquigarrow \emptyset$.

These very simple ideas are all that are involved in acceptances. But we do need to state the axioms which are all rather obvious given familiarity with the notation. We write the set of behaviours defining a process P as $\mathcal{B}\,P$, and functions are written in Curried form.

H1: $$\forall b \in \mathcal{B}\,P \bullet \quad \langle\rangle \in traces(b)$$

H2: $$\forall b \in \mathcal{B}\,P \bullet \forall s \in \Sigma^* \bullet \forall x \in \Sigma \bullet \\ s^\frown\langle x\rangle \in traces(b) \Leftrightarrow s \in traces(b) \wedge (\exists X \subseteq \Sigma \bullet x \in bsX)$$

H3: $$\forall b \in \mathcal{B}\,P \bullet \forall s \in traces(b) \bullet \forall X \subseteq \Sigma \bullet \quad bsX \subseteq X^{\checkmark✗}$$

H4: $$\forall b \in \mathcal{B}\,P \bullet \forall s \in traces(b) \bullet \forall X, Y \subseteq \Sigma \bullet \\ bsX = \emptyset \wedge Y \subseteq X \Rightarrow bsY = \emptyset \\ \wedge \\ bsX \cap Y \neq \emptyset \Rightarrow bsY \neq \emptyset$$

H5: $$\forall b \in \mathcal{B}\,P \bullet \forall s \in traces(b) \bullet \forall X, Y \subseteq \Sigma \bullet \\ bsX \cap Y^{\checkmark,✗} \neq \emptyset \wedge Y \subseteq X \Rightarrow bsY = bsX \cap Y^{\checkmark,✗}\,.$$

Properties from these axioms are captured in

(C1)	$bsX = \emptyset$	$\wedge\ Y \subseteq X$	$\Rightarrow$	$bsY = \emptyset$	
(C2)	$bsX \cap Y \neq \emptyset$		$\Rightarrow$	$bsY \neq \emptyset$	
(C3)	$bsX \cap Y^{\checkmark✗} \neq \emptyset$	$\wedge\ Y \subseteq X$	$\Rightarrow$	$bsY = bsX \cap Y^{\checkmark✗}$	,

for a typical trace s where X and $Y \subseteq \Sigma$. We will need the following in our definition of hiding later:

Definition 3.1 behave(b) *is an abbreviation for*

$$b : \Sigma^* \nrightarrow \left(\mathbb{P}\,\Sigma \rightarrow \mathbb{P}\,\Sigma^{\checkmark✗}\right) \\ \wedge \\ \langle\rangle \in \operatorname{dom} b \\ \wedge \\ \forall s \in \operatorname{dom}(b) \bullet \forall X \subseteq \Sigma \bullet\ bsX \subseteq X^{\checkmark✗} \\ \wedge \\ \forall s \in \Sigma^* \bullet \forall x \in \Sigma \bullet \\ s^\frown\langle x\rangle \in \operatorname{dom}(b) \Leftrightarrow s \in \operatorname{dom}(b) \wedge (\exists X \subseteq \Sigma \bullet x \in bsX) \\ \wedge \\ \forall s \in \operatorname{dom} b \bullet \forall X, Y \subseteq \Sigma \bullet \\ bsX = \emptyset \quad \wedge\ Y \subseteq X \Rightarrow \quad bsY = \emptyset \quad \wedge \\ bsX \cap Y \neq \emptyset \quad \Rightarrow \quad bsY \neq \emptyset \quad \wedge \\ bs(X) \cap Y^{\checkmark✗} \neq \emptyset \ \wedge\ Y \subseteq X \Rightarrow bsY = bsX \cap Y^{\checkmark✗}\quad .$$

4 Overtures and Hesitant Offers

Consider

$$\Big((a \rightarrow \mathit{Stop}) \overleftarrow{\Box} (b \rightarrow \mathit{Stop})\Big) \setminus \{b\}$$

Suppose that it is offered $\{a\}$. Then we expect that a will be performed because it has higher priority. If an offer which does not include a is made, we expect the internal hidden b to happen, so the offer is refused: none of the offered external events occurs. So at first sight, one might expect $\Big((a \rightarrow \mathit{Stop}) \overleftarrow{\Box} (b \rightarrow \mathit{Stop})\Big) \setminus \{b\} = a \rightarrow \mathit{Stop}$. In fact, there is another possibility: an offer of $\{a\}$ can be refused. In general

$$\Big((a \rightarrow \mathit{Stop}) \overleftarrow{\Box} (b \rightarrow \mathit{Stop})\Big) \setminus \{b\} = (a \rightarrow \mathit{Stop}) \sqcap \mathit{Stop} \qquad (2)$$

This shows that we need to take care in the definition of hiding.

The purpose of this section is to explain how the nondeterminism in hiding, not all of which is immediately obvious, arises. To do this we introduce *overtures* and *hesitant offers* which are tools built on the foundation of acceptances. It is particularly important to use precise instruments since $\mathcal{CSPP}$ explores new territory: hiding in the presence of priority.

To fix ideas, suppose that $P = (a \rightarrow \mathit{Stop}) \overleftarrow{\Box} (b \rightarrow \mathit{Stop})$ is implemented as a synchronous state machine. Hiding b means that it is continuously available. Suppose this is implemented on an input wire *b_ready*: it will be continuously active. If a is offered to the circuit, this means that the corresponding *a_ready* is similarly active. If both signals are sampled as active when the circuit is first clocked, then a will happen and the circuit behaves as $a \rightarrow \mathit{Stop}$.

However, the *a_ready* signal will typically come from another circuit, and may not be ready on the first clock edge. On some later clock edge, the other circuit will be ready to communicate, and only then will *a_ready* become active. But meanwhile, our circuit will have performed the hidden event b on the first clock edge. And will have ceased activity since it now behaves as *Stop*. Effectively our circuit has been made the external offer $\emptyset$ before a later offer of $\{a\}$. We call a pair of offers of the form $\langle \emptyset, X \rangle$ a *hesitant offer*. So it is clear that P will behave as *Stop* when it receives an hesitant offer: $\mathit{Stop} \sqsupseteq P$. And equation 2 is explained.

In more general situations, especially when software rather than circuits are the target, 'tentative offers' can arise. Consider the case of two processes P and Q attempting to communicate. These might have conflicting priorities. A scheduler must resolve the conflict and determine which events can be agreed by the two processes. The scheduler may need to query each process to determine which events are acceptable in response to various offers. Thus offers may be made in order to determine the response, rather than to actually permit the event at this preliminary stage. Thus a process may be subjected to an *overture* of offers of the form $\langle X_1, \ldots, X_n \rangle$. Only the last offer X_n may actually result in an event being performed, even through earlier tentative offers may have signaled acceptance of at least one event. Such a scheduler strategy is not advocated here. We would expect practical language implementations to apply usage rules to exclude most if not all cases of priority conflict in the interests of efficiency. But $\mathcal{CSPP}$ is a closed algebra and so defines the outcome of such conflicts. And in the interests of abstraction, it must permit all possible implementations. Hiding must cover such situations.

The conclusion from the above is that an offer X should always be regarded as *the last element of an overture* $\langle X_1, \ldots, X \rangle$. But hiding is the only case in which this matters. For example, tentative offers made before a final offer to $a \rightarrow \mathit{Stop}$ make no difference. The response to X is always $X \cap \{a\}$. But not so for $\Big((a \rightarrow \mathit{Stop}) \overleftarrow{\Box} (b \rightarrow \mathit{Stop})\Big) \setminus \{b\}$. The overture $\langle\{c,d\},\{e,f\},\{a\}\rangle$ results in a refusal, $\{a\} \rightsquigarrow \emptyset$, while $\langle\{a,d\},\{a,c\},\{a\}\rangle$ gives $\{a\} \rightsquigarrow \{a\}$. It is important here to realize that we are here using the ideas of acceptances in a dynamic way for understanding and to guide intuition. Our lack of knowledge of the particular overture leading up to an offer is modelled in static nondeterminism. Equation 2 just gives two ways of writing exactly the same process: neither has any trace of hidden state — that resides only in our intuitive understanding of the left hand side. But the dynamic intuition does explain *why* we equate the two expressions. This rather obvious point has occasionally been misunderstood.

Overtures seem to model all the sorts of implementation that arise, including asynchronous circuits. We use them to motivate the formal definition of hiding in $\mathcal{CSPP}$ below.

The reason that we pick out certain special overtures of the shape $\langle\emptyset, X\rangle$ and call them hesitant offers is that these are maximally permissive with respect to hidden events. Consider $P \setminus H$ and suppose that p is a behaviour of P. Suppose now that X is offered externally: the events of H are always available internally. So effectively, the internal offer is $X \cup H$. If some hidden $h \in H$ can be performed, that is $h \in pt(X \cup H)$ for some trace t of p, then the hidden trace t and event h can also occur when the offer is $\emptyset$. This is from **C3** which gives

$$pt(X \cup H) \cap H \neq \emptyset \;\Rightarrow\; ptH = pt(X \cup H) \cap H \;.$$

So hesitant offers uncover all the possible initial states. Thus we normally need only bother with the two special overtures $\langle X\rangle$ and $\langle\emptyset, X\rangle$ when deciding on the possible responses to the offer X.

5 Defining Hiding

In this section we build the general definition of hiding in $\mathcal{CSPP}$. A process is identified with a set of behaviours, and these behaviours must obey the axioms. So we will only admit functions q which obey the abbreviation $\mathsf{behave}(q)$. When we define $P \setminus H$, each behaviour $q \in \mathcal{B}(P \setminus H)$ arises from some matching behaviour p of P.

To define a behaviour q we need to specify what happens on one of its traces s. This trace is externally visible, and will match a hidden trace $t \in \mathit{traces}(p)$. s will just be t with any hidden events from H filtered out: $s = t \setminus H$. And there will be a most recent visible event of t.We capture the position in t where that occurs by an abbreviation:

Definition 5.1
$\mathsf{LastExposure}(t, H)$ *is an abbreviation for* $\max(\{n \in \operatorname{dom} t \mid t(n) \notin H\} \cup \{0\})$.

So $\mathsf{LastExposure}(\langle h_1, h_2\rangle, \{h_1, h_2, h_3\}) = 0$ and $\mathsf{LastExposure}(\langle h_2, a, h_3\rangle, \{h_1, h_2, h_3\}) = 2$.

The next abbreviation identifies all the possibilities for hidden progress during overtures. The last visible event, if any, is the last element of s. This matches the position $\mathsf{LastExposure}(t, H)$ of t. We admit any t which involve further trailing events drawn from H under any conceivable overture. This is all accessible members of H except any eliminated by priorities within H itself. All such t give rise to the same s. We will see below that some extensions of t may also contribute to the acceptances at s, and for that reason we define a more general abbreviation which filters the 'tail' of t to permit only events available under an offer X:

Definition 5.2 *Let p be a behaviour and t a trace.* $\mathsf{Accessible}(p, t, X, H)$ *is an abbreviation for*

$$\begin{array}{c} \forall\, 1 \le n \le \#t \bullet \exists\, Y \subseteq \Sigma \bullet t(n) \in p(t \downharpoonright (n-1))(Y \cup H) \\ \wedge \\ \forall\, \mathsf{LastExposure}(t, H) \le n < \#t \bullet t(n) \in p\,(t \downharpoonright (n-1))\,(X \cup H) \end{array} \qquad (3)$$

The notation $t \downharpoonright n$ is trace truncation: $t \downharpoonright 0 = \langle\rangle$ and $\langle t_1, \ldots, t_n\rangle \downharpoonright i = \langle t_1, \ldots, t_i\rangle$ at least when $i \leqslant n$. In the light of our earlier result from **C3**, setting $X = \emptyset$ above then identifies all the possible t matching s as above. These will be the possible 'starting points' for s: a shortest trace underlying s. When there are several possibilities they constitute nondeterminism. In the case of equation 2, take $s = \langle\rangle$. Then both $t = \langle\rangle$ and $t = \langle b\rangle$ are possible starting points. The two possibilities match the two components on the right hand side. That is $\mathsf{Accessible}(p, \langle\rangle, \emptyset, \{b\})$ and $\mathsf{Accessible}(p, \langle b\rangle, \emptyset, \{b\})$ are true for the only behaviour p of the hidden process on the right hand side.

The introduction noted that divergence can happen when an unbounded stream of events is hidden: external control of these events is lost, and the process can 'run wild':

Definition 5.3 $\mathsf{Wild}(p, t, X, H)$ *is an abbreviation for*

$$\exists w \in H^{\omega} \bullet \forall n > 0 \bullet (t^\frown(w \downharpoonright n)) \in traces\,(p) \wedge w(n) \in p(t^\frown(w \downharpoonright (n-1)))(X \cup H)\ , \quad (4)$$

where $n \in \mathbb{N}$ is understood.

w above is any infinite trace of events from H: $w = \langle h_1, h_2, h_1, h_2, \ldots\rangle$ perhaps when $H = \{h_1, h_2, h_3\}$. That provides a simple way of saying that there are arbitrary hidden extensions of the trace t when X is being offered. Notice that if a process goes into an internal loop during a tentative offer in an overture, it might be pulled back out of the loop when a later different offer is made. What matters is whether it is in an infinite loop under the final offer. Of course, in most situations, the external offer is irrelevant. But

$$\mu P \bullet \left((a \to Stop) \overleftarrow{\Box} (b \to P)\right) \setminus \{b\} = (a \to Stop) \overleftarrow{\Box} \mathsf{div}$$

is a process that can eventually be pulled out of a loop by an offer of a. Processes of this type are sometimes used in hardware, so the extra divergence information retained in $\mathcal{CSPP}$ compared to standard CSP is useful. It is also realistic since $\mathcal{CSPP}$ handles infinite behaviour properly, and divergence is inherently unbounded.

The next abbreviation is the crux of the definition of hiding: it specifies the responses

Definition 5.4 $\mathsf{Slide}(q, p, s, t, X, H)$ *is an abbreviation for*

$$qsX = \left(\begin{array}{c} \bigcup \left\{ (pt'(X \cup H)) \setminus H \;\middle|\; \begin{array}{c} t' \geqslant t \wedge t' \in traces\,(p) \wedge s = t' \setminus H \\ \wedge \\ \mathsf{Accessible}(p, t', X, H) \end{array} \right\} \\ \cup \\ (\{\text{✗} \end{array} \right) \quad (5)$$

✗ is the token representing livelock (divergence). Here t is the 'starting ' point for the trace s: as we have seen there are often several choices for that, and much of the nondeterminism in hiding arises in this way. But we will see that the response to an offer X can involve 'sliding' along trace extensions t'. The best way to understand that is with an example:

$$\left((a \to Stop) \overleftarrow{\Box} (h \to b \to Stop)\right) \setminus \{h\} = (a \to Stop) \overleftarrow{\Box} (b \to Stop) \quad \sqcap \quad b \to Stop$$

First of all the process might perform h internally before an effective offer is made: then it behaves like $b \to Stop$. That disposes of the second component on the right above. But suppose that an offer X is made before the hidden process has performed b. If $a \in X$, then

a occurs. However, if a is not available, then h happens: the processes 'slides' down an extended trace and then behaves like $b \to Stop$. It is this phenomena that Slide captures.

We are now almost equipped to define hiding in general, but there is one extra step. We need to make sure that we allocate 'starting traces' t to externally visible traces s in consistent way. We do that very simply with a function $\theta : traces\,(q) \to traces\,(p)$. A function ensures that there is a unique t for each s. We call this a 'locater' and it must be monotone:

Definition 5.5 $\mathsf{locater}(\theta, S, H)$ *is an abbreviation for*

$$\forall\, s, s' \in S \bullet \left(\begin{array}{c} s > s' \;\Rightarrow\; \theta(s) > \theta(s') \\ \wedge \\ s = \theta(s) \setminus H \end{array} \right)$$

Armed with these abbreviations, we can now give the definition which is rather simple and hopefully now transparent:

$$\mathcal{B}\,(P \setminus H) = \left\{ q \;\middle|\; \begin{array}{c} \mathsf{behave}(q) \;\wedge\; \exists\, p \in \mathcal{B}\,P \bullet \mathrm{dom}\, q \subseteq (\mathrm{dom}\, p) \setminus H \\ \wedge \\ \exists\, \theta : \mathrm{dom}\, q \to \mathrm{dom}\, p \bullet \mathsf{locater}(\theta, \mathrm{dom}\, q, H) \\ \wedge \\ \forall\, s \in traces\,(q) \bullet \\ \mathsf{Accessible}(p, \theta(s), \emptyset, H) \\ \wedge \\ \forall\, X \subseteq \Sigma \bullet \mathsf{Slide}(q, p, s, \theta(s), X, H) \end{array} \right\}. \qquad (6)$$

Trying this on one of the simplest of compliant examples gives

$$\Big((a \to Stop) \overset{\leftrightarrow}{\Box} (b \to Stop)\Big) \setminus \{b\} = (a \to Stop) \;\sqcap\; Stop$$

We mentioned in section 2 that one way of embedding standard CSP in $\mathcal{CSPP}$ was to identify standard operators with the compliant versions. With that identification, we have reproduced the standard result $((a \to Stop) \Box (b \to Stop)) \setminus \{b\} = (a \to Stop) \;\sqcap\; Stop$.

In the more general embedding of standard CSP, we have

$$\begin{aligned}
((a \to Stop) \Box (b \to Stop)) \setminus \{b\} \;&=\; \left(\begin{array}{c} (a \to Stop) \overset{\leftarrow}{\Box} (b \to Stop) \\ \sqcap \\ (a \to Stop) \overset{\leftrightarrow}{\Box} (b \to Stop) \\ \sqcap \\ (a \to Stop) \overset{\rightarrow}{\Box} (b \to Stop) \end{array} \right) \setminus \{b\} \\
&= \left(\begin{array}{c} \Big((a \to Stop) \overset{\leftarrow}{\Box} (b \to Stop)\Big) \setminus \{b\} \\ \sqcap \\ \Big((a \to Stop) \overset{\leftrightarrow}{\Box} (b \to Stop)\Big) \setminus \{b\} \\ \sqcap \\ \Big((a \to Stop) \overset{\rightarrow}{\Box} (b \to Stop)\Big) \setminus \{b\} \end{array} \right) \\
&= \left(\begin{array}{c} (a \to Stop) \sqcap Stop \\ \sqcap \\ (a \to Stop) \sqcap Stop \\ \sqcap \\ Stop \end{array} \right) \\
&= (a \to Stop) \sqcap Stop
\end{aligned}$$

so we reproduce the standard result once again. Revisiting the more general case in equation 1 on page 104, we have

$$\begin{aligned}
\Big((a \to P) \overleftarrow{\Box} (b \to Q)\Big) \setminus \{b\} &= (a \to P') \overleftarrow{\Box} Q' \sqcap Q' \\
\Big((a \to P) \overleftrightarrow{\Box} (b \to Q)\Big) \setminus \{b\} &= (a \to P') \overleftrightarrow{\Box} Q' \sqcap Q' \\
\Big((a \to P) \overrightarrow{\Box} (b \to Q)\Big) \setminus \{b\} &= Q'
\end{aligned}$$

When we identify standard CSP with the compliant subset, the second equation above is identified with equation 1. But in the more general embedding, we have

$$\begin{aligned}
((a \to Stop) \Box (b \to Stop)) \setminus \{b\} &= (a \to P') \overleftarrow{\Box} Q' \sqcap (a \to P') \overleftrightarrow{\Box} Q' \sqcap Q' \\
&\sqsupseteq ((a \to P') \Box Q') \sqcap Q'
\end{aligned}$$

That is we have a *refinement* of the standard result. This is entirely to be expected: we can draw finer distinctions in $\mathcal{CSPP}$ than we can in CSP.

A more interesting case of hiding a compliant process is

$$\Big((a \to Stop) \overleftrightarrow{\Box} (b \to c \to Stop)\Big) \setminus \{b\} = (a \to Stop) \overleftrightarrow{\Box} (c \to Stop) \sqcap c \to Stop$$

6 Hiding and Recursion

A proper semantics for any variant of CSP must define recursion properly. In the case of a denotational semantics, it lies at the heart of the theory. Recursion is defined by some sort of fixed point theory. This in turn is initimately connected with infinite behaviour, and acceptance semantics handles this in a novel and powerful way.

CSP and its variants have a much wider scope than just finite state machines that can be implemented in software and hardware. It is a much more powerful mathematical tool. It is typical of such mathematical theories that they can describe some rather exotic beasts. Thus there are some processes, usually involving infinite behaviour in the form of unbounded nondeterminism, which have been awkward to handle in previous semantics. This is not a defect of CSP, but a sign of its power. The acceptance semantics of $\mathcal{CSPP}$ completely solves all these problems in a simple way.

There are two complementary ways of locating fixed points in the acceptance approach to $\mathcal{CSPP}$:

1. $\mathcal{CSPP}$ is a complete lattice under refinement.

2. $\mathcal{CSPP}$ is a complete pseudometric space.

The approach based on refinement uses the classic Knaster-Tarski fixed point theorem which shows that any monotone function f on a complete lattice has a least fixed point. It is given by $\mu f = \bigwedge\{x \mid x \sqsupseteq f(x)\}$ which we would write here as $\mu f = \bigsqcap\{x \mid x \sqsupseteq f(x)\}$. If f is also *continuous* with respect to the partial order, there is a more constructive form for the fixed point, but we will omit that here.

Refinement is very simple in acceptance semantics: it is just set inclusion in the behaviours: $P \sqsupseteq Q \;\Leftrightarrow\; \mathcal{B}P \subseteq \mathcal{B}Q$. That in turn means means that hiding is monotone: $P \sqsupseteq Q \Rightarrow (P \setminus H) \sqsupseteq (Q \setminus H)$ which follows by inspection of equation 6. Since all the other $\mathcal{CSPP}$ operations are also monotone, we find that recursions involving hiding are meaningful: they represent the least fixed point. A rather extreme example is $\mu P \bullet P \setminus H = \bot \setminus H$, where $\perp$ is the most nondeterministic of all processes. Another example which will be of interest below is $P = a \rightarrow P \setminus \{a\}$. It is easy to see that the most nondeterministic solution is $a \rightarrow \perp \setminus \{a\}$. Other examples appeared in the last section.

The complementary way to define fixed points is to use a metric, or in the case of $\mathcal{CSPP}$, a pseudometric d. A *pseudo*metric because there are certain distinct processes are not distinguished by d. When functions are contracting with respect to d, then a fixed point exists. Matching recursions are called *constructive*. Because d is a pseudometric, there are cases where the fixed points are not unique, but those are well defined, and one can always identify the cluster of 'matching' processes. In such cases, we just take the most nondeterministic such process as the meaning of the recursion which matches the least fixed point with respect to refinement. This is an application of Janes maximum entropy principle, so we would have made this choice even in the absence of a refinement order. Most applications of recursion use constructive functions, and most $\mathcal{CSPP}$ operators are contracting. However, hiding is an exception. Our earlier example of $P = a \rightarrow P \setminus \{a\}$ is a guarded recursion, yet is not contracting. This is a standard phenomena and indeed a standard example, but we mention it here for completeness. This does not necessarily mean that the presence of hiding in a recursion prevents it from contracting, but it does mean that each recursion involving hiding must be examined individually.

7 Conclusions

Hiding has been shown to be a fundamental part of CSP and so also of $\mathcal{CSPP}$. Indeed it dictates the presence of $\sqcap$ and div, and accounts for much of the flavour of the theory.

The nondeterminism that is asociated with hiding has been carefully investigated and explained, and the ideas of *hesitant offer* and *overture* formulated as precise tools for the investigation.

Hiding has been extended to handle priority properly, and the formal definition in $\mathcal{CSPP}$ acceptance semantics presented with careful justification and explanation, all with a full treatment of unbounded behaviour.

References

[1] C.A.R Hoare. *Communicating Sequential Processes*. Prentice Hall International, 1985.

[2] A.W. Roscoe. *The Theory and Practice of Concurrency*. Prentice Hall, 1998.

[3] Davide Sangiorgi and David Walker. *The π-calculus*. Cambridge University Press, 2001.

[4] Robin Milner. *Communicating and Mobile Systems: the π-calculus*. Cambridge University Press, 1999.

[5] P.H.Welch. Process Oriented Design for Java: Concurrency for All. In H.R.Arabnia, editor, *Proceedings of the International Conference on Parallel and Distributed Processing Techniques and Applications (PDPTA'2000)*, volume 1, pages 51–57. CSREA, CSREA Press, June 2000.

[6] P.H.Welch, J.R.Aldous, and J.Foster. CSP networking for java (JCSP.net). In P.M.A.Sloot, C.J.K.Tan, J.J.Dongarra, and A.G.Hoekstra, editors, *Computational Science - ICCS 2002*, volume 2330 of *Lecture Notes in Computer Science*, pages 695–708. Springer-Verlag, April 2002.

[7] Peter H. Welch and Jeremy M.R. Martin. Formal analysis of concurrent java systems. In *Communicating Process Architectures – 2000*, Concurrent Systems Engineering, pages 275–301, Amsterdam, Sept 2000. IOS Press.

[8] A.E. Lawrence. Extending CSP: Denotational semantics. *IEE Proceedings: Software*, 150(1):51–60, Feb 2003.

[9] A. E. Lawrence. Acceptances, Behaviours and infinite activity in CSPP. In *Communicating Process Architectures – 2002*, Concurrent Systems Engineering, pages 17–38, Amsterdam, Sept 2002. IOS Press.

[10] A.E. Lawrence. CSP extended: imperative state and true concurrency. *IEE Proceedings: Software*, 150(1):61–69, Feb 2003.

[11] A. E. Lawrence. HCSP, imperative state and true concurrency. In *Communicating Process Architectures – 2002*, Concurrent Systems Engineering, pages 39–55, Amsterdam, Sept 2002. IOS Press.

[12] A.W. Roscoe, editor. *A Classical Mind*. Prentice Hall Series in Computer Science. Prentice Hall, 1994. Essays in Honour of C.A.R. Hoare.

Communicating Process Architectures 2003
Jan F. Broenink and Gerald H. Hilderink (Eds.)
IOS Press, 2003

Sampling and Timing: A Task for the Environmental Process†

Gerald H. HILDERINK and Jan F. BROENINK
Twente Embedded Systems Initiative,
Drebbel Institute for Mechatronics and Control Engineering,
Faculty of EE-Math-CS, University of Twente,
P.O.Box 217, 7500 AE, Enschede, the Netherlands
g.h.hilderink@utwente.nl

Abstract. Sampling and timing is considered a responsibility of the environment of controller software. In this paper we will illustrate a concept whereby an environmental process and multi-way events play an important role in applying timing for untimed CSP software architectures. We use this timing concept for building our control applications based on CSP concepts and with our CSP for C++ (CTC++) library. We present a concept of sampling of control applications that is orthogonal to the application. This implies global timing on the basis of timed events. We also support traditional local timing on the based of timed processes,

1. Introduction

At Control Engineering at the University of Twente, we use concepts from the theory of *Communicating Sequential Processes* (CSP) [1; 2] to design real-time control software architectures and to reason about concurrency and their reactive/real-time behaviours [3]. The CSP concepts contribute in managing complexities and provide guidelines for a clean and prescribed software development approach. The CSP concepts provide an excellent separation of concerns at a high level of abstraction in terms of processes and their interrelationships. The result is that nothing will be designed or implemented in an ad-hoc manner and students easily learn these abstract concepts of designing and implementing real-time software.

The real-time control software is the software part of a mechatronic system that aims at improving the dynamics of the system or to automate it. The dynamic behaviour of the plant, i.e. the 'machine'-part of the mechatronic system, imposes timing requirements on the software. The control loop from obtaining sensor data via processing to sending actuator signals must be completely processed within every sampling period *Ts*. This sample period *Ts* is determined by the dynamic behaviour of the plant and lies in our domain typically between 0.1 to 10 ms. Furthermore, all sensors must be sampled at the same moment causing as little jitter as possible (sampling). The actuators too must set their new value at the same moment (actuation). In practice it is often required that sampling is done before actuation in order to avoid disturbance of the measurements by the actuation. The distance between sampling and actuation is fixed within the sampling period. See Figure 1. The inputs can be done in parallel and the outputs can be done in parallel, or a

† This research is supported by PROGRESS, the embedded system research program of the Dutch organization for Scientific Research, NWO, the Dutch Ministry of Economic Affairs and the Technology Foundation STW.

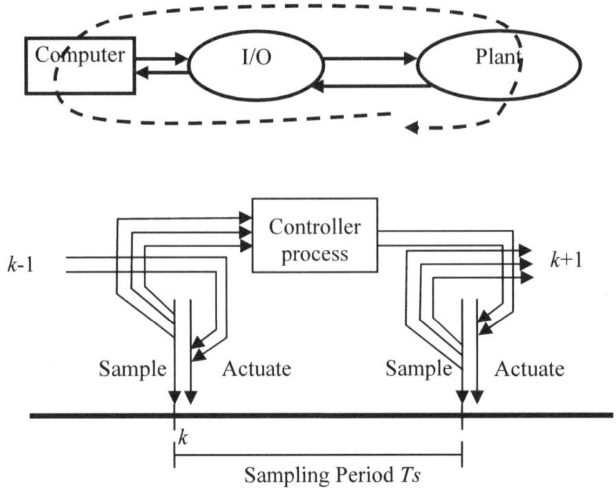

Figure 1 Real-time control loop (above) and timing of one sample–process–actuate–loop where the kth sample is shown (below)

fixed sequence can sometimes be more practical when converters are under control of the software.

Real-time implies on the one hand that the response is expected before a deadline or precisely on a specified moment (neither before nor after). Its timeliness implies correctness. On the other hand real-time systems are intrinsically concurrent because they are coupled to the real world. Discrete control systems require a constant and precise sampling rate of their inputs and outputs to the real world they control. With precise we mean a predefined moment in time within a narrow range of variation that is acceptable to ensure a reliable and stable control system. Furthermore, the processes between these inputs and outputs should guarantee that they perform their tasks and deliver their outputs for actuation before their deadlines expire. Each control loop deals with a parallel aspect of the system and therefore control loops are naturally parallel to each other.

Sensors and actuators require signal conversions between the continuous-time and real world representation and discrete-time and digital representation. Common converters are Analog to Digital Converters (ADCs), Digital to Analog Converters (DACs), and counters. These converters require accurate timing with as less jitter as possible between the input sampling and output actuation, and with a fixed distance between the successive sampling events.

A shared clock between the converters is the fastest and therefore the most accurate solution, but not every interface card has timers providing a shared external timer among all peripheral devices. Often configurations require the system timer; the internal timer on the computer board where the software (e.g. the interrupt service routine) has control over the converters. The correspondence between these different configurations of external and internal timers is that timing is part of the environment of the application and orthogonal to the application. This separation of concerns can be used to our advantage when we make an untimed design timed through encapsulation within its environment orthogonal to the design.

We use a graphical CSP notation by Hilderink [4] to illustrate the examples in the form of CSP diagrams.

In this paper we propose a conceptual approach to incorporate timing for accurate sampling on external CSP channels in software. This approach differs from the solution found in the programming language occam [5] and the CSP for Java libraries JCSP [6] and

CTJ [7]. This proposal may become a basic concept behind the timing framework of CT developed by Hilderink [7]. CT stands for *Communicating Threads* which provides concurrent programming where multithreading in execution is encapsulated in CSP-like abstractions of processes and channels. Currently, this timing framework has been successfully implemented in CT for C++ (CTC++).

Section 2 discusses the problem with timed processes as a result of local timing. The conceptual background to this proposal from which our solution is derived is discussed in Section 3. The solution in the form of environmental services is presented in Section 4. Section 5 applies these services in two examples.

2. Sampling problems with timed processes

In occam and on the transputer, time was provided by a TIMER object [5]. The TIMER object is a kind of channel on which processes can read the time or on which processes get blocked until the specified time has expired. The same syntax is used to specify a timeout-guard in the alternative construct. A similar object, called CSTimer, is defined in JCSP [6]. CTJ implements a Java-like static sleep method that will block a process for a specified time and CTJ provided separate timeout-guards for timeouts on alting. These timing concepts in occam, JCSP and CTJ are encapsulated in processes, called *timed processes*. The problem is that these processes cannot guarantee precise sampling on channel communication. In particular, external channels are concerned with data conversion between the digital computer and the analog world at specified moments in time. The timed processes do not provide an adequate solution to sampling and timing issues on channels.

In this example, we illustrate a few problems with timed processes concerning sampling. Consider two independent controller processes, LController and HController, as depicted in Figure 2. The channels c_1, c_2, and d_1 are connected to sensors and channels c_3 and d_2 are connected to actuators. In this example, input-channels c_1, c_2, and d_1 communicate with an AD converter each and output-channels c_3 and d_2 communicate with a DA converter each. Of course, processes see channels and *not* the converters. In this sense they are hardware independent. They only depend on the integrity of the values. The HController process repeatedly reads values from channel d_1 and after computation the result is sent to channel d_2. The LController process repeatedly reads values from channel c_1 and c_2 and after computation the result is send over channel c_3.

At the open-ends of the arrows one can imagine processes in hardware to which processes in software communicate. These processes in hardware are not rendered in the diagram, because they are out of the context of the control software.

The sensor data on channels c_1, c_2 should be sampled at the same time. Channel c_3 should be actuated shortly after c_1 and c_2 are sampled, in order to not disturb the measurements at c_1 and c_2. This is similar for the channels d_1 and d_2 of HController but this can be done at the same or a different sampling interval than for LController.

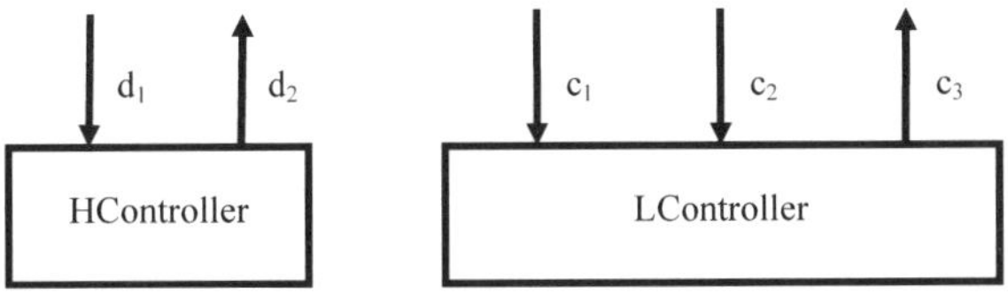

Figure 2 Example: control application consisting of a higher-priority controller process and a lower-priority controller process

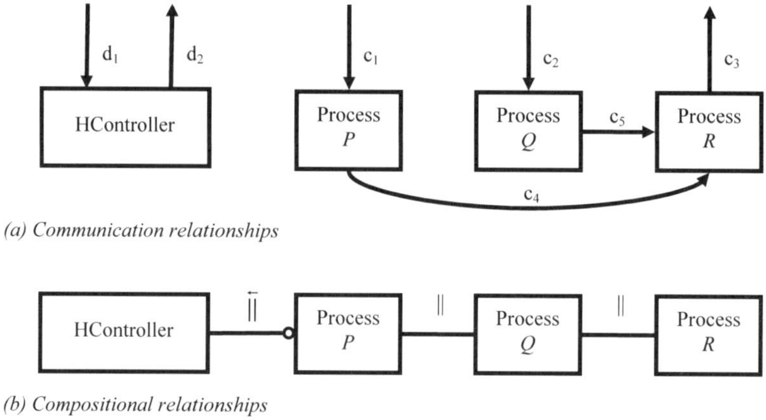

Figure 3 Lower-priority controller process divided into sub-processes

In this example we will assume that HController operates at a higher sampling frequency than LController. This requires that HController executes at a higher priority then LController. The problems that rise in this situation are common problems with timed processes. We will illustrate this by the following example in which LController consists of three sub-processes each carrying out a computational task, see Figure 3a. We will omit their functional descriptions.

Process *P* receives information from channel c_1, process *Q* receives information from channel c_2, and process *R* determines the output and sends the results to channel c_3. The processes *P* and *Q* send the results of their calculations to process *R* via the internal channels c_4 and c_5. All processes are part of the same control loop and therefore the channels c_1 and c_2 should be sampled and c_3 should also be actuated at the same sampling frequency. As mentioned in Section 1, the order of reading and writing should be fixed and reads and writes should be very close to each other: first input on c_1 then input on c_2 then output on c_3. Any varying delay between the inputs and outputs can cause significant jitter that may bring the system into an unstable state. Figure 3b shows that all three processes are meant to execute in parallel, $P \parallel Q \parallel R$. This example suffers from the following problem when we apply the timed processes for sampling.

Sampling cannot be based solely on reading and writing on channels. Pre-emption and arbitrary scheduling of processes can cause a varying delay between reading and writing on channels. Thus, sampling on reading and actuation on writing on channels is not an adequate solution. The resulting variation in sampling intervals or jitter can make plant unstable and unpredictable. It is important that jitter is negligibly small which makes any delay approximately constant compared to the sample time Ts. Thus, scheduling of the processes does not guarantee a fixed order of inputs and outputs. Any order is possible. Serializing these processes does not really solve the problem. This is because the computations that are performed by the processes cause a distance (delay) between the inputs and outputs. To overcome this problem we need to make sure that sampling and actuation is done fast and in direct sequence. Disassembling the processes and moving the inputs and outputs to another sequential process is not an option because we want to specify three parallel tasks for some reason that is not addressed here. Serializing requires breaking the internal loops of the processes into a common outer loop. The problem increases when multiple frequencies are involved. This approach has tremendous effect on architectural changes and prevents reusing processes.

A common problem with the sleep/after method of the timer object is that after its wakeup *the process can be delayed by a higher priority process*. This is obvious the case when two controllers at two different sampling frequencies are performed. In order to fulfil the real-time requirements we apply the rate-monotonic priority ordering; the process with the higher frequency gets higher priority than a process with a lower frequency. Atomic coherency between conversions during sampling and actuation in a control loop is required, which a (timed) process cannot provide. This particular order of sampling and actuation should be orthogonal to the software architecture otherwise this would have a tremendous impact on the design.

❑

In order to achieve sampling, special precautions in hardware or in software have to be taken. In hardware, a common external timer on which peripheral devices are triggered and an atomic sequence of inputs and outputs is guaranteed. In software, this implies some sort of ceiling priority for each sequence of pairs of inputs and outputs to guarantee adequate sampling and actuation; the processor interrupt handling can be used for this.

After conversion the devices usually generate an external interrupt that reflects a timed event on which values are read from peripherals into memory or values are copied from memory to the device. Unfortunately, not all devices are supplied with a common timer whereby devices are triggered at the same time. More often devices are triggered by a timed interrupt service routine of low-level instructions using the internal timer of the computer. Often, this is a flexible solution whereby the software stays in control over the devices.

In this paper we bring this concept of internal and external timing to a higher level of abstraction conform the CSP concepts on which we build our applications. We will apply timed events instead of timed processes, which proposal conceptually resembles with timing in hardware *and* timing in software. With the latter we mean that we can also create timed processes based on the same proposal. If this proposal is carefully applied then the software architecture becomes independent of external timing such as sampling and actuation.

3. Timing concept

3.1 Timing in CSP

Timing in CSP primarily concerns events, because *an event is an occurrence in time and space*. An event happens at a certain moment in time (e.g. periodic or sporadic) and somewhere in the system. We can observe the real-time behaviour of processes by measuring the time between events in which the processes engage. This is a simplified and a more pragmatic approach than determining the exact execution time of processes. Traces of events help determining the possible patterns of events in time. This takes into account the compositional relationships between processes. The worst case time can help in determining that processes can guarantee their deadlines. Timing analysis is not discussed in this paper.

The CSP concepts we use are based on untimed CSP [1; 2] which theory does not deal with intrinsic timing concerns. Timed CSP [8] does deal with timing, but it does not offer a pragmatic approach to be useful in software. The problem is that the timeout operator in timed CSP is far too complex and inefficient to build in software. Furthermore, the timeout operator requires semantically changes to original CSP operators.

We will address a different timing concept for untimed CSP designs, which does not need any semantically changes to the original CSP operators (i.e. ;, ||, and □) and to extended CSP operators (i.e. $\overleftarrow{\|}$, $\overleftarrow{\Box}$, and $\overrightarrow{\Delta}$). In this approach, exceptions are involved but these exceptions are orthogonal to the original semantics—the original semantics of these CSP constructs remain intact. The exact exception handling is not discussed in this paper.

CSP defines that processes communicate with each other through shared channel objects. Processes that are willing to communicate on a channel will be blocked until the other side is willing to communicate. When a reader process and a writer process engage in communication then data is passed from writer to reader. At that moment both processes will continue in parallel. This moment is called *the communication event*. For sampling it is important that those communication events occur at periodic moments in time. This involves external channels that connect the controller processes in software to processes in hardware such as signal converters to which sensors and actuators are connected.

Considering the notion of events, we believe that the time stamp of occurrence is a property of an event (see previous definition of event) and not a property of a channel. However, timing can be performed by channel communication when some sort of process at a peer-end of a channel is willing to communicate at a precise moment in time.

3.2 Environmental process and timing

In CSP, an event requires two or more processes to engage in—an event is multi-way. However, a communication event with channels is considered to be strictly two-way—one producer process and one consumer process must be willing to communicate in order to let the communication event happen. Another form of communication between multiple processes is a barrier object as described in the theory of BSP [9]. When all participating processes synchronize (block) on a barrier object then data exchange between the processes takes place and successively the barrier releases the processes to continue in parallel. The rendezvous is the communication event. With a barrier object a communication event is two-way or multi-way which value is equal to the number of participating processing ($\# \geq 2$).

In CSP textbooks, the environment is often mentioned as an external or *environmental process* that also must be willing to accept events. We will assume that this includes communication events. In this approach we take the environmental process into account which also participates in communication events. An environmental process is basically another CSP process that synchronizes on internal events. In this case a communication event can be considered to be $(n+1)$-way, where $n=2$ for channel synchronization and $n \geq 2$ for barrier synchronization. This is exactly what this approach is about.

The environment can influence the behaviour of concurrent software in several ways. This concerns the acceptance of events. In an ideal environment all events are accepted in the application. This idealness is usually the starting-point of a design. In reality, things can go wrong whereby the environment can become partially disabled. In this case the system may not engage in certain events. For example, consider a plant with embedded systems and field-busses whereby a communication cable is accidentally cut or plugged-out. When that happens no communication events will happen over a disconnected channel. The reverse is also possible when a disconnect channel becomes connected. These disabilities and abilities can rise in time. The non-idealness of the environment is considered during the refinement of the design.

More generally, when a producer process and a consumer process are willing to communicate then the environment should also accept the event in order to let the event happen. It is possible that the environment cannot accept the event.

The environment is time-variant (depending on time) and memory-based (non-anticipating) and can contain multiple timelines. The timelines can be divided in *continuous time*, *discrete time*, or a hybrid of both. For example, a processor performs its tasks in discrete time. The entire system, consisting of the computer (discrete time) connected to the real world (continuous time), is called a hybrid system. Therefore, we consider time a property of the environment being not pre-emptable. Time and the environment are orthogonal to the application. Of course, the sampling period could be a constant parameter used by the control laws that are performed by the control application, but this is simply and solely a value derived from the dynamics of the system.

In our proposal, the environmental process can perform services for the application processes that are running in the environment. These processes can call upon these services through call-channels. In this paper we introduce such an environmental process with services that provides us means for sampling. The environmental process is invisible to the application and is not part of any compositional construct in the application. This environmental process can be very complex and we do not always have to know what this environmental process exactly is. We only need to know in what way the environment can influence the application. This may or may not be in control of the application. In a distributed system, every computer system has a separate environmental process.

Figure 4 illustrates the presence of an environmental process in a CSP diagram. Consider the three processes *P*, *Q*, and *R* communicating over channels *c* and *d* in Figure 4a. For completeness, Figure 4b specifies that *P*, *Q*, and *R* are in parallel and that the objects *c* and *d* are rendezvous channels on which the processes synchronize on communication. For internal channels we assume that *P* and *Q* engage in the communication event when they both are willing to communicate over channel *c* whereby *ENV* always accepts that communication event. Also we assume that no exceptions occur. This is similar for the processes *Q* and *R* with channel *d*. This assumption is not general, because we forget the facts that events can only occur in an environment that accepts them, i.e. the environmental process *ENV* must also be willing to engage in these events. The environmental process *ENV* is the highest-priority process that is never pre-empted by the application processes. Usually, process *ENV* is hidden from the design because it is not explicit part of the design;

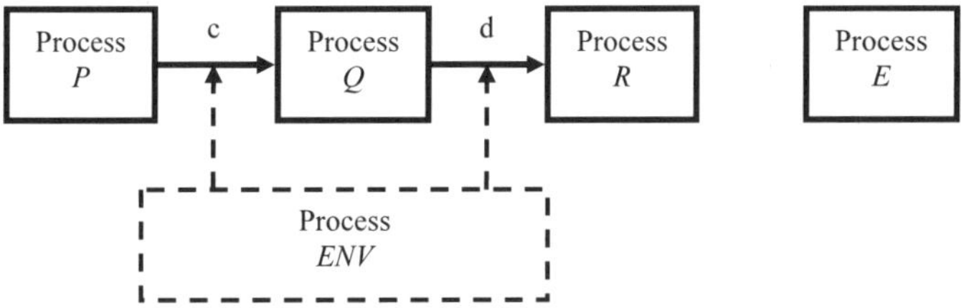

(a) Communication relationships with environmental process ENV

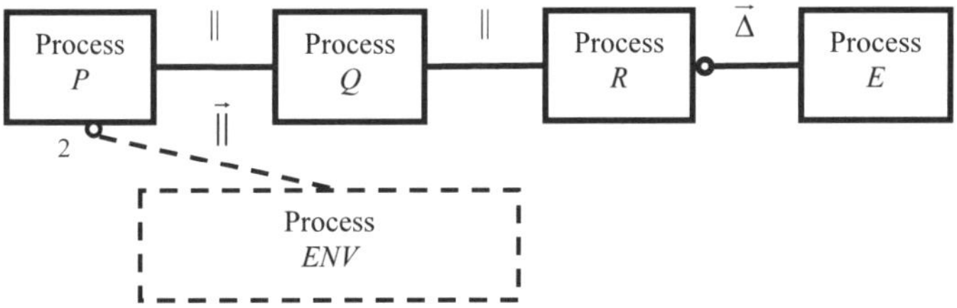

(b) Compositional relationship with environmental process ENV

Figure 4 Example with visualized environmental process

therefore the process and relationships are dashed. It is clear to see that *ENV* is connected to each channel in the design. Process *ENV* does not send any data. On error, process *ENV* can throw an exception to pair *P-Q* or pair *Q-R* with an error message that specifies the reason why the communication event cannot take place. This exception can be handled by the exception handling construct in the application, see process *E* and the $\vec{\Delta}$-relationship.

4. Environmental services

The following methods are services that can be carried out by the environmental process. These methods are called by the control application and they are served when the environmental process is willing to accept the call. We developed an `Environment` class that provides a global static call-channel which service can be invoked by any process at any time. All methods, except for the `time()` method, act upon a communication event and require a channel or a barrier. This includes timed-guards in the alternative construct. The call-channel approach makes these methods automatically thread safe.

We apply these methods systematically by accommodating these to the network building process that creates the network of processes and channels. Thus, the sampling frequencies are set at the top-level of the application.

- **Accept communication event `at` specified time**

```
Environment::at(channel, time);
Environment::at(channel, time, Time interval_time);
```

The producer and consumer processes need to be willing to communicate before the environmental process is willing to accept the communication event on the specified channel or barrier at the specified time (in microseconds) or period. If this is not the case and the processes engage in the event after the specified time expires then any blocked process will be released and an exception is thrown at the producer and at the consumer. Hence, the real-time requirement has not been met.

❑

```
Environment::at(barrier, time);                   // single-shot
Environment::at(barrier, time, interval_time);    // periodical
```

The environmental process will participate in the barrier synchronization and will commit to the synchronization at the specified time. If one participant does not sync before the environmental process then exceptions will be thrown to all processes and all processes will be released. Hence, the real-time requirement has not been met.

❑

- **Accept communication event `after` specified time**

```
Environment::after(channel, time);                // single-shot
Environment::after(channel, time, interval_time); // periodical
```

The communication between the producer and consumer processes will be delayed until specified time. Any communication after the specified time will be accepted and they both immediately continue. No exceptions are thrown. If an interval time is specified then the next waiting time will be incremented with the interval time.

❑

```
Environment::after(barrier, time);                   // single-shot
Environment::after(barrier, time, interval_time);    // periodical
```

The environmental process will participate in the barrier synchronization and will commit to the synchronization at specified time. No exceptions are thrown. If an interval time is specified then the next waiting time will be incremented with the interval time.

```
Environment::after(guard, time);                     // single-shot
Environment::after(guard, time, interval_time);      // periodical
```

If the alternative construct is waiting and the alting process at the other end is willing to communicate before the specified time then the guard will become ready at the specified time. Otherwise the guard will be ready when the alting process at the other end is willing to communicate. This guard is called a *timed-guard*. No exceptions are thrown. A timed *skip-guard* is used for specifying a *timeout-guard*. The skip-guard will be ready at the specified time and no exceptions are thrown at timeout. As with channels and barriers the guard can be periodically timed. The specified interval time increments time each period. Since guards are local to a process, implies that `after(guard,time,..)` is used locally and no other process can alter time on a local guard.

The timed-guard can be used with `at(channel,time,..)` and `after(channel,time,..)`.

The method `after(guard,time1,..)` is independent of the time as specified with `at(channel,time2,..)` or `after(channel,time2,..)`. The method `after(guard,time1,..)` on `channel` could very well endanger the deadline as specified by `at(channel,time2,..)` which results in an exception when `time1` > `time2`. Although this is a natural behaviour, this combination is not very applicable. Therefore, `after(guard,time1,..)` and `at(channel,time2,..)` should be used with care to avoid timeout exceptions

We do not support guards for barriers, because we are uncertain about multiple alting on a barrier; this introduces a conflict that is similar to two-way alting on a channel (i.e. input-guarding and output-guarding on one channel) at the same time as described by Jones [10].

❑

The (periodic) timing stops when `accept(channel)` or `refuse(channel)` are used.

- **`Accept` communication**

```
Environment::accept(channel);
Environment::accept(barrier);
```

The environmental process will accept any communication event on the specified channel or barrier. This will cancel any timing as specified with `at(..)` or `after(..)`, or any refusal that was specified with `refuse(..)`. Also after a refusal of the channel one can perform accept to cancel the refusal. If the channel or barrier cannot be accepted then an exception it thrown. The `accept(..)` method is synchronized. The method will block when it was called before `at(..)`, `after(..)`, or `refuse(..)`. This prevents any race hazards between the methods. A guard is not influenced by the environmental process.

The `accept(..)` methods can be used to undo any timing on channels or to let the environment accept an event that it refused. Furthermore, if the environment does not want to accept the event then the application has no control over it and a `UnacceptableException` results.

❑

- **`Refuse` communication and (optionally) throw exception**

```
Environment::refuse(channel, exception_message);
Environment::refuse(barrier, exception_message);
```

This method will let the environmental process refuse the acceptance of the communication event on the specified channel or barrier. If an exception message is specified then it will let the channel or barrier throw the exception message to the participating processes. The exception message is passed to the producer and consumer processes. If no exception message is specified then the channel or barrier will block processes until the environment is willing to accept the events.

The `refuse(..)` method can be used to command the environment that an artificial refusal should be carried out. This method can be used for two main reasons:

1. The application can be tested on possible failures on communication. This way the robustness of the application can be tested.
2. In case the application deadlocks or livelocks then there is no way to terminate the program or a particular part of the program. If an deadlock or livelock is detected then with `refuse(..)` one can throw exceptions to channels or barriers that will release synchronization and exception handling must gracefully terminate the program.

A `UnacceptableException` results if the environmental process cannot refuse events on the specified channel or barrier.

❑

- **Get the actual `time`**

```
Environment.time() : Time
```

Returns the absolute time read by the environmental timer.

❑

- **Let a process `sleep` or `sleepUntil` a specified time**

Besides the `Environment` class CTJ supports the `Thread` class that is similar as the Java Thread class in order to manipulate the thread of control in processes. Since no channel or barrier is directly involved we cannot use the environmental process. Thus we need the help of the `Thread` class instead of the `Environment` class. The services by the Thread class are local to the process and thread-oriented rather than communication event oriented.

```
Thread.sleep(relative_time)
```

Let the thread of control in a process sleep for the specified relative time.

❑

```
Thread.sleepUntil(absolute_time)
```

Let the thread of control in a process sleep until the specified absolute time.

❑

These methods may internally use the `Environment` methods to perform temporarily blocking of the process on some hidden timed event. For sampling and steering we do not use these methods.

5. Example

In this example we will illustrate the use and mechanism for sampling. Consider the two controller processes HController and LController depicted in Figure 2. The external channels d_1 and d_2 are timed on sample interval T_{s1} and external channels c_1, c_2 and c_3 on sample interval T_{s2}. In the following code we create the external channels and assign them to the environmental process with specified start time and sampling interval. The sampling rate for HController is 1 kHz and the sampling rate for LController is 0.1 kHz. The start time is specified such that sampling starts when everything is constructed otherwise deadlines may be passed on start-up.

```
//--- create external channels
Channel<int> d1 = new ADC(0);
Channel<int> d2 = new DAC(0);
Channel<int> c1 = new ADC(1);
Channel<int> c2 = new IncCounter(0);
Channel<int> c3 = new DAC(1);

//--- set up sampling timing and register channels to environment
long Ts1 =  1000; // in usec
long Ts2 = 10000; // in usec

long starttime = Environment::time() + 100000;

// firstly the inputs
Environment::at(d1, starttime, Ts1);
Environment::at(c1, starttime, Ts2);
Environment::at(c2, starttime, Ts2);

// secondly the outputs
Environment::at(d2, starttime, Ts1);
Environment::at(c3, starttime, Ts2);

//--- create processes and compositional relationships
...
```

Although processes can read and write on these channels in parallel, the conversions will be in some sequence. This is because the environment uses the interrupt service routine that demands an atomic sequence.

Every registration with the same start time and sampling interval belong to the same atomic group and its order of execution is determined by the sequence of registration. Therefore, the sequence of registration can be important. The sequence of inputs and outputs will be sorted by its time stamp and when the time stamps are equal then the sequence is determined by the sequence of registration. Due to this constraint, the programmer can minimize conversion latencies be choosing a most optimal order of registration. We will not discuss the implementation of how this is achieved in this paper.

We will illustrate this mechanism in Figure 5 using CSP diagram notations. Figure 5a shows the communication relationships of both controllers with their input-output counterparts in hardware (the input/output bubbles in the grey rectangle). In Figure 5b-d, the compositional relationships between these the hardware inputs/outputs are rendered for

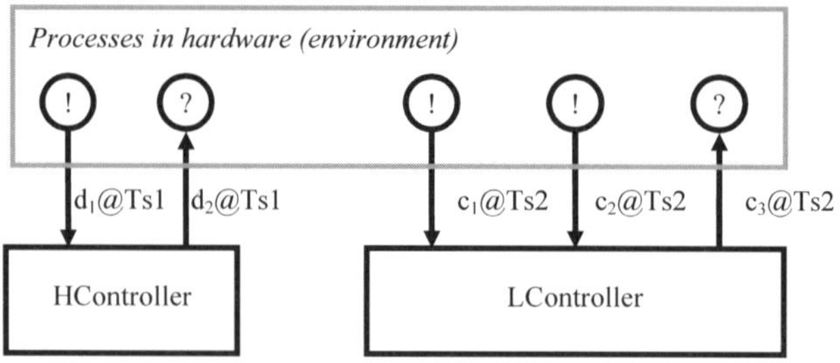

(a) Communication relationships

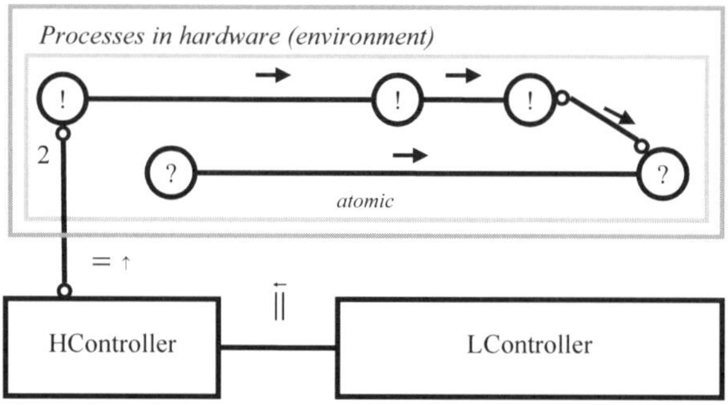

(b) Compositional relationships on $t = t_{s1} = t_{s2}$

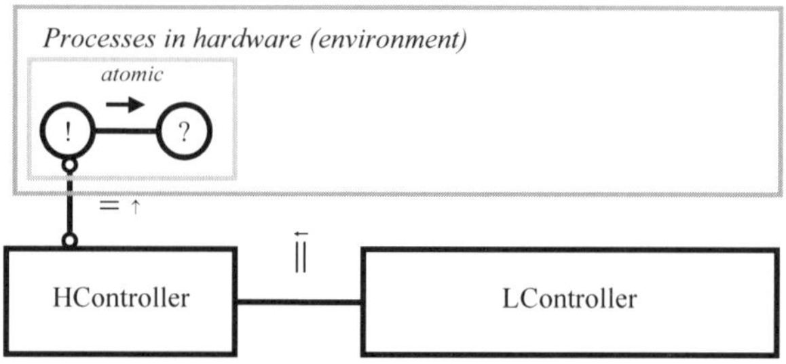

(c) Compositional relationships on $t = t_{s1}$ *and* $t \neq t_{s2}$

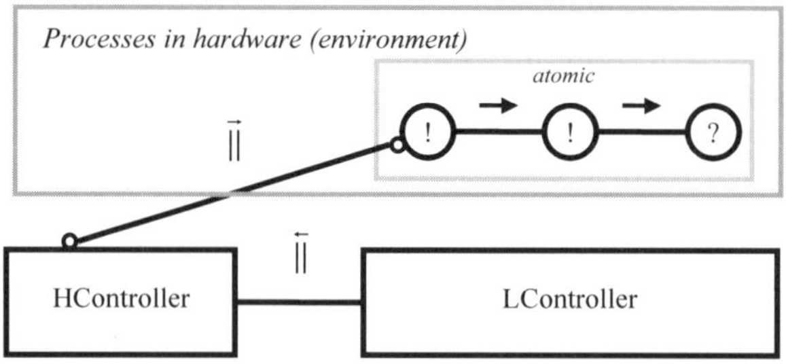

(d) Compositional relationships on $t = t_{s2}$ *and* $t \neq t_{s1}$

Figure 5 Atomic sequence of inputs and outputs by the environmental process

different scenarios. This is the solution for using interrupt handling on the internal timer.

Process LController has a lower sampling frequency ($1/T_{s2}$) than the sampling frequency

($1/T_{s1}$) of process HController, with $T_{s1} < T_{s2}$. Thus, we specify that LController gets a lower priority than HController.

The conversions of the devices are done atomically. That is, they cannot be interrupted by the application. This is depicted by the atomic rectangles in grey. Processor interrupt mechanisms are sequential and mostly priority or pre-emption based. Therefore, the sequential compositions and the prioritized parallel composition with the hardware input/output processes are enforced by this environment.

This mechanism adapt to three scenarios, where:

1. $t = t_{s1} = t_{s2}$ $\Leftrightarrow t_{s1} = t_{s2}$
2. $t = t_{s1}$ and $t \neq t_{s2}$ $\Rightarrow t_{s1} \neq t_{s2}$
3. $t = t_{s2}$ and $t \neq t_{s1}$ $\Rightarrow t_{s1} \neq t_{s2}$

with variable t be the actual time. The sampling times are defined as:

$$t_{s1} \in \{starttime + k_1 * T_{s1} | \text{ for } k_1 = 0,1,2,3,\ldots\}$$

$$t_{s2} \in \{starttime + k_2 * T_{s2} | \text{ for } k_2 = 0,1,2,3,\ldots\}$$

Scenario 1: Figure 5b shows two equal time stamps. Sampling and actuation of all channels will be done in a predefined sequence. Firstly, all sampling processes must be performed and secondly all actuation processes must be performed. This introduces some jitter between sampling and a varying distance between sampling and actuation. Usually this jitter is so small that it does not endanger the stability or the accuracy of the controllers.

If it is really required that this variety is not allowed then all conversions should be done at the same time at a fixed sequence. This can also be specified with a sequence of `at(…,starttime,Ts)` registrations. In this case scenario 2 and 3 do not exist. The problems can be that this solution consumes more energy and the quality and costs of all components depend on the highest sampling frequency in the system. This may require more expensive converters. **Scenario 2**: Figure 5c shows the situation when only time stamp t_{s1} is reached. **Scenario 3**: Figure 5d shows the situation when only time stamp t_{s2} is reached. However, in the practical case presented in this paper, Scenario 3 will never occur when frequencies are multiplicities like in our JIWY case study of 1 kHz and 0.1 kHz. It is included here for completeness.

We have applied the environmental process concept successfully for a mechatronic system, called JIWY [11]. The control software is programmed in C++ with CTC++. A small variation of jitter between sampling and actuation is allowed, and therefore both scenario 1 and 2 apply for JIWY.

6. Conclusions

In this paper we described a concept of timing for untimed CSP architectures which involves an environmental process that participates in communication events. An environmental process is basically another CSP process that synchronizes on internal events. This approach allows adequate timing on external channel communication that provides sampling for control systems; something that was not possible with a timer object or a sleep method. Proposing multi-way events contributed in an orthogonal concept that does not require changes to an untimed CSP architecture. This concept may increase the extendibility, portability, and maintainability of CSP based software designs.

References

[1] C.A.R. Hoare. Communicating Sequential Processes. Prentice-Hall. London, UK, 1985.

[2] A.W. Roscoe. The Theory and Practice of Concurrency. C. A. R. Hoare and R. Bird *Series in Computer Sciences*, Prentice-Hall. 1998. 0-13-674409-5.

[3] G.H. Hilderink, A.W.P. Bakkers and J.F. Broenink. A Distributed Real-Time Java System Based on CSP. In *The Third IEEE International Symposium on Object-Oriented Real-Time Distributed Computing (ISORC-2000)*, IEEE Computer Society. Newport Beach, California: pp. 400-407, 2000.

[4] G.H. Hilderink. A Graphical Modelling Language for Specifying Concurrency based on CSP. In *IEE Proceedings Software*, IEE. **150:** 108-120, 2002. ISSN 1462-5970.

[5] INMOS. occam 2 Reference Manual. C. A. R. Hoare *International Series in Computer Science*, Prentice Hall. 1988. ISBN 0-13-629312-3.

[6] P.H. Welch and P.D. Austin. The JCSP home page, http://www.cs.ukc.ac.uk/projects/ofa/jcsp, 1999.

[7] G.H. Hilderink. Communicating Threads for Java (CTJ) home page, http://www.rt.el.utwente.nl/javapp, 2002.

[8] S. Schneider. Concurrent and Real-time Systems. S. U. U. D. Barron and B. U. U. P. Wegner *Worldwide series in computer science*, John Wiley & Sons. Chichester, UK, 2000. 0-471-62373-3.

[9] U.o. Oxford. Better Ways to Program Parallel Processors, http://oldwww.comlab.ox.ac.ul/oucl/ocpara/methods.html, (UK).

[10] G. Jones. On Guards. In E. T. Muntean *7th Occam User Group & International Workshop on Parallel Programming of Transputer based Machines*, LGI-IMAG. Grenoble, 1987.

[11] D.S. Jovanovic, G.H. Hilderink and J.F. Broenink. A Communicating Threads (CT) Case Study: JIWY. In J. S. Pascoe, P. H. Welch, R. J. Loader and V. S. Sunderam *Communicating Process Architecture 2002*, IOS Press. University of Reading, UK. **60:** pp. 311-320, 2002. ISSN 1383-7575.

Communicating Process Architectures – 2003
Jan F. Broenink and Gerald H. Hilderink (Eds.)
IOS Press, 2003

Parallel Processing — the picoChip way!

Andrew DULLER Gajinder PANESAR Daniel TOWNER
picoChip Designs Ltd., Bath, UK

Abstract. This paper describes a new approach to parallel processing within the well targetted application domain of wireless communications systems, using the picoArray™. The picoArray™ is a tiled-processor architecture, containing 430 heterogeneous processors, connected through a novel, compile-time scheduled interconnect. We show how the features of the picoArray™ allow deterministic processing to be achieved, and how the tool chain allows programming to be performed effectively in a combination of high level assembly language and C. By handling a wide variety of types of processing within the picoArray™ a single design flow can be used to produce complex communications systems. The effectiveness of this approach is demonstrated through the use of the picoArray™ to build a working 3G base-station.

1 Introduction

Within the applications space of wireless communications, systems are typically designed using a mixture of DSPs, FPGAs and custom ASICs, resulting in systems that are awkwardly parallel in nature. In addition, the tool chain forces the user to regard the systems from several different perspectives, introducing problems with specification, implementation and verification. All these factors result in solutions that are complex, non-deterministic, and difficult to debug.

The picoArray™ architecture [1] was developed to create a massively parallel system which did not suffer from many of the problems of conventional general purpose parallel systems. There was no intention to create a general purpose parallel processor but one which provided an alternative to creating an ASIC in a range of applications within the wireless communications domain. In some senses it can be thought of as a "programmable ASIC". The fundamental attributes of the approach are that communications between processes have to be fixed at compile time and therefore no support for dynamic communications is required and problem partitioning is largely done by the user since this is what would occur during an ASIC design process.

The design of both the picoArray™ and the tool chain strongly reflect the application domain whose attributes are:

- application partitioning is traditionally done by the user and is well understood,
- high degree of replicated code due to multi-user and multi-antenna systems,
- considerable use of stream based processing [1],
- hand coding of assembly language is often used to achieve compactness of code and efficiency.

[1] Stream based processing is where data is passed at high rate through a chain of processes each of which do relatively simple operations on the data before passing them on to the next stage.

This paper consists of an overview of the picoArray^{TM} architecture, an introduction to the applications domain and a summary of the tool chain. Finally, the current status of the work is described which helps to show the effectiveness of the approach through the goals that have been achieved.

2 Overview of the picoArray^{TM} architecture

The picoArray^{TM} is a tiled processor architecture in which 430 heterogeneous processors are connected together using a deterministic interconnect as shown in figure 1. The interconnect consists of bus switches joined by picoBus^{TM} connections. To simplify the diagram the interconnect is only drawn between the switches, each processor is actually connected directly to the picoBus^{TM} above and below it (an enlarged view of part of the interconnect is shown in figure 2, again to simplify the diagram only two of the four vertical bus connections are shown).

The level of parallelism is relatively fine grained with each processor having a small amount of local memory. There are four RISC processor variants which share a common core instruction set, but have varying amounts of memory and additional instructions to implement certain wireless baseband control and digital signal processing functions. Each processor runs a single process in its own memory space and they use "signals" to synchronise and communicate. Multiple picoArray^{TM} devices may be connected together to form systems containing thousands of processors using on-chip peripherals which effectively extend the on-chip bus structure. A brief description of the four processor variants and a breakdown of the internal memory distribution is given in table 1.

The initial version of the picoArray^{TM}, the PC101, is a large chip and it is therefore necessary to use redundancy within the array to improve the yield.

The routing strategy used was determined largely by the real time nature of the intended applications where the indeterminate latency due to bus arbitration would be unacceptable. All of the communications are determined during the "compilation" of the system which means that the communications bandwidth can be guaranteed.

2.1 Interconnect

Within the picoArray^{TM} , processors are organised in a two dimensional grid, and communicate over a network of 32-bit unidirectional buses (the picoBus^{TM}) and programmable bus switches. The physical interconnect structure is shown in figure 2. The processors are connected to the picoBus^{TM} by ports which contain internal buffering for signal data. These act as nodes on the picoBus^{TM} and provide a simple processor interface to the bus based on *put* and *get* commands. The processors are essentially independent of the ports unless they specifically use a *put* or a *get* instruction.

The inter-processor communication protocol implemented by the picoBus^{TM} is based on a time division multiplexing (TDM) scheme. There is no run-time bus arbitration, so communication bandwidth is guaranteed. Data transfers between processor ports occur during specific time slots, scheduled in software, and controlled using the bus switches. Figure 2 shows an example in which the switches have been set to form two different signals between processors. Signals may be point-to-point, or point-to-multi-point. In the latter case, the data transfer will not take place until all the processor ports involved in the transfer are ready.

Communication time slots throughout the picoBus^{TM} architecture are allocated according to the bandwidth required. Faster signals are allocated time-slots more frequently than slower signals. The user specifies the required bandwidth for a signal by giving a rate at which the

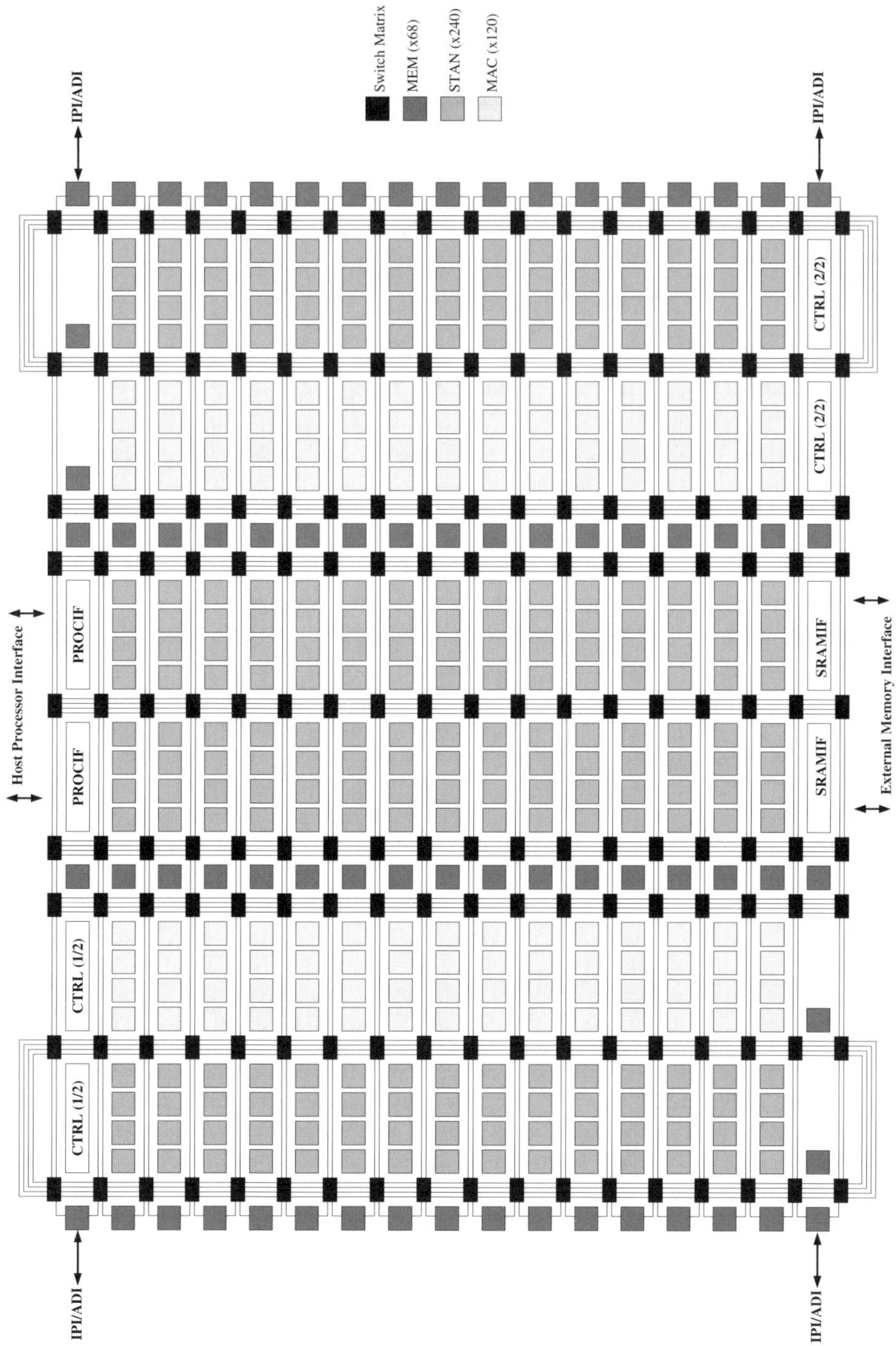

Figure 1: Top-level Diagram showing processors and interconnect

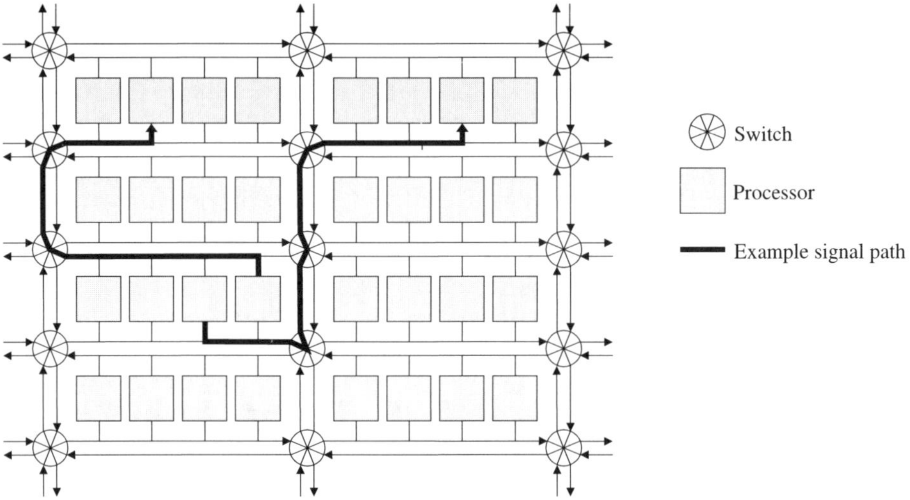

Figure 2: Interconnect

signal must communicate data. For example, a transfer rate might be described as @4, which means that every fourth time-slot has been allocated to that transfer.

The default signal transfer mode is synchronous; data is not transfered until both the sender and receiver ports are ready for the transfer. If either is ready before the other then the transfer will be retried during the next available time slot. If, during a *put* instruction no buffer space is available then the processor will sleep (hence reducing power consumption) until space becomes available. In the same way, if during a *get* instruction there is no data available in the buffers then the processor will also sleep. Using this protocol ensures that no data can be lost.

There is also an asynchronous signal mode where transfer of data is not handshaken and in consequence data can be lost by being overwritten in the buffers without being read.

2.2 Processors

All of the processors in the picoArray™ are 16-bit, and use 3-way VLIW scheduling. The basic structure of the processors is shown in figure 3. Each processor has its own small memory (between 1KB and 32KB), which is organised as separate data and instruction banks (i.e. a Harvard architecture). The processor contains a number of communication ports, which allow access to the interconnect buses through which it can communicate with other processors. Each processor is programmed and initialised using a special configuration bus. The processors have a very short pipeline which helps programming, particularly at the assembly language level. The architecture of the four processor variants are shown in figure 4.

In addition to the general purpose processors, there are a number of special peripherals, including a host interface, an SRAM interface, asynchronous data and inter-picoArray interfaces. These peripherals are connected to the bus structure through ports, which enables them to be treated as though they are special purpose processes. The overall distribution of processors and peripherals is shown in figure 1 with the peripherals being placed in the top and bottom rows of the array.

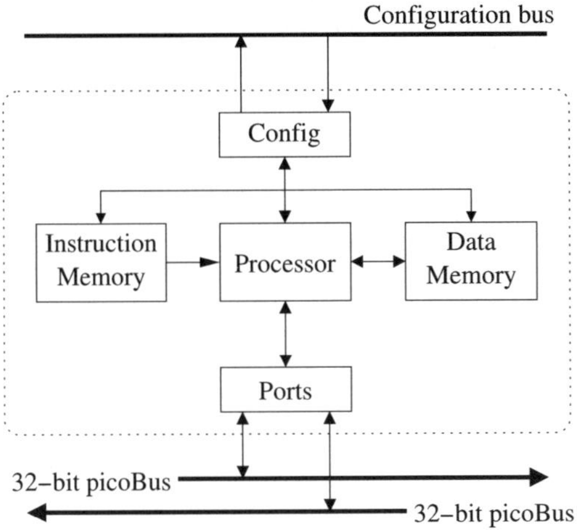

Figure 3: Processor Structure

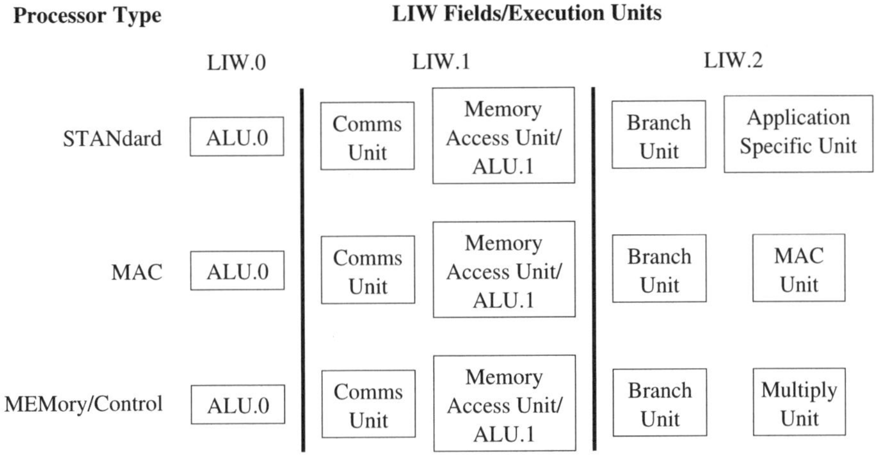

Figure 4: VLIW and execution unit structure in each processor

Type	Description	Number	Memory (bytes)
STAN	*Standard*: A standard processor optimized for CDMA spread and de-spread and other wireless base station signal processing functions	240	768
MAC	*Multiply accumulate*: A processor that includes a multiply-accumulate unit and which supports additional multiply-accumulate instructions	120	768
MEM	*Memory*: A processor having a multiply unit and additional data memory	68	8,704
CTRL	*Control*: A processor with a multiply unit and larger amounts of data and instruction memory optimized for the implementation of control functionality	2	32,768

Table 1: PC101 processor variants and memory distribution

2.3 Host interface

The Host or microprocessor interface is used to configure the picoArray^{TM} device and to transfer data to and from the picoArray^{TM} device using either a register transfer method or a DMA mechanism. The DMA memory-mapped interface has a number of ports mapped into the external microprocessor memory area. Two ports are connected to the configuration bus within the PC101 and the others are connected to the internal picoBus^{TM}. These enable the external microprocessor to communicate with the internal array elements using signals.

2.4 SRAM interface

Each picoArray^{TM} has an amount of memory distributed amongst the processors for data and instruction storage. However, an external SRAM interface is provided to supplement the on-chip memory. This interface allows processors within the core of the picoArray^{TM} to access external SRAM across the internal picoBus^{TM}.

2.5 Asynchronous data/Inter-picoArray interfaces

There are four interfaces on each device which can be configured in one of two modes: either the inter picoArray interface (IPI) mode or the asynchronous data interface (ADI) mode. The choice of interface mode is made for each interface separately during device configuration.

2.5.1 Inter picoArray interface

The four IPI interfaces are bidirectional and designed to allow each picoArray^{TM} to exchange data with up to four others. Using this feature, a grid of picoArray^{TM} devices can be constructed to implement highly complex and computationally intensive signal processing systems. The IPI interface operates in full duplex, sending and receiving 32-bit words. The 32-bit words on the internal picoBus^{TM} are multiplexed as two 16-bit data on the interface itself.

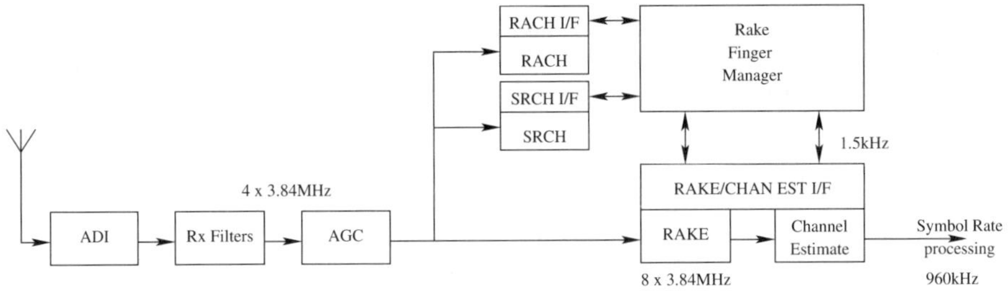

Figure 5: Typical data rates

2.5.2 *Asynchronous data interface*

The asynchronous data interface (ADI) allows data to be exchanged between the internal picoBus™ and external asynchronous data streams such as those input and output by data converters or control signals between the base band processor and the RF section of a wireless base station.

2.6 *Low-power considerations*

Potentially, a device such as the PC101, which contains 430 processors and a TDM interconnect, could use a lot of power. A number of methods have been used to enable the power consumption of the picoArray™ to be reduced. For example, individual processors are able to 'sleep' when they are waiting for communications events, thus consuming minimal power, and parts of the picoArray™ which are not used in a particular design are switched off.

2.7 *Array layout*

As can be seen from section 3 the target applications are fairly varied although they have many common attributes. The layout of the array in terms of quantities and locations of processors was determined to match these attributes as far as possible. Since any realistic system will make use of many picoArray™ devices, 4, 8, 16 or more, the layout has to be a compromise between the different types of processing that are required within a wireless infrastructure system. At the input to the system the data rates are very high but the processing is simple, as data flows through the system the data rates reduce, the control becomes more complex and the operations become more complex.

Figure 5 shows a typical piece of processing from the front end of the receive chain in a base station. At the input to the system, through the ADI, the data rate is high during which the incoming signals are filtered. This is then transformed into a symbol rate stream of data at a much lower rate, this rate can vary between the 960kHz given, down to 15kHz depending on the type of signal. It should be noted that this rate is for each user of the base station and typically there will be 64 users or more. In addition, multiple antennas will be used making the initial input data rate even higher.

The majority of the array, 360 processors (STAN and MAC processors), are designed for stream based processing and therefore have small amounts of local memory. The split between STAN and MAC was determined largely by the requirement for the specialist spread and de-spread operations (present in the STAN processor) required by CDMA. The target applications tend to have a smaller requirement for block based processing and this is supported by the 68 MEM processors which have more local storage and can be used in conjunction with the SRAM interface. The low level control requirements of application systems are handled by the provision of 2 CTRL processors. Each processor has a performance comparable to an ARM 9 for control type functionality, or a TI C55XX for DSP tasks.

3 Target applications

The picoArray™ is targetted at providing the processing solution for a wide range of communications systems, specifically those related to wireless infrastructure. These include the UMTS FDD/TD-SCDMA [2], IEEE 802.16 [3] and CDMA2000 [4] standards. While the details of these standards vary considerably the style of processing and algorithmic requirements have many similarities. The typical functions are:

- Signal filtering and conditioning for both transmit and receive paths (adjacent channel rejection, digital pulse shaping, Inter-Symbol-Interference minimisation)
- Multiple access processing: including both mux and demux, and its management (CDMA/OFDMA/TDMA)
- Signal synchronisation (timing acquisition and tracking and local oscillator offset compensation)
- Equalisation: compensation for channel distortion (terrestrial mobile radio channel is probably the most severe of any communications channel)
- Forward error correction: Encode and decode. CRC, Maximum Likelihood Viterbi and Turbo (the many flavours!). Interleaving and de-interleaving to make distribution of raw channel errors amenable to FEC.
- Control functions: Cellular systems have a huge amount of event-driven signal processing and adaptation of that signal processing (Call setup/teardown, closed loop control of transmission time/power/frequency/code and channel equalisation in dynamic conditions for example)

These applications typically consist of a mixture of DSP functionality which is both stream based and block based. In addition there is a requirement for distributed control of the DSP operations. The picoArray™ architecture provides stream based support by having a large number of processors which have small amounts of local instructions and data. Block based support is provided by having a number of the processors which have increased local data and if this is insufficient then there is also access to an external memory. Conventional architectures (and some newer ones) force designers to fit the algorithm to the architecture rather than choose the algorithm implementation to suit the performance required (e.g., block-based approaches incur latency overheads unpalatable in some systems but conventional DSPs are only efficient when operating block based).

Previous approaches to these applications have used ASIC, DSP and FPGA to produce a solution. Often they require a combination of these technologies which in itself produces problems with co-design and co-simulation. In addition, the most cost effective method, ASIC, is extremely inflexible, which causes problems when wireless standards change or algorithmic improvements need to be made. The DSP/FPGA solution can be flexible but tends to be expensive, power hungry and combining the two technologies appears never to be simple. The picoArray™ produces a reprogrammable solution to a wide range of these applications using a single tool chain, rather than having to combine DSP tool chains with FPGA and perhaps ASIC.

To provide some idea of the complexity of this type of application figure 6 shows the major software components in a typical UMTS baseband system. Many of the blocks in the diagram are complex systems in their own right and the blocks may need to be replicated for each user of a base-station.

4 Programming the picoArray™ - the tool chain

The design of the tool chain strongly reflects the application domain of the picoArray™ and this has been used to simplify the tool flow. For example, the C compiler deals with the code for a single processor and no attempt is currently made to automatically parallelise code across multiple processors. In addition there is considerable use of assembly language coding and experience has shown that this can be achieved efficiently (from the user's perspective) due to the high degree of replication of the code.

The picoArray™ is programmed using a mixture of VHDL [5], ANSI/ISO C and assembly language. The VHDL is used to describe the structure of the overall system, including the relationship between processes, and the signals which connect them together. Each individual process is programmed in conventional C (albeit with additional communication functions), or in assembly language. A simple example is given in figure 7.

A comprehensive tool chain exists to convert the input VHDL into a form suitable for execution on one or more picoArray™ chips. The tool chain includes a compiler, an assembler, a VHDL parser, a cycle-accurate simulator, debuggers, a place-and-switch tool, design partition tools and verification tools. The relationship between these tools is shown in figure 8. The following sections examine each of these tools in turn.

4.1 VhdlParser

The VHDL parser is the main entry point for the user's source code. A complete VHDL design is given to the parser, which coordinates the compilation and assembly of the code for each of the individual processes. An internal representation of the machine code for each processor and its signals is created.

4.2 C Compiler

The C compiler is a port of the GNU Compiler Collection (GCC) [6]. A few simple extensions have been provided to support communication, but the compiler otherwise supports conventional ANSI/ISO C. GCC is designed primarily for 32-bit general purpose processors capable of using large amounts of memory, making it a challenge to support 16-bit embedded processors with just a few kilobytes of memory. The compiler uses a Deterministic Finite Automata scheduling algorithm [7] to generate efficient VLIW schedules.

In addition to being called by the VhdlParser to compile code for processes, the compiler may be used stand-alone to generate library files. These libraries can be linked with the processes in a VHDL design, allowing effective code reuse.

4.3 Cycle accurate simulation

The cycle accurate simulator can be constructed directly from the output of the VhdlParser, since there is no need to determine how a design must be partitioned between chips, or how processes are allocated to processors. The simulator can be used in two modes, functional or cycle accurate, depending on the requirements of the user.

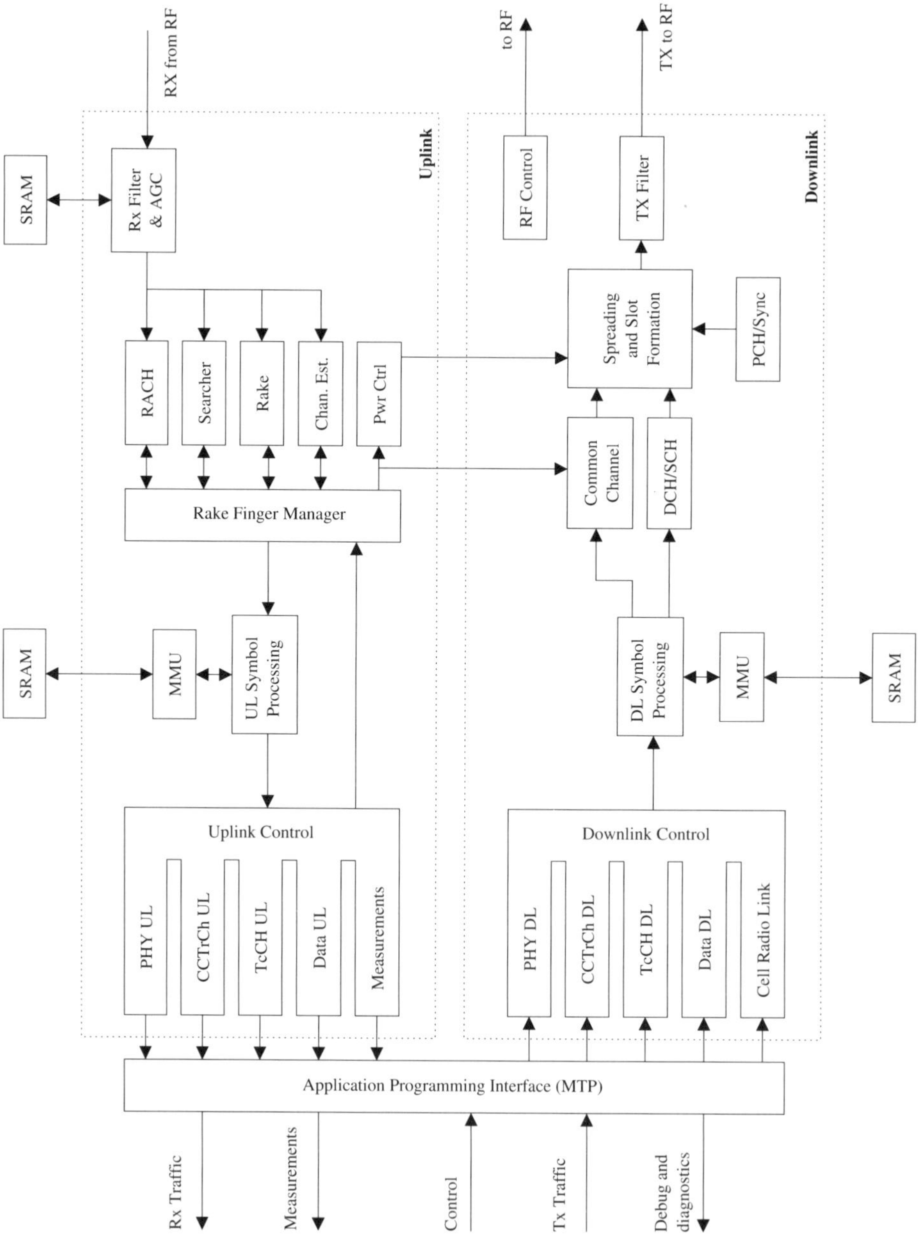

Figure 6: Diagram showing software components in a typical UMTS baseband system.

```
------------------------------------------------------------------------
entity Producer is                       -- Declare a producer
  port (channel:out integer32@8);        -- 32-bit output signal
end entity Producer;                     --   with @8 rate

architecture ASM of Producer is          -- Define the 'Producer' in ASM
begin MEM                                -- use a 'MEM' processor type
  CODE                                   -- Start code block
    COPY.0 0,R0 \ COPY.1 1,R1            -- Note use of  VLIW
  loopStart:
    PUT R[0,1],channel \ ADD.0 R0,1,R0   -- Note communication
    BRA loopStart
  ENDCODE;
end;                                     -- End Producer definition.

------------------------------------------------------------------------

entity Consumer is                       -- Declare a consumer
  port (channel:in integer32@8);         -- 32-bit input signal
end;

architecture C of Consumer is            -- Define the 'Consumer' in C
begin STAN                               -- Use a 'STAN' processor
  CODE
  long array[10];                        -- Normal C code

  int main() {                           -- 'main' function - provides
    int i = 0;                           --    entry point

    while (1) {
      array[i] = getchannel();           -- Note use of communication.
      i = (i + 1) % 10;
    }

    return 0;
  }
  ENDCODE;
end Consumer;                            -- End Consumer definition

------------------------------------------------------------------------

use work.all;                            -- Use previous declarations

entity Example is                        -- Declare overall system
end;

architecture STRUCTURAL of Example is    -- Structural definition
  signal valueChannel: integer32@8;      -- One 32-bit signal...
begin
    producerObject: entity Producer      -- ...connects Producer
      port map (channel=>valueChannel);
    consumerObject: entity Consumer      -- ...to Consumer
      port map (channel=>valueChannel);
end;
------------------------------------------------------------------------
```

Figure 7: Example source program

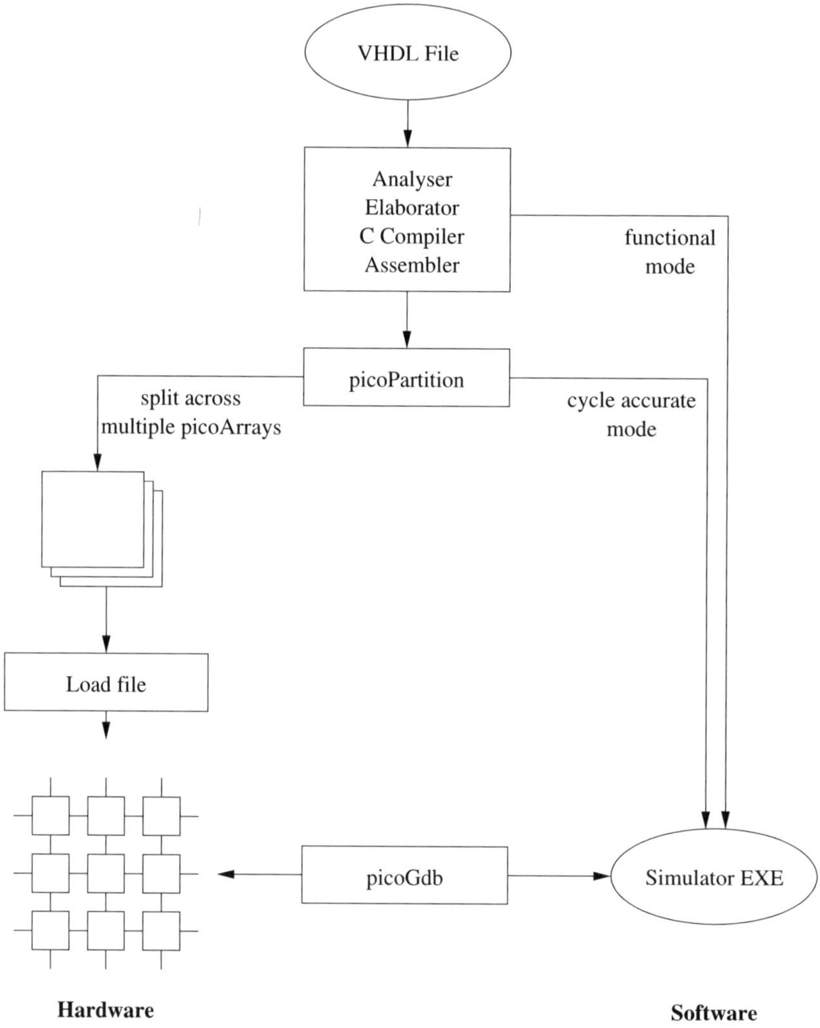

Figure 8: Tool Structure

In functional mode it can model the logical functionality of a design with cycle accuracy only provided within each individual processor, the timing of the communications is not modelled accurately. This can be used early in the design process as it is not necessary to determine where processes are to be placed on the hardware and even which chip they will reside on.

In cycle accurate mode the design must be partitioned between chips (see Section 4.4), processes must be placed onto specific processors and the signals must be routed on the picoBus™ (see Section 4.5). This allows inter-chip timing information and signal timing information to be back annotated into the simulation to produce a fully cycle accurate simulation.

4.4 Chip partitioning

If a design requires more processors than are available in a single picoArray™, the design must be partitioned across multiple chips. This process is currently manual, with the user specifying which processes map to which chip, although the splitting of signals between the chips is automated.

4.5 Place and Switch

Once a design has been partitioned between chips a process akin to place and route in ASIC designs has to be done for each chip. This assigns a specific processor to each entity in the design and routes all of the signals which link entities together. The routing must use the given bandwidth requirements of signals while routing. The routing algorithm should also address the power requirements of a design, by reducing the number of bus segments that signals have to traverse, enabling unused bus segments to be switched off. This process is performed using the Plastic (PLace And Switch To IC) tool.

When a successful place and switch has been achieved a "load file" can be produced which can be loaded directly onto the hardware.

4.6 Debugging

The debugging tools allow an entire design to be easily debugged, either as a simulation, or using real hardware. The tools support common debugging operations such as setting breakpoints, single and multi-step execution, halt-on-error, status display, and memory/register tracing. A port of the GNU Debugger (GDB) [8] allows source-level debugging, using either C or assembly language. For flexibility, both graphical and command-line interfaces are provided.

Debugging parallel processes is notoriously difficult, so some special debugging operations are provided to aid the user, including "spies" and traffic analysis. Spies are special processes which are inserted in the middle of a signal, and allow the flow of data through that signal to be monitored. Traffic analysis is used to identify how much bandwidth is being used in different parts of the picoArray™, and can find hot-spots in the bandwidth usage, or detect deadlock.

5 Current Status and Performance

The current implementation of the picoArray™, the PC101, runs at 160MHz and is implemented using 130nm technology. A PC101 device running at 160 MHz can execute 206

GIPS, excluding acceleration from the application-specific instructions. A demonstrator system consisting of 16 devices is capable of approximately 3,300 GIPS. Comparison of performance with other architectures is notoriously difficult but a Texas Instruments TMS320-C6416 running at 600 MHz can execute approximately 9.6 GIPS.

The main application area that has been chosen to demonstrate the technology is a UMTS base-station and much of this has been coded using PC101 hardware and the tool chain described above. First silicon for PC101 was received in December 2002 and the first telephone call through a demonstration base-station using this silicon was made in May 2003. This uses a system consisting of 4 PC101's with 740 processors being used.

6 Conclusion

The basic ideas behind the picoArray™ concept together with the links to the application domain have been described. A pragmatic approach has been demonstrated to the tool chain in which the expertise of the user and the specific application domain has been used where possible rather than trying to create "magic" tools which would be complex, time consuming to create and may never materialise. This is not to say that approaches to automatic partitioning and automatic code scheduling are not being investigated but picoArray™ technology does not stand or fall on the basis of the existence of such tools.

The creation of a major application, a UMTS base-station, in only a few months using the technology demonstrates the power of the approach taken in terms of ease of use due to the single design flow.

7 Acknowledgements

Thanks in particular to Peter Claydon and Doug Pulley who co-founded picoChip Designs Ltd. Thanks to all who have worked on the huge variety of tasks that have made PC101 systems a reality.

References

[1] P. Claydon. Picoarray Switch Matrix. Patent number GB 0030993.0, 2002.

[2] 3GPP. *3GPP TS25 Series (FDD + TD-SCDMA).*

[3] IEEE. *802.16 IEEE Standard for Local and metropolitan area networks.*

[4] TIA/EIA. *TIA/EIA-IS-2000 series.*

[5] Peter Ashenden. The Designer's Guide to VHDL. Morgan Kaufmann, ISBN 1-55860-270-4, 1996.

[6] Richard Stallman. Using and porting the GNU compiler collection. ISBN 059510035X, `http://gcc.gnu.org/onlinedocs/gcc/`, 2000.

[7] Vladimir Makarov. The finite state automaton based pipeline hazard recognizer and instruction scheduler in GCC. The 2003 GCC Developers' Summit Conference Proceedings, `http://www.linux.org.uk/~ajh/gcc/gccsummit-2003-proceedings.pdf`, May 2003.

[8] Richard Stallman, Roland Pesch, and Stan Shebs. Debugging with GDB. ISBN 1882114884, `http://sources.redhat.com/gdb/current/onlinedocs/gdb.html`, 2002.

Communicating Process Architectures – 2003
Jan F. Broenink and Gerald H. Hilderink (Eds.)
IOS Press, 2003

An Introduction to the Kent C++CSP Library

Neil BROWN and Peter WELCH

Computing Laboratory, University of Kent, Canterbury, Kent, CT2 7NF, England.
neil@twistedsquare.com, P.H.Welch@kent.ac.uk

Abstract. This paper reports on a CSP library for C++, developed over the past year at the University of Kent. It is based on the OO-design and API of JCSP and the lightweight algorithms of KRoC **occam**, with extensions to exploit specific C++ capabilities (such as templates). Both user-level and operating system threads are used to provide a range of implementation options and semantics (e.g. for managing blocking system calls intelligently) that run efficiently under either Windows or Linux. The library is presented from the user's point of view, mainly by way of a tutorial. Implementation details are also outlined and some benchmark results given. The performance of C++CSP is between that of KRoC **occam** and JCSP — fairly close to KRoC.

1 Background

This year sees the 25th anniversary of C.A.R. Hoare's paper describing Communicating Sequential Processes (CSP) [1, 2], a new programming model for dealing with parallelism and concurrency that has a strong mathematical basis. Supported by a range of direct implementation technologies, CSP avoids most of the fundamental problems inherent in concurrent design – allowing it to be used with confidence by systems engineers with only a modest amount of training and no specialist mathematical knowledge.

Despite this, no mainstream operating systems or programming languages have built-in support for CSP mechanisms. For concurrency, they offer only unstructured varieties of threads and optional locking mechanisms (such as semaphores or monitors). They provide no support to enforce correct and efficient use of those locks – and cannot be made to provide it. Received wisdom[3] is that this concurrency is hard, should only be used as a last resort (for performance reasons) and only by very experienced designers and programmers. That received wisdom is correct. But CSP turns this around.

Of course, **occam**[4, 5] provides the standard for CSP implementation against which others must be measured – but **occam** is not (yet) a mainstream language. For C, CCSP[6] is a library and run-time kernel that provides the occam model. JCSP[7] and CTJ[8] are CSP/**occam** libraries for Java. This paper reports on C++CSP: a CSP library for C++ that combines an Object Oriented API (derived from JCSP, but with modifications to take advantage of specific C++ capabilities, such as templates) with fast kernel support (along the lines of CCSP and **occam**).

C++CSP runs native under either Windows or Linux/Unix. Applications, therefore, have direct access to all standard Windows – or Linux/Unix – libraries.

This paper provides a user's introduction to the library, describes some of the internal workings and provides early performance figures. It assumes readers have familiarity with the basic ideas and mechanisms of CSP, **occam**, JCSP and C++.

2 Library Details

At the heart of the CSP paradigm are processes and channels – processes being pieces of code that can be executed in sequence or in parallel and channels being the means by which they communicate.

Like JCSP before it, C++CSP adopts an object-oriented approach to implementing CSP. A process is a subclass of `CSProcess`, and channels are classes, such as the standard one-to-one unbuffered channel `One2OneChannel`. Channels are templated classes (the mechanism for generic programming in C++) that can be used to communicate any assignable type, be it a primitive, a "pod" class or a more complex class. In a typical scenario, a parent process creates a channel and passes a reading end (also a class – `Chanin`) to one process and a writing end (again, a class – `Chanout`) to another process. That way, neither child process can access the channel directly to try and obtain a class representing the opposite end of the channel.

As in JCSP, a new process is implemented by declaring a new class as a child of the `CSProcess` class and implementing the `void run()` method (though in C++CSP this is protected rather than public[1]). Unfortunately, it is necessary to pass to the constructor of `CSProcess` the size of the stack needed for the process. This will be explained later.

The C++CSP library provides:

- processes (run either in sequence or in parallel);
- channels (either or both of the reading and writing ends may be *shared*; also, they may be *buffered* using a variety of library or user-defined buffers);
- parallel channel communications (a special case of parallelism included for convenience);
- `ALT`ernatives;
- *buckets* and *barriers*[9];
- *mobile* objects with the same semantics as their (modern) **occam** counterparts;
- easy and secure channel *poisoning* for clean shutdown;
- support for Microsoft Windows (98 and later);
- support for Linux and most other Unix-variants.

3 Tutorial

3.1 The Semantics of Assignment and Communication

Channel communication is just a distributed form of assignment – a value is computed in one process and assigned to a variable in another (after synchronising the two processes of course). In occam, there is the equivalence:

```
FOO x, y:                                FOO x, y:
...                                      ...
x := y              equals               CHAN FOO c:
                                         PAR
                                           c ! y
                                           c ? x
```

[1] This means that processes can only be run by the library itself and cannot be run directly by the user, due to problems with stacks that will be explored later in section 4.2.

Therefore, the semantics of communication must be compatible with the semantics of assignment.

In C++, the meaning of the assignment operator, =, can be overridden to mean whatever the programmer wishes it to mean... and this will be reflected in the meaning of communication in the C++CSP library. So, if a class overrides assignment to mean "shallow copying", a shallow copy of an object will be sent across a C++CSP channel. If a class has not overridden its assignment operator, then a "member-wise" copy (the default for assignment of C++ objects) is received by the reading process. If the channel carries array types, only the pointers will be communicated, since in C++ an array is treated as a pointer. If the channel carries primitive types (int, float etc.), only their values will be communicated.

In this paper, most of the examples work with `int`-carrying channels. If `Bar` is a class, then `Bar`-carrying channels are handled with equal ease - examples are referred to in Sections 5.3 and 6.3.

3.2 Our First Program

We will now look at creating a simple program using the C++CSP library. There will be a writer process that writes 100 integers to a channel then exits, and a reader process that reads 100 integers from a channel and exits. These processes are declared as follows[2]:

```
class Writer : public CSProcess
{
private:
    Chanout<int> out; //writing end of the channel
protected:
    virtual void run(); //overrides the method in CSProcess
public:
    Writer(const Chanout<int>& o); //constructor
};

class Reader : public CSProcess
{
private:
    Chanin<int> in; //reading end of the channel
protected:
    virtual void run(); //overrides the method in CSProcess
public:
    Reader(const Chanin<int>& i); //constructor
};
```

The constructors are as follows (we will use an arbitrarily-sized large stack):

```
Writer::Writer(const Chanout<int>& o)
    :    CSProcess(65536),out(o)
{
}

Reader::Reader(const Chanin<int>& i)
    :    CSProcess(65536),in(i)
{
}
```

The above pieces of code are a general template for any process. Each process is a child of `CSProcess` and must implement the run method. They will have as class members

[2]In all sample code, it is assumed that the namespace of the library, `csp`, is in scope.

any channel ends they require; these are passed to the class constructor. As can be seen above, channel ends can be assigned for ease of use. Following the object-oriented paradigm, channel ends are the same regardless of what type of channel (buffered, shared, or both) they are attached to, allowing any channel type to be used with any process (bar a few exceptions regarding `ALT`ernatives and shared channels). The run methods are given below:

```
void Writer::run()
{
    for (int i = 0;i < 100;i++)
    {
        out << i; //see note below
    }
}

void Reader::run()
{
    int n;
    for (int i = 0;i < 100;i++)
    {
        in >> n; //see note below
    }
}
```

Note: the << and >> operators are overriden so that the indicated above lines expand to `out.write(&i);` and `in.read(&n);` respectively. The ampersands are required because the implementation of our generic channels expects pointers to the source and destination for the communication. However only the integer *values* (not the pointers to the values) will be communicated in the above example.

These functions could hardly be simpler. Internally, input and output calls pass the source and destination respectively as pointers to avoid copying large data structures (pointers are chosen over references to help emphasise that point). Now all that is required is a parent function to glue the components together:

```
void function()
{
    Start_CSP();
    One2OneChannel<int> channel;
    Parallel(
        new Writer(channel.writer()),
        new Reader(channel.reader()),
    NULL);
    End_CSP();
}
```

The `Start_CSP()` and `End_CSP()` calls must be made at the start and end of the section of code in a program that uses CSP to properly initialise and shutdown the library. A channel is declared and its `writer()` and `reader()` methods are called to obtain writing and reading ends of the channel. These are passed to the constructors of the processes which in turn are passed to the vararg `Parallel` function which takes in a varying-size `NULL`-terminated list of processes to be run in parallel. The library takes care of deleting the processes when they have finished executing.

So there is a simple working CSP program. It doesn't print out greetings for the globe but it does what it is intended to do.

3.3 The Poisoned Chalice

At the moment our program is very rigid – the reader expects precisely 100 numbers before it exits. To make our program more flexible we might like to allow any number of inputs before it exits. One way of accomplishing this would be to use a special terminator value but this is a messy solution and not always viable if all integer values might be used in the communication. It can also cause massive deadlock headaches in the design of a shutdown poison path. For this frequently occuring need the library provides a poison mechanism.

Poisoning in other CSP frameworks such as **occam** usually involves poison protocols [10] – a dressed up terminator value solution. C++CSP allows you to poison a channel, a form of "stateful" poisoning. A process can set any channel that it has an end of to be poisoned. Subsequently, any attempts by any process to access this channel will cause a poison exception to be thrown. This can then be caught by the process that is attempting to access the channel, and dealt with by poisoning all channels it can access before exiting gracefully. So we can redesign our writer process as follows:

```
void Writer::run()
{
    for (int i = 0;i < 100;i++)
    {
        out << i;
    }
    out.poison(); //<-- added line
}
```

If we wanted to be cautious then we would include a try-catch block to check for a poison exception in the writer but for our small example it is unnecessary. So now we must modify the reader to keep inputting numbers until the channel becomes poisoned. In fact there is a "common process" supplied with the library that does what we require – the BlackHole process. The code is given here:

```
template <class DATA_TYPE>
class BlackHole : public CSProcess
{
    DATA_TYPE t;
    const Chanin<DATA_TYPE> channelIn;
protected:
    void run()
    {
        try
        {
            while(true)
                channelIn >> t;
        }
        catch(PoisonException)
        {
            channelIn.poison(); //You never know!
        }
    };
public:
    inline BlackHole(const Chanin<DATA_TYPE>& chIn)
        :    CSProcess(65536),channelIn(chIn)
    {
    };
};
```

This code is fairly self-explanatory, fitting the pattern of processes that has already been seen. When the black-hole gets poisoned it will drop to catching the poison exception. At this point (as referenced by the cryptic "you never know!" comment) it poisons all the channels it can. This is technically unnecessary because that must have been the channel that threw the poison exception originally, but it shows what is good practice – had this been an id process then it would had to have poisoned both its channels because there is no way to tell which channel threw the poison exception. Poisoning an already poisoned channel has no effect on the channel (it does not throw a poison exception either).

Replacing our reader process with this black-hole process will produce a program that has the same semantics but is more flexible. The change to the code for `function` is trivial:

```
void function()
{
    Start_CSP();
    One2OneChannel<int> channel;
    Parallel(
        new Writer(channel.writer()),
        new BlackHole<int>(channel.reader()),
    NULL);
    End_CSP();
}
```

3.4 Shared/Buffered Channels

Using a shared channel is simply a matter of declaring `Any2OneChannel` etc instead of the normal `One2OneChannel` and then giving reading and writing ends to the appropriate processes. This can be demonstrated by changing our program so that there are two black-hole processes listening on a one-to-any channel:

```
void function()
{
    Start_CSP();
    One2AnyChannel<int> channel; //change the channel type
    Parallel(
        new Writer(channel.writer()),
        new BlackHole<int>(channel.reader()),
        new BlackHole<int>(channel.reader()),
        // ^^^ - add an extra black hole
    NULL);
    End_CSP();
}
```

Likewise, buffered channels are used by declaring `One2OneChannelX` (note the X) or `One2AnyChannelX` etc and passing to the constructor a buffer object to specify how to buffer the channel (for example limited/unlimited FIFO buffering or overwriting buffering). This follows the mechanisms of JCSP. Making our shared channel buffered requires only the following change to its declaration:

```
    Buffer<int> buffer(10);    // declare a buffer of the required type
    One2AnyChannelX<int> channel(&buffer);
```

The channel takes its own copy of the buffer to use, to avoid the user passing the same buffer to multiple channels and many channels trying to use the same buffer.

3.5 Parallel Communication

In **occam**, processes may be run in parallel at granularities down to single statements. Here, as in JCSP and CCSP, real classes must be set up before processes may be run in parallel – which can be tedious and expensive. A common CSP/**occam** idiom where this happens is for performing primitive inputs/outputs in parallel, e.g.

```
PROC delta (CHAN INT in?, out.0!, out.1!)
  WHILE TRUE
    INT x:
    SEQ
      in ? x
      PAR
        out.0 ! x
        out.1 ! x
:
```

To support this idiom, C++CSP provides a special `ParallelComm` class. To demonstrate its use, here is a *generic* `Delta` class for doing the same as the above **occam** process, but which can be applied for channels of any data type (using C++ templates), and for any user-defined number of parallel output channels:

```
template <class DATA_TYPE,unsigned OUTPUT_CHAN_AMOUNT = 2>
class Delta : public CSProcess
{
private:
    const Chanin<DATA_TYPE> channelIn;
    DATA_TYPE t;
    ParallelComm* parComm;
protected:
    void run()
    {
        while (true)
        {
            channelIn >> t;
            parComm->communicate();     // outputs t on all channels
                                        // (see below in the constructor)
        };
    };
public:
    inline Delta
        (
            const Chanin<DATA_TYPE>& chIn,
            const Chanout<DATA_TYPE>* const chOut
        )
        :
            CSProcess(65536),channelIn(chIn)
    {
        ParCommItem* pars[OUTPUT_CHAN_AMOUNT];          // the size of
        for (int i = 0;i < OUTPUT_CHAN_AMOUNT;i++)      // the output array
        {                                               // is determined
            pars[i] = chOut[i].parOut(&t);              // by the template
        }
        parComm = CSP_NEW ParallelComm(pars,OUTPUT_CHAN_AMOUNT);
    };
};
```

The poison-related portions of the run method have been omitted for the sake of simplicity. The run method is simple – continually input a data item and perform the parallel communication. This communication is set up in the constructor. An array of (pointers to) `ParCommItem` data types is created and initialised by calling the `parOut` method of each channel writing end to obtain a `ParCommItem` (that will output from the variable `t`).

To construct an instance of this process for integer channels and with, say, 10 outputs:

```
Chanin<int> in;
Chanout<int> out[10];

...  initialise in and out

Delta<int,10> myDelta(in,out);
```

Parallel communication is encapsulated in a class for reasons of expected efficiency and typing problems with implementing it using functions.

This mechanism is currently being developed to a more generalised system that will allow other common actions, such as syncing on a barrier or "falling into" a bucket to be used in a similar way, and this development is likely to lead to a name change to reflect this expansion from pure communications.

3.6 *Alternatives*

Being able to choose between channels or other events (Alternatives) is one of the fundamental primitives of CSP. As with **occam** and JCSP, C++CSP restricts this to choosing between channel inputs, timeouts and skips. Here is an example from our test suite – a process that repeatedly `ALT`s between two channels and a timeout:

```
class Alter : public CSProcess
{
    const Time t;
    Chanin<char> in0,in1;
protected:
    void run()
    {
        char c;
        Alternative alt(
            in0.inputGuard(&c),
            in1.inputGuard(&c),
            new RelTimeoutGuard(t),
        NULL);

        while (true)
        {
            switch(alt.priSelect())
            {
                case 0:                 //input has taken place on in0
                    ...  deal with it
                    break;
                case 1 :                //input has taken place on in1
                    ...  deal with it
                    break;
                case 2:                 //timeout expired
                    ...  perform action for timeout
                    break;
            }
```

```
            }
        };
    public:
        Alter(
            const Time& time,
            const Chanin<char>& c0,
            const Chanin<char>& c1
        )
            :   CSProcess(65536),t(time),in0(c0),in1(c1)
        {
        };
    };
```

The `ALT` is set up in the run method. The constructor takes a list of varying size of pointers to Guard objects that is `NULL`-terminated (as per the `Parallel` function). The `inputGuard` method of the channel reading end fetches an input guard (that will write into `c`). A "relative" timeout guard is constructed that will timeout after the time length `t` has elapsed since the beginning of the `ALT` (absolute timeout guards that expire after time `t` are also available). The Alternative class is executed using the `priSelect` method and then its return value (the zero-based index of the guard that fired) can be acted on. Note that there is a difference between this API and the JCSP version: in C++CSP, the guard action (e.g. input) is performed automatically by the `select` method of the `Alternative` (whereas, in JCSP, programming that action is the user's responsibilty). There are arguments for both approaches – the current C++CSP mechanism is open for discussion.

C++CSP offers "`PRI ALT`", "fair `ALT`" and "same `ALT`" capabilities. The difference between the three occurs when multiple guards are ready; `PRI ALT`s always pick the earliest guard in the list in this situation. A fair `ALT` evens out the priority over the course of many calls to the select method by least favouring the last activated guard, and a same `ALT` favours the last activated guard – useful for guards that will activate infrequently in short bursts. C++CSP also offers an *arbitrary* `ALT`, where the decision between multiple ready guards is non-deterministic.

4 Implementation Details

4.1 Concurrency/Scheduling

One of the main implementation details of any CSP framework is how the essential element of concurrency is provided. JCSP uses Java's threads to run processes in parallel. KRoC[5, 11], Kent's **occam** compiler, uses assembly language to switch workspaces (by altering the stack register). This means that JCSP does not have control over the context switches that occur (the JVM does), but KRoC does have control over the context switches. In other words KRoC is co-operatively multitasking.

C++CSP uses user-space threads to follow the KRoC model of being co-operatively multitasked. Since the library then has control over exactly when context switches occur, many optimisations can be made regarding the need for mutexes and similar synchronisation mechanisms for internal library data because only one process will ever be scheduled at any one time.

If this was the only mechanism used then C++CSP would not be able to take advantage of multiple processors at all as it would only use a single thread (besides which, the optimisations would all be invalid if multiple processors were in use).

One possibility for altering this would have been to create an OS-thread version along the lines of Vella et al. [12, 13]. C++CSP instead uses a many-to-many threading model. The library contains a function to spawn a process in a new thread. This new thread is initialised with a separate user-space kernel as described above and the specified process is run using this kernel in the new thread. Special channels for communicating between threads are available in the library, as is an inter-thread barrier that is used to synchronise processes in different threads (especially at the time of termination). Their algorithms are not discussed in this paper, only the algorithms for the user-space threading.

C++CSP runs under Microsoft Windows by utilising the little-known "fibers" lightweight user-space thread capabilities. For Unix variants, C++CSP uses the standard ANSI C function `setjmp` / `longjmp` on supported processors. This uses a single line of assembly language (the only one in the whole library) to manipulate the stack. It can also use the highly-portable (if a little slow) GNU `pth`[14] library to function under virtually every known Unix-variant.

The scheduler can also be ported to use other scheduling mechanisms (for example an assembly language based scheduler in an embedded device). This is because at its core the scheduler only needs three primitive operations – start process, stop process, and switch to specified process.

4.2 Stacks

Using user-level threads means that either the C++CSP library or a threading library that it uses must allocate the stack from the heap as a normal piece of memory. While Microsoft offer automatic stack-growing in their fibers library, such mechanisms are not provided under the other scheduling methods. This leads to the unfortunate situation where the user must guess the size of stack that a process needs (in a portable library!) and if it is too low, stack overflow and hence memory access violations will occur. Equally, setting too large a size can cause the program's memory requirements to bloat, especially if one process must be run thousands of times in parallel. Profiling tools could potentially be used to aid in getting the stack size right. A possible solution to this could be to automatically check for nearing the end of the stack area in memory and making the library expand the stack, but this would only be possible using the `setjmp` mechanism as the `pth` library does not allow for direct stack manipulation such as this. CCSP has this same problem (although **occam** does not – even for the latest version[5, 15] that includes run-time creation of processes and mobiles).

4.3 Channel Algorithms

The channel algorithms are much simplified by the fact that only one process can actually be running at any one time and that the library is in control of switching between processes. This removes the need for mutexes or monitors to be used with the channel data – it can be accessed directly.

The core of the channel algorithms work on the assumption that the channel is one-to-one – shared channels wrap these functions up with mechanisms to ensure that only one writer and one reader will actually be attempting to access the channel at any time (the other readers and writers must form a queue).

At any point in time, a one-to-one channel that is only being used for normal inputs and outputs can be in one of three states – empty, reader waiting or writer waiting. There will never be a situation where both are waiting because the second party to arrive will see the other waiting and perform the communication. From the point of view of one of the participating parties a channel can only be in two states – empty or another process waiting (since the channel is one-to-one they must be a reader for our writer or vice versa).

So the main data member of a channel is a handle to a process. It can either be a null handle indicating that the channel is empty/unused, or it will point to a waiting process. There should also be a data member to store the destination/source of the communication (for a waiting reader/writer respectively) so that the second process to arrive at the channel can perform the communication itself. So the pseudo-code of either process consists of:

- If the channel is empty:
  - Put our pointer in the channel, and our destination/source pointer
  - Suspend ourselves from the run queue and schedule another process
- Else:
  - Perform the transfer using the waiting process's destination/source pointer
  - Put the waiting process back on the run queue

A simple and elegant algorithm. Of course more complexities are required for the finished library due to the variety of different communication types (shared channels, buffered channels, `ALT`s, extended inputs, parallel communication, and various permissible combinations of these).

One particular problem with the above function is that if the waiting party is performing a parallel communication then they will be "woken up" (put back on the run queue) as soon as one of them completes, regardless of whether the remaining channel communications have completed. So we can use the concept of an "occupied count". Every process will have an occupied count (that basically counts the number of actions it is occupied with). During normal operation it is irrelevant but when the process is waiting on a channel (and therefore not on the run queue) it sets its occupied count to be the amount of communications that it is waiting to finish.

Instead of putting a waiting process straight back on the run queue, the occupied count of the process should be decremented and compared to zero – if it is zero then the process is put back on the run queue, otherwise it is waiting for other transfers and should be left off the run queue for now. Processes wishing to perform normal communications simply set their occupied count to 1 so they will be woken up as soon as the transfer completes, as before.

The last channel procedure that will be examined here is that of `ALT`ing. If a process wants to `PRI ALT` over many channels then it should first check the channels in order to see if a writer is waiting to output on any of them. If there is a waiting writer then the `ALT`ing process can complete the transfer, wake the writer and complete the `ALT`. Complications arise however if there are no waiting writer.

Clearly the `ALT`ing reader should wait on all these channels. If it uses the usual algorithm however then as soon as a writer finds the reader, it will complete the transfer and wake the reader. Should two writers arrive in quick succession (that is, the second arrives before the reader has run again), they will both complete the transfer – clearly an error as an `ALT`ing reader should only ever complete one transfer per `ALT`.

The solution to this is to have the reader place its handle in the channel as normal, but put a null pointer into the data destination address in the channel. Then any writer that finds this null pointer must not perform the transfer – instead it should overwrite the process handle in the channel with its own, set the source address and suspend itself as if the channel were empty, but before it reschedules, it should place the reader back on the run queue.

When the reader wakes up, it knows that *at least* one writer must be ready. It should check the channels (in reverse order, from least to highest priority), checking the process handles in the channels. If it finds its own handle then it should remove it (to make the channel unused again). If it finds a writers handle then it should record the index of the channel, as this will be the highest priority ready writer it has found. When all the channels have been checked, it can perform an input on the appropriate channel and then return. Any writer that it did not input from can be left as it is set-up ready for the next input on the channel.

This is the standard `ALT`ing algorithm, first applied by INMOS for the transputer's microcoded scheduler[16] and subsequently developed for non-transputer **occam** kernels[11] and JCSP[17].

5 Advanced Features

5.1 Extended Rendezvous

C++CSP has some features that extend the core CSP functionality. One idea borrowed from KRoC is that of the extended rendezvous[15]. This is the term coined for when a channel communication takes place between two processes, but the writing process is not free to continue until the reading process orders it (rather than free as soon as the communication has finished). This allows for interesting new functionality – for example, a semantically invisible *snooping* process, inserted into any channel (where the writer is not free until the *snooper* has successfully inspected the data and forwarded it to the reader). C++CSP allows these extended inputs to be performed both stand alone and inside an `ALT`. An example is given in section 6.3.

5.2 Blocking System Calls

As mentioned in the concurrency section, C++CSP allows processes to be placed in separate threads. On a single-processor machine using operating-system threads will be slower to perform normal computations than using the normal user-level threads. The difference arises in relation to blocking system calls (the most common of which are input/output operations).

If a normal process makes a blocking system call, it will cause the entire CSP kernel of that thread to freeze for an indefinite period of time. This could be tolerable in some cases but often this is totally unacceptable. The solution in C++CSP is fairly simple – write the code in a process with no regard for blocking, and use the functionality of the library to create this process in another thread. The process can block without freezing the original thread, and can still communicate with the processes in the original thread by using the special inter-thread channels.

It is hoped that features such as this and the fact that most GUIs and many toolkits are written in C/C++ will allow C++CSP to be used for GUIs – it has been noted before that CSP has a natural fit to graphical interfaces. Unfortunately most GUIs are not thread-safe, a classic example being Java's Swing[3]. Those built on C++CSP, however, will be easy to make thead-safe – simply by conforming to the usual (**occam**) *parallel useage* rules.

5.3 Mobiles

Mobiles [18] are a relatively new addition to the KRoC compiler. They are a compound data structure that has zero-aliasing enforced by the compiler, but they are still referred to by pointers. Whenever the mobile is send down a channel or involved in an assignment, the compiler transfers the pointer but changes the value of the source of the operation to be undefined, thereby ensuring that only one reference to the mobile exists at any one time.

C++CSP translates these semantics into a so-called "smart pointer" class. It holds a single pointer, and whenever an assignment is performed, the pointer in the source mobile is set to

NULL. The library uses the assignment operator inside its communications, so no special case is needed for sending mobiles down channels, they are used as every other object and the mobile semantics are inherently used.

5.4 KRoC Inter-operability

KRoC is an **occam** compiler developed by the University of Kent. It provides ultra-fast context switches using its own user-level kernel. One relatively new feature is its user-defined channel mechanism that allows custom channels to be provided to communicate almost seamlessly with the **occam** code. C++CSP uses this, along with its inter-program channels (that can communicate between any two programs on the same machine run by the same user), to allow C++CSP processes to communicate with **occam**. Both programs can use their own native channels, allowing for very easy inter-operability.

The inter-program channels only allow for one-to-one communication (with no ALTs) but 'id' processes can be used at either end to allow for any variety of communication to be performed.

6 Performance Measures And Tests

6.1 CommsTime

One benchmark[19] that is commonly used for CSP frameworks is the commstime benchmark. This consists of building a simple network with a prefix, successor, delta and custom recorder process to measure how fast it takes the network to produce each number. The results recorded on a Celeron 667 MHz. (256 MBytes SDR RAM) were:

CSP Framework	Time Per Iteration
occam (KRoC 1.3.3)	1.3 microseconds
C++CSP (setjmp)	5 microseconds
C++CSP (GNU pth)	75 microseconds
JCSP (JDK 1.4)	230 microseconds

Unsurprisingly, KRoC is the fastest of all the CSP implementations, due to its heavily optimised and CSP-centred nature. C++CSP is not far behind however – only a factor of 4, despite being high-level and portable. This is a very encouraging result that shows it can be within an order of magnitude of KRoC. The results for GNU pth are included here to show the time differences between the scheduling methods. It is disappointingly slow, given that the library is built using the same setjmp/longjmp functions that C++CSP uses for its own scheduling. Therefore the native C++CSP setjmp implementation is recommended for use over GNU pth wherever possible.

JCSP is two orders of magnitude behind KRoC and C++CSP. This is because it sits on top of whatever OS/JVM thread mechanisms are implementing Java's native threads. On our test platforms, these are based on non-cooperative multitasking with relatively large-grained time slicing. Whilst not particularly fast, this does bring other benefits – namely that blocking system calls do not block the rest of the parallel system.

6.2 Stressed ALT

Another test, from JCSP, is the stressed ALT test. This involves connecting p writing processes to n shared channels and getting one process to fair ALT over the n channels, meaning it will take input from a total of $p \times n$ processes. The library has successfully run this test with 200 processes on each of 100 shared channels, yielding a total of 20,000 processes trying to write to one (very stressed!) reader. This test succeeded, with the fair ALT performing correctly. Subsequently, timings were performed to compare with both KRoC and JCSP:

CSP Framework	Time for test A	Time for test B	Time for test C
occam (KRoC 1.3.3)	0.6 microseconds	0.7 microseconds	1 microsecond
C++CSP (setjmp)	3 microseconds	7 microseconds	10 microseconds
JCSP (JDK 1.4.3)	130 microseconds	200 microseconds	—

Test A ran with 10 channels (n), 10 writers on each (p). Test B ran with 20 channels, each with 100 writers. Test C used 100 channels, each with 200 writers. The times recorded are the average time for each successful input by the reading process. The machine used was the same Celeron 667 MHz.

KRoC's *fair ALTing* mechanism delivers unit time for each choice, regardless of the number of choices being checked — so long as the system is under stress (i.e. messages are always pending on each channel)[3]. The same applies for C++CSP and JCSP. The increase in timing from test A to C is the result of cache misses — the **occam** version of test A fitting entirely on to primary cache!

The times for KRoC and C++CSP are one order of magnitude apart, but it is pleasing to see that C++CSP appears to perform proportionally better on the small test (A).

The results for JCSP are, perhaps, unfairly penalised by the very small cache on the Celeron processor. For example, running Test A on a 233 MHz. Pentium 2, gave a timing of 80 microseconds. Unfortunately, that machine had only 64 MBytes of main memory, which was simply not enough for the 2001 Java threads required for Test B — the system thrashed its disc heavily and timings fluctuated wildly (above 300 microseconds). Test C requires 20,000 Java threads and the consequent disc thrashing brought both these machines to their knees.

6.3 Mandelbrot Fractal

To further demonstrate the abilities of the library, here is an abbreviated version of a system to produce a Mandelbrot fractal. The layout of the network will be very naive and simple – one process for each pixel of the Mandelbrot, with each feeding (via a shared channel) into a single "pixel drawer" process.

The pixel-drawer process could be implemented simply by inputting a pixel (coordinates and colour) and drawing it. The feeder channel could be made buffered but that would make little difference to this uni-processor demonstration. An alternative would be for the drawer process to use extended inputs. This means that outputting a pixel down the channel has the semantics of an imperative procedure call – once the output has completed, the pixel has been drawn. This alternative method is the one that is used below.

The pixel-drawer process listed below includes some synchronisation code to synchronise with the locking/unlocking of a frame buffer. Whenever the process receives an input on the sync channel (apart from the initial one), it waits for a second input; the first message

[3]This was not the case for the *transputer* implementation, which always delivered an $O(n)$ overhead for each choice, where n was the number of choices.

indicates that the frame buffer has been locked, the second that the frame buffer is now unlocked:

```
class PixelDrawer : public CSProcess
{
private:
    Chanin<Pixel> in;
    Chanin<bool> sync;
    Pixel p;
protected:

    static void DrawPixel(csp::CSProcess* process)
    {
        PixelDrawer* t = (PixelDrawer*)process;
        Pixel p = t->p;
        ... draw the pixel
    };

    void run()
    {
        try
        {
            bool b;
            sync >> b;

            csp::Alternative alt(
                in.extInputGuard(DrawPixel,&p),
                sync.inputGuard(&b),
            NULL);

            while (true)
            {
                if (alt.priSelect() == 1)
                {
                    sync >> b;    // wait for screen to be unlocked
                };
            }
        }
        catch (PoisonException)
        {
            in.poison();
            sync.poison();
        }
    };
public:
    PixelDrawer(
        const csp::Chanin<Pixel>& i,
        const csp::Chanin<bool>& sy
    )
        :   CSProcess(65536),in(i),sync(sy),screen(scr)
    {
    };
};
```

The syntax for extended inputs in alternatives sets up the `extInputGuard` method of a `Chanin` with two parameters – the first is a function that takes a single parameter of type `CSProcess` and returns void, the second must be the destination address for the input. This function is executed for the extended part of the input (before the outputter is freed) and is passed the pointer to the process that is doing the input. In this example, no further action needs to be taken in response to the extended input guard – action is only needed in response to a sync input. The code for the main process is given here:

```
class Mandelbrot : public CSProcess
{
protected:
    void run()
    {
        One2OneChannel<Pixel> draw;
        One2OneChannel<bool> sync;

        Barrier barrier(1);
        // ^^^ - initialise with (this) one process already registered

        ...  initialise graphics

        spawnProcess(
            new PixelDrawer(draw.reader(),sync.reader()),
            &barrier
        );

        const int X = 160, Y = 120;

        CSProcess* mandels[X * Y];
        for (int i = 0;i < X;i++)
        {
            for (int j = 0;j < Y;j++)
            {
                mandels[(i * Y) + j] =
                    new MandelbrotPixel(draw.writer(),i,j);
            }
        }

        sync.writer() << true;     // lock the frame buffer

        Parallel(mandels,X*Y);
        // ^^^ - will only return when all the pixels have been drawn

        draw.writer().poison();
        // ^^^ - slight cheat, to poison the channel

        barrier.sync();            // unlock the frame buffer

        sleepFor(Seconds(30));
        // ^^^ - Simplistic way of displaying the image for a while

        ...  de-initialise graphics
    };
public:
    Mandelbrot() : CSProcess(65536*8)
    {
    };
};
```

The `MandelbrotPixel` process simply calculates the pixel colour for its given coordinates, sends the pixel down its channel and exits – omitted here for brevity. Graphics initialisation etc. code has also been omitted. In this example, the `PixelDrawer` process never actually recieves a `sync` message associated with the unlock – it is poisoned instead. The `spawnProcess` function forks off the specified process, taking as its second parameter a pointer to a `Barrier` (C++CSP's implementation of the KRoC/JCSP barrier primitive) that can then be `sync()`ed with to wait for the termination of the forked process.

7 Conclusion

C++CSP is a new but stable library that implements all common CSP paradigms and is highly portable. Possible future developments include more integrated network support, some GUI support and possibly other tools to aid in the development (such as an automatic process template generator). Following JCSP.net[20] and KRoC.net[21], C++CSP.net seems almost required!

The library is still in its infancy however and needs to cultivate a user base while ensuring (and proving) its stability and providing new functionality – without succumbing to so-called "feature-bloat". It is hoped that this library will prove useful to existing CSP users, and hopefully even introduce some C++ users to the useful concept of Communicating Sequential Processes.

The library will be made available on the web by the University of Kent shortly. Enquiries can be made to either author.

References

[1] C.A.R.Hoare. Communicating Sequential Processes. In *CACM*, volume 21, pages 666–677, August 1978.

[2] C.A.R.Hoare. *Communicating Sequential Processes*. Prentice-Hall, 1985.

[3] Hans Muller and Kathy Walrath. Threads and Swing. `http://java.sun.com/products/jfc/tsc/articles/threads/threads1.html`.

[4] INMOS LIMITED. Occam 2.1 Reference Manual. Technical Report, May 1995. Available at: `http://www.wotug.org/occam`.

[5] University of Kent at Canterbury. Kent Retargetable occam Compiler. Available at: `http://www.cs.ukc.ac.uk/projects/ofa/kroc/`.

[6] J.Moores. CCSP – a Portable CSP-based Run-time System Supporting C and occam. In B.M.Cook, editor, *Architectures, Languages and Techniques for Concurrent Systems*, volume 57 of *Concurrent Systems Engineering series*, pages 147–168, Amsterdam, the Netherlands, April 1999. WoTUG, IOS Press.

[7] University of Kent at Canterbury. Java Communicating Sequential Processes. Available at: `http://www.cs.ukc.ac.uk/projects/ofa/jcsp/`.

[8] Gerald H. Hilderink. CTJ (Communicating Threads for Java). `http://www.ce.utwente.nl/javapp/`.

[9] Peter H. Welch and David C. Wood. Higher Levels of Process Synchronisation. In A. Bakkers, editor, *Parallel Programming and Java, Proceedings of WoTUG 20*, volume 50 of *Concurrent Systems Engineering*, pages 104–129, Amsterdam, The Netherlands, April 1997. (WoTUG), IOS Press. ISBN: 90-5199-336-6.

[10] P.H.Welch. Graceful Termination – Graceful Resetting. In *Applying Transputer-Based Parallel Machines, Proceedings of OUG 10*, pages 310–317, Enschede, Netherlands, April 1989. Occam User Group, IOS Press, Netherlands. ISBN 90 5199 007 3.

[11] P.H.Welch and D.C.Wood. The Kent Retargetable occam Compiler. In *Proceedings of WoTUG 19*, volume 47, pages 143–166, March 1996.

[12] K.Vella and P.H.Welch. CSP/occam on Shared Memory Multiprocessor Workstations. In B.M.Cook, editor, *Architectures, Languages and Techniques for Concurrent Systems*, volume 57 of *Concurrent Systems Engineering Series*, pages 87–119, Amsterdam, the Netherlands, April 1999. WoTUG, IOS Press.

[13] Kevin Vella, Kurt Debattista, and Joseph Cordina. Cache-Affinity Scheduling for Fine Grain Multithreading. In James Pascoe, Peter Welch, Roger Loader, and Vaidy Sunderam, editors, *Communicating Process Architectures 2002*, volume 60 of *Concurrent Systems Engineering*, pages 321–352, IOS Press, Amsterdam, The Neterlands, September 2002. IOS Press.

[14] GNU. GNU `pth`. `http://www.gnu.org/software/pth/`.

[15] F.R.M.Barnes and P.H.Welch. Prioritised Dynamic Communicating and Mobile Processes. *IEE Proceedings-Software*, 150(2):121–136, April 2003.

[16] M.D.May, P.W.Thompson, and P.H.Welch. *Networks, Routers and Transputers*, volume 32 of *Transputer and occam Engineering Series*. IOS Press, 1993.

[17] Peter H. Welch and Jeremy M.R. Martin. Formal Analysis of Concurrent Java Systems. In Peter H. Welch and Andr W.P. Bakkers, editors, *Communicating Process Architectures 2000*, volume 58 of *Concurrent Systems Engineering series*, pages 275–301, Amsterdam, the Netherlands, 2000. IOS Press.

[18] F.R.M.Barnes and P.H.Welch. Mobile Data, Dynamic Allocation and Zero Aliasing: an occam Experiment. In *Communicating Process Architectures*. IOS Press, 2001.

[19] Roger M.A. Peel. A Reconfigurable Host Interconnection Scheme for Occam-Based Field Programmable Gate Arrays. In Alan G. Chalmers, Henk Muller, and Majid Mirmehdi, editors, *Communicating Process Architectures 2001*, volume 59 of *Concurrent Systems Engineering*, pages 179–192, IOS Press, Amsterdam, The Neterlands, September 2001. IOS Press.

[20] P.H.Welch, J.R.Aldous, and J.Foster. CSP networking for java (JCSP.net). In P.M.A.Sloot, C.J.K.Tan, J.J.Dongarra, and A.G.Hoekstra, editors, *Computational Science - ICCS 2002*, volume 2330 of *Lecture Notes in Computer Science*, pages 695–708. Springer-Verlag, April 2002.

[21] Mario Schweigler, Fred M.R. Barnes, and Peter H. Welch. Flexible, Transparent and Dynamic occam Networking With KRoC.net. In J.F.Broenink, editor, *Communicating Process Architectures – 2003*, volume 61 of *Concurrent Systems Engineering*, Amsterdam, The Netherlands, September 2003. (WoTUG), IOS Press.

Communicating Process Architectures 2003
Jan F. Broenink and Gerald H. Hilderink (Eds.)
IOS Press, 2003

Agents for Concurrent Programming

Enrique GONZALEZ, César BUSTACARA, Jamir AVILA
Research Groups SIDRE-SIRP
Pontificia Universidad Javeriana - Facultad de Ingeniería
Bogotá, Colombia
egonzal@javeriana.edu.co

Abstract. This paper aims to demonstrate that concepts from Distributed Artificial Intelligence are very useful to design concurrent systems. The BESA framework and the AOPOA methodology are introduced as tools to achieve this goal. The Behavior-oriented, Event-driven and Social-based Agent (BESA) framework combines the concepts of MultiAgent Systems with the design of concurrent systems: an agent can be constructed as a set of behaviors; the notion of behaviors can be directly applied to concurrent systems design using the Agent Oriented Programming paradigm. The internal architecture of a BESA agent integrates two important features: a modular composition of behaviors and an event dispatcher based in a *select* like mechanism. The Agent Oriented Programming based in an Organizational Approach (AOPOA) methodology provides a systematic procedure to build complex system based in three concepts: a hierarchical recursive decomposition of the system, a goal-oriented role identification, and an evolution of the cooperation relationships linking the system components.

1 Introduction

In the last decade, the field of MultiAgent Systems (MAS) has evolved quickly. Applications of MAS based systems include many diverse domains such as information retrieval, robotics, education and training, social simulation, e-commerce, and many others [5][2][3][7]. FIPA, the Foundation for Intelligent Physical Agents, is an organisation aimed at producing standards for the interoperation of heterogeneous software agents [9]. The FIPA architecture includes basic communication services and specifies the Agent Communication Language (ACL). Several FIPA compliant platforms have emerged, some of the most significant Java based ones are JADE, ZEUS, FIPA-OS and April [10]. As MAS based systems are inherently parallel, MAS platforms can fusion elements from the artificial intelligence and from the concurrent system fields; the Behavior-oriented Event-driven and Social-based Agent (BESA) framework is an attempt to accomplish this fusion [13].

This paper aims to demonstrate that concepts from distributed artificial intelligence are very useful to design concurrent systems. In particular, MAS provide a framework inherently concurrent, as they are composed of a set of autonomous entities called agents, working in parallel. In a MAS, agents cooperate to achieve the system goals, the global behavior of the system emerges from the interactions between agents [6]. An agent is an entity that encapsulates data and behavior, in a similar way objects from the OOP paradigm do. Objects are passive, agents are active, agents are continuously acting to achieve their goals in a proactive and autonomous way. An agent is situated in an environment, owns the means to sense and modify its environment, thus agent behavior is closely related to it [3].

Agent concepts can be used to build parallel entities with a high degree of abstraction.

The BESA architecture [13] is the basis of an agent framework that aims to combine the concepts of MAS, as presented in the next section, with the design of concurrent systems. In BESA, an agent is composed of concurrent modules, called behaviors, and includes a channel that supports a *select* like mechanism (cf. section 2.4). Concurrence is present, not only at system level, but also inside agents. In order to design complex systems supported by the BESA platform and following a distributed artificial intelligence approach, an Agent Oriented Programming based in an Organizational Approach (AOPOA) methodology is proposed [12] [11].

This paper is focused in the concurrent design aspects; it presents how the concepts from MAS are implemented and used for concurrent system design. Once the relation between artificial intelligence and concurrent design is analyzed, the fundamental concepts of the three layers of the BESA architecture are introduced. Then a closer look of the concurrent internal architecture of a BESA agent is presented, followed by the main aspects of the AOPOA methodology and their implications for concurrent programming. Finally, some implementation issues and results are discussed.

2 Artificial Intelligence and Concurrent Design

The AI notions can be used to build concurrent systems based in a higher abstraction approach. In particular, agents could be seen as the basic concurrent entities composing the system; interaction between agents provide a clear semantic to the communication between entities; an organizational decomposition allows to deal with complexity; the behavior concept is directly mapped to the *select* mechanism used in concurrent systems.

2.1 Identification of Concurrent Entities

When designing a concurrent system, one of the critical points is to identify entities or modules that can run in parallel to achieve a task in an efficient way. For instance, the methodology presented by Foster performs a four steps process: partition, communication, agglomeration and mapping [1]; the obtained entities when applying this process are meaningless in the context of the task. In opposition, based in the MAS paradigm, a system can be seen as a set of agents, each of them having a well-defined role. A role defines the responsibilities of the agent in terms of its goals, and also identifies unambiguously its relations with the environment and other agents. Several instances of a role can run in parallel in a natural fashion; each of them owning a clear semantics in the context of the problem that the system is dealing with.

From the point of view of software design, it is very valuable to build a system composed of entities with well-defined semantics. The system can be developed and maintained more easily, and also the obtained components can be reused as building blocks in other applications or expanded to enlarge their functionality.

2.2 Semantics of the Interactions

Rationality is one of the most important concerns when dealing with agents from the artificial intelligence perspective [4]. However, when dealing with the design of concurrent systems, what is interesting is to study how agents can interact in a well-defined fashion. The relationships between agents have a high degree of abstraction, thus the model of interactions among the entities of the system posses a well-defined semantics. In fact, there are multiple MAS techniques designed to foster the cooperation between agents.

Cooperation in MAS is achieved by incorporating three essential components: collaboration, which is in charge of task allocation; coordination, that includes synchronization and planning; and conflict resolution which is needed when concurrent resource access or incompatible objectives appear. Ferber [6] and Huhns [2] compile some of the more important techniques to deal with these three problems.

When a cooperation technique is applied, an interaction protocol is established between the cooperating agents. Interaction protocols are well-formed conversations that define communication patterns between the participants; the exchanged messages are considered as speech acts; they contain not only data, but also an intentional semantics [8]. From the perspective of concurrent design, it is very valuable to build a system where the relationships between its concurrent components have a clear semantics: it is easier to construct the system and to verify its correctness.

2.3 Complexity Management

Dealing with complexity is one of the major concerns of software system design; actual systems are large and their requirements are huge, so the relationships among its components become very difficult to manage. The n-tier architectures and the use of software patterns have permitted to construct large systems in a more systematic and robust way [24][25]. The MAS approach offers further concepts and tools to deal with complexity. As it was pointed out before, MAS design provides techniques to identify the set of entities that compose a system. In particular, a MAS can be studied as being an organization [6]; thus, it can be analyzed in a hierarchical recursive way: an organization is decomposed into simpler ones.

This approach is very useful to deal with complexity; if there is a very complex entity, it can be decomposed into simpler ones. The goals and relations of the original organization are delegated to the simpler ones; additional interactions between the new entities can be needed to make them work together. Building complex systems based in a decomposition strategy is systematic and produces well-structured and modular software.

2.4 Non Determinism Management

Minsky [14] proposed an agency model to build intelligent systems: the idea that a mind can be constructed from simple concurrent modules was introduced. This notion of modular components was exploited, mainly in the robotic research field, and complemented with biological inspired principles. The result of this evolution is the concept of behavior as the fundamental building block of intelligence. A behavior is a mapping of sensory inputs to a pattern of motor actions, which are used to achieve a task [15]. Thus, an agent can be constructed as a set of behaviours and the overall behavior of the agent emerges from them; the recursive organizational decomposition is applied again inside the agent.

The notion of behavior can be directly applied to concurrent systems design. A concurrent entity can be seen as being composed of several modules, similar to agent behaviors, each module having well-defined semantics, purpose and responsibility. An entity can be built using a *select* like mechanism, similar to the guard concept implemented in Joyce [22] and the CSP alting-channel [23]. Each module is associated to a different option/guard of the *select* mechanism. When a new stimulus arrives to the entity, it is redirected to the appropriate module by the *select* mechanism. Once again artificial intelligence notions facilitate parallel systems design; in fact, the abstraction provided by the behavior approach makes it easier to develop systems applying the traditional mechanisms for dealing with non-determinism.

3 BESA Model

The abstract model of BESA is based in three fundamental concepts: a modular behavior-oriented agent architecture, an event-driven control approach implementing a *select* like mechanism, and a social-based support for cooperation between agents [13]. In this section, the concept of event used and the layers of the BESA architecture are introduced.

3.1 Event-Driven Control

When building agents it is fundamental to consider the external information coming from the environment, agent behavior derives from it. Normally, the agent reacts to the stimulus provided by the environment. From the concurrent system perspective, stimulus can be modelled as incoming events; in BESA all interactions between agents are modeled by events. Since the agent is cooperating with others, messages coming from them must be also considered as events. In this case, the event includes also the data associated to the message, and represents an speech act. A speech act is more than a communication exchange because it includes a well-defined intention that yields to an action; it constitutes a way to achieve a goal.

In the BESA model, an agent is activated only when a fact, concerning it, occurs. Thus, this event-driven control model avoids CPU busy-wait, agents do not consume processor while waiting for events.

3.2 BESA Architecture

The BESA architecture is composed of three levels: agent level, social level and system level.

Agent Level: an agent is composed by a set of behaviors, a channel acting as an event selector mechanism, and a state. A detailed explanation of the agent architecture will be presented in the next section.

Social Level: it is intended to define and manage hierarchical organizations. These organizations are composed by a set of agents with a common goal. In the BESA model, a facilitator agent could be included in the organization to provide cooperative support to the other agents. The services provided by the facilitator agent include task and resource allocation, conflict resolution, distributed planning and synchronization. This level offers tools to design semantic interactions between the components (agents) of the concurrent systems.

System Level: it comprises the execution environment for agents conformed by BESA containers that could reside in different processing units. In order to comply with the FIPA standard, this level includes a unique entry port that receives ACL request coming from other external MAS [9]. In figure 1, the components of the system level are presented. The container manages the life cycle of the agents; the local administrator supports the agent communication; directory facilitators provided naming and lookup services. Local administrators are synchronized and include automatic replication facilities; it also includes the coordination mechanism to achieve agent mobility.

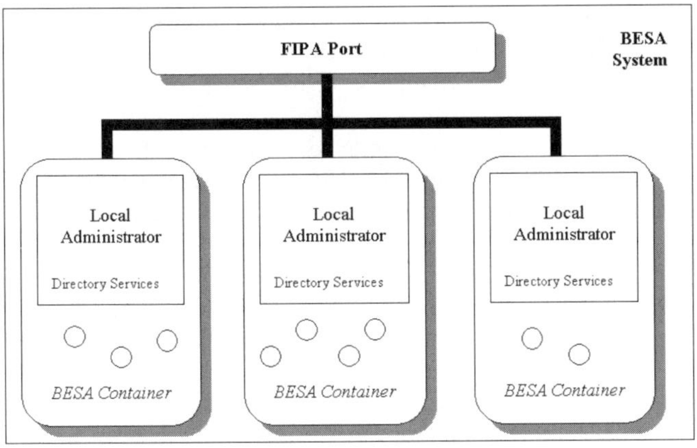

Figure 1. BESA System Level Model

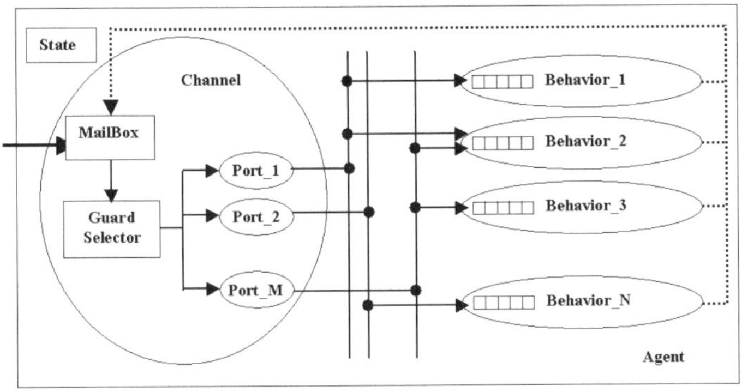

Figure 2. Internal Architecture of an Agent

4 Architecture of a BESA Agent

The internal architecture of an agent integrates three important features: a modular composition of behaviors, a channel implementing an event selector mechanism and an agent internal state. In figure 2 the relations between components are illustrated. In BESA, channel and behaviors run in parallel providing high level of concurrency. This approach allows to easily manage the natural non-determinism present in complex systems. In fact, the channel includes a set of ports to provide a *select* like mechanism. BESA is implemented in the Java language in order to exploit features such as portability, object oriented approach, support for multi-platform and multi-threading, and communication facilities.

4.1 Channel Model

The channel receives all events directed to the agent from the environment or other agents. The channel represents the only entry point to the agent, ensuring that the agents are always ready to receive events in asynchronous way. The incoming events are transferred to the concerned behaviors using the selector mechanism. The channel components are a mailbox, a guard selector and a set of ports.

The mailbox is a temporary mechanism used to store events; it is implemented using a queue. As the channel includes an unique entry point, while an event is picked out and processed, it is possible that new concurrent events arrive; the mailbox assure that these events are not missed.

Once an event arrives, the guard selector receives and processes it. An event includes a tag that defines its semantics; the selector uses this tag to place the incoming event in the appropriated port. There is a unique port for each defined tag in order to avoid ambiguous behavior activation.

The abstract model of a *select* mechanism includes a set of guards that can be concurrently fired in a non-deterministic pattern. A guard is fired when a specified type of event arrives and a logical condition is satisfied. Once the guard is fired, the appropriated actions are executed in response to the incoming event. In the BESA channel, for each guard exists an associated port. When an event is placed in a port, if the respective logical condition is satisfied, the event is transferred to the interested behaviors; otherwise, the event is held in the port. When the logical condition becomes true, if there is any event in the port, the first one must be transferred immediately; in order to assure this requirement, the logical condition is associated to a predefined object; when this object is modified, an attempt to fire the guard is performed automatically.

A channel is implemented as a thread that performs four steps:

- The channel *setup* method is automatically called in order to initialize the local data structures of the agent and to create its behaviors.
- A barrier synchronization of the behaviors is performed in order to assure that all the components of the agent are completely operational before processing any event.
- The heart of the channel run method is an infinite loop that picks out events from the mailbox, in a blocking receive statement, transfer them to the appropriate port, and tries to fire the corresponding guard.
- Finally, when the *kill* method is called, the infinite loop is terminated and the channel *setdown* procedure is executed; it includes behavior destruction and house keeping tasks.

This technique to handle events allows the agent to be ready to react to multiple simultaneous incoming events. As the run method is fixed in advance, the programmer must provide only the methods that are called automatically; which assures the channel robustness and allows to implement the *select* mechanism in a proper way.

4.2 Behavior Model

In the BESA model, an agent is composed by a set of behaviors. Behaviors are concurrent processes that encapsulate the actions that must be executed in response to the received events. A behavior handles at least one type of event; in order to obtain automatically the events of interest, it must bind itself to the corresponding guard. When analyzed from the abstract view, a behavior encapsulates the data and the code required to manage a set of guards.

The behavior has a priority queue, where events are transferred by the channel. All external events have the normal priority level; only some special system events have a higher priority. This design avoids starvation problems, while allowing to deal with special situations.

A behavior is implemented as a thread that performs three steps:

- The behavior *setup* method is automatically called in order to initialize the local data structures of the behavior and to create and bind its associated guards.

- The main part of the behavior run method is an infinite loop that picks out events from its queue, in a blocking receive statement, and depending on the event tag calls the method corresponding to the associated guard.
- Finally, when the *kill* method is called, the infinite loop is terminated and the behavior *setdown* procedure is executed, which includes guard destruction and house keeping tasks.

This technique to build an agent allows it to process several events in parallel; however, notice that events associated to the same behavior are processed sequentially. Since the run method is fixed in advance, the programmer must provide only the methods that are called automatically; which assures the behavior robustness and the execution of actions in a proper way. These actions include changes on the state, generation of events, or modifications of the environment.

4.3 Internal Communication and Mobility

The agent state serves to store its internal state or the model of the environment, which may include its beliefs concerning other agents. The state is unique and shared by all the behaviors, thus a mutual exclusion mechanism is implemented using the monitor supplied by the Java language. The state can be used as a communication mean between behaviors.

In order to respect the semantics of the *select* abstract model, this shared memory mechanism does not allow to directly exchange events between behaviors. Therefore, when a behavior needs to synchronize with another one, it must send an event to the channel mailbox.

Agent mobility is one important feature that can be implemented using the BESA agent architecture. In fact, as the behavior and channel run methods are pre-coded, a synchronization of all the active components can be achieved; thus, allowing to safely stop and restart an agent in order to be moved.

When it is required to move an agent, the agent's channel sends a high priority special event to its behaviors, causing them to terminate its infinite loop without calling the *setdown* method and to stop in a well-defined and stable state. Once this point is reached, the behavior informs the channel, which using a synchronization barrier mechanism detects when the agent is ready to be moved. At this point, the actual BESA container transfers the agent instance to the desired destination container, where a clone of the moving agent is created and started in the adequate point to continue its normal operation. The local administrator retains the events arriving to the agent while the agent moving procedure is performed and sends them to the agent once it has been restarted.

5 Agent Oriented Programming

Shoham introduced the agent oriented programming (AOP) approach [20]. The AOP can be seen as an extension of the object oriented programming (OOP) model; in fact, AOP is a new paradigm to analyze and design engineering systems. The world is inherently concurrent, the agent abstraction allows to map it in a more realistic fashion: the model is closer to the natural parallelism of the problem being analyzed. AOP can incorporate into unique framework notions from the OOP approach and from the concurrent programming techniques, enriched by the artificial intelligence concepts of rational agents. AOP methodologies must be used in order to develop modular, flexible and extensible systems, which can be implemented on agent platforms like the BESA framework.

The construction of flexible systems exhibiting complex behaviors, emerging form the interaction of modular components, can be achieved in a structured fashion [17]. Most of the known AOP methodologies include, sometimes with different denominations, the following steps: requirements analysis, role identification and task allocation, agent type specification, agent system architecture design, agent internal architecture design, and final system integration. The artifacts produced and the way these steps are performed are particular for each methodology. For instance, Boer argues that five models must be constructed: agent/role, skills, knowledge, organization and interaction [16]. Tropos is supported by two basic ideas: the notion of modeling an agent based in its mental states and a well-defined set of environment requirements [18]. The Gaiga approach analyzes the system in a macro level, the agent society, and in a micro level, the individual agent [21]. Miles approach is centered in the agent interaction analysis, roles as means to achieve goals allow representing the social aspects of the system [19]. Ferber proposes that a MAS can be seen as an organization, which has qualities not present in its individual components [6]. An agent-oriented programming based in an organizational approach (AOPOA) methodology is proposed by Ahogado [12].

AOPOA aims to provide a systematic and general procedure to build complex systems based in three concepts: a hierarchical recursive decomposition of the system, a goal-oriented role identification, and a evolution of the cooperation relationships linking the system components. The resulting MAS is composed by a set of active entities that aim to accomplish a well-defined set of objectives. The organizational approach makes it easier to perform an iterative and recursive decomposition of complex entities into simpler ones based in a goal split and merging procedure; and at the same time to identify the interactions between the entities composing the system. The design process is iterative, an organization level is developed at each iteration. In the analysis phase, tasks and roles are detected; during the design phase, the interactions between roles are specified and managed by cooperation links. At the final iteration, the role parameterization is performed, which allows to specify the events and actions associated to each agent. The abstraction level of the cooperation links evolve: first they represent abstract relations, then a cooperative mechanism is associated to deal with the specific interaction situation, a well-formed interaction protocol is used to implement it, finally an appropriate communication pattern is obtained. Figure 3 illustrates the iterative role decomposition process; the new identified roles must achieve the goals of the original ones; the cooperative links present in *i-th* iteration remain the same; notice that these links are assumed by the new inner roles.

The AOPOA final model includes a set of artifacts that specify and describe in a complete and structured way the roles, the cooperation links and the agent parameterization of the MAS. This model is independent of the platform where the system will be implemented; it contains a set of entities linked by well-defined message exchanges (interaction protocols), and a statement of what actions must be taken in response to each message.

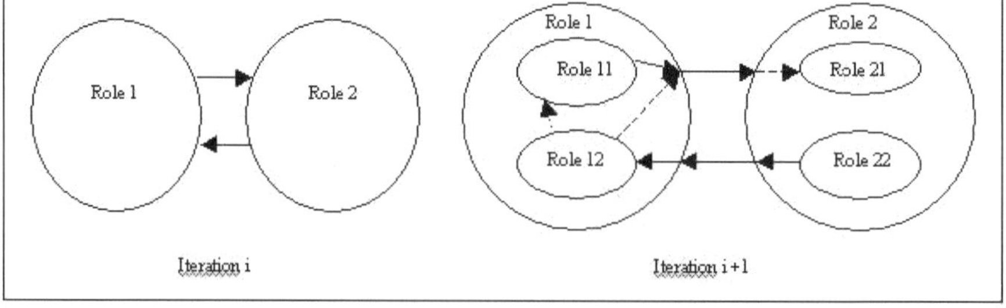

Figure 3. AOPOA role decomposition and cooperative link inheritance.

The BESA platform is very well suited to implement this kind of model. The messages are seen as events; for each event an associated guard is in charge of performing the specified actions. The guards of an agent that aim to assure the same goal and that can run sequentially are grouped into the same behavior. Finally, the obtained agents are deployed in one or several BESA containers.

6 Results and Discussion

The agent and system levels of the BESA architecture have been implemented successfully using the Java programming language; this choice aims to run BESA systems in a heterogeneous, hardware and software, platform. The BESA architecture was evaluated by the implementation of two applications, which include most of the challenging situations that MAS must deal with. The first application case concerns the cooperative robotics field; a multirobot simulator was built, and then extended to control a physical multirobot system; the simulator has been used intensively in artificial intelligence courses. The second study case was used for refining the AOPOA methodology; the simulation of a restaurant was developed following the steps of AOPOA. The obtained model was easily implanted in BESA. In both cases, the operation of the systems was satisfactory, a large number of agents interact concurrently in a correct fashion. In the near future a more detailed evaluation of BESA, including quantitative results, will be performed.

The BESA framework and the AOPOA methodology complement each other. AOPOA provides an iterative systematic procedure to generate a MAS abstract model that fulfills the specific application requirements. BESA incorporates concepts and mechanism from the artificial intelligence and concurrent systems fields to supply a framework that is well suited to implement MAS. When both are combined, a complete development tool for concurrent system design is obtained, which includes identification of concurrent entities, specification of semantic interaction patterns, organization of system complexity, and management of non-determinism. Besides, the system is composed of modular components with increased cohesion and reduced coupling; the inherent high level of abstraction of the agent approach makes the system easy to extend, scale and maintain.

BESA makes easy to develop concurrent applications because the framework hides the low-level details to the programmer, such as thread management and communications. You have to code only the business logic and map it to guards that are grouped in behaviors; the concurrent aspects of programming are automatically provided. However, in order to obtain a proper system, a coherent agent approach must be applied to analyze the specific problem and to design an adequte solution.

The MAS approach is not only the basis of a promising technology, but also it emerges as a new way of thinking: a paradigm for analyzing and designing inherently concurrent systems. In this paper, it has been demonstrated that concepts from distributed artificial intelligence are very useful to design concurrent systems. In particular, the synergy between BESA and AOPOA provides a set of interesting agent based tools to achieve this goal.

Acknowledgements

This work is part of the ASMA project supported by the Pontificia Universidad Javeriana. The authors gratefully acknowledge the contributions of the students working in ASMA related projects.

References

[1] Foster I., Designing and Building Parallel Programs. Addison-Wesley, 1995.
[2] Huhns M. et al., "Readings in Agents", Morgan Kaufmann, 1998.
[3] Weiss G. (Editor),"Multiagent Systems", MIT Press, cap. 1-2, 1999.
[4] Russell S., Norving P., "Inteligencia Artificial: Un Enfoque Moderno". Prentice Hall, 1996.
[5] CMU, "The MultiRobot Lab". http://www-2.cs.cmu.edu/~multirobotlab.
[6] Ferber J., MultiAgent Systems: an Introduction to Distributed Artificial Intelligence, Addison Wesley, 1999.
[7] Knapik M. et al., "Developing Intelligent Agents for Distributed Systems". McGraw Hill, 1998.
[8] Labrou Yannis, Finin Tim, Agent Communication Languages: The Current Landscape, IEEE Intelligent Systems, March/April 1999.
[9] FIPA Organization, "FIPA - Foundation for Intelligent Physical Agents". http://www.fipa.org.
[10] FIPA, "Publicly Available Implementations". http://www.fipa.org/resources/livesystems.html.
[11] Pontificia Universidad Javeriana, "PUJ_Ingeniería – Proyecto de Investigación ASMA". http://ainsuca.javeriana.edu.co/asma.
[12] Ahogado D., Reinemer A.M., González E., "AOPOA: Aproximación Organizacional para Programación Orientada a Agentes". Submitted to CLEI'03, 2003.
[13] González E., Avila J., Bustacara C., "BESA: Behavior-oriented, Event-driven and Social-based Agent Framework". To appear in PDPTA'03, 2003.
[14] Minsky M., The Soceity of Mind, Touchstone Book, 1986.
[15] Murphy R.R., Introduction to AI Robotics, MIT Press, 2000.
[16] De Boer F., "Methodology for Agent-Oriented Software Design", NOAG-i project, http://www.cwi.nl/~frb/.
[17] Caire G., Leal F., Evans R., "Agent Oriented Analysis using MESSAGE/UML".
[18] Castro J., Kolp M., Mylopoulos J., "Towards Requirements-Driven Information Systems Engineering: The Tropos Project", 2001.
[19] Miles S., Joy M. y Luck M., "Designing Agent-Oriented Systems by Analysing Agent Interactions", Department of Computer Science, University of Warwick Coventry, CV4 7AL, United Kingdom.
[20] Shoham Y.,"Agent-Oriented Programming", Artificial Intelligence, 60-1, pp. 51-92, 1993.
[21] Wooldridge M., Jennings M. Kinny D., "The Gaia Methodology for Agent-Oriented Analysis and Design", Autonomous Agents and Multi-Agent Systems 3, Kluwer Academic Publishers, pp. 285-312, 2000.
[22] Hansen B., "A JOYCE Implementation", Software - Practice and Experience, Vol. 17 Nº 4, April 1987.
[23] Hoare C., "Communicating Sequential Processes", Comm. ACM, Vol. 21 Nº 8, Aug. 1978.
[24] Alur D. et al., "Core J2EE Patterns: Best Practices and Design Strategies". Prentice Hall, 2001.
[25] Gamma E., Helm R., Jonson R., Vlissides. "Design Patterns: Elements of Reusable Object-Oriented Software". Addison-Wesley. 1999.

Communicating Process Architectures 2003
Jan F. Broenink and Gerald H. Hilderink (Eds.)
IOS Press, 2003

occam for reliable embedded systems: lightweight runtimes and model checking

Denis NICOLE
Department of Electronics and Computer Science, University of Southampton, Southampton, SO17 1BJ, UK.
dan@ecs.soton.ac.uk

Sam ELLIS
ARM Ltd.,110 Fulbourn Road
Cambridge, CB1 9NJ, UK
sam.ellis@arm.com

Simon HANCOCK
Airbus, Filton, Bristol, BS99 7AR, UK
simon.hancock@airbus.com

Abstract. We describe some more recent developments of the SPoC system. We describe a new module in the occam compiler which performs substantial simplifications of the run-time demands made by the compiled code. This has been used successfully both to target a simple PIC microcontroller and to generate input for the SMV model checker.

1 Introduction

The initial development of the SPoC occam compilation system was reported[1] at a previous conference of this series. This is a robust programming environment, suitable for commercial code development. In its original form, it accepted the occam 2.5 programming language and converted a complete parallel occam program into a single program in the C programming language. A simple scheduler implemented within the C code allows the execution of the parallel occam program by the cooperative sequencing of the various program fragments. No operating system support for multithreading or concurrency is required. The resultant code is highly portable and has run on big- and little-endian processors, UNIX and Microsoft operating systems and dedicated signal processors. Macros for the `gdb` debugger permit the debugging of running programs at the occam source level. Additional experimental versions of SPoC have supported distributed message-passing architectures, allcache shared memory systems, dynamic remote process creation and channel migration.

Despite its wide application, there were two key application areas for which SPoC was not entirely suitable: microcontrollers and model checking. In both cases, this was because the domain required that the run-time support demands be even "lighter" than for SPoC. In the case of the microcontrollers, the burning need was to minimise the amount of run-time state and thus of RAM required for an occam program; a 16F84[2] has only 36 bytes of data storage and 1024 words of instruction storage. For model checking, not only did the amount of run-time state have to be minimised to reduce the amount of checking to a reasonable level, but the restricted execution model of mainstream model checkers had to be targeted. Newer version of the PIC, such as the 16F628, relax the memory constraints considerably; the increase in data memory to 224 bytes is particularly welcome.

We believe we have achieved a reasonable degree of success in tackling both these domains; perhaps one of our key achievements has been to eliminate the learning curve for both targets. The occam programmer is able to perform interesting model checking entirely at the occam level, with no need to learn either a FSM description language or the handling of temporal logic expressions.

2 The occam simplifier

In addition to their limited memory, the PIC microcontrollers have only a restricted stack depth for subroutine return addresses. A module for SPoC was developed which performed a variety of standard compile-time optimizations:

1. Elimination of `SKIP` processes.
2. Constant propagation and elimination of unreachable `IF` and `CASE` processes.
3. Replicator unrolling. In occam, `PAR` replicators are always constant, as are many `ALT`, `IF` and `SEQ` counts. Elimination of these explicit replicators greatly simplifies the run-time and allows many new opportunities for constant propagation. This unrolling is essential to reduce the amount of run-time state exposed to the PIC.
4. Function and procedure inlining. Occam has a very clean copy-in copy-out semantics for procedure parameters and thus inlining, even in the presence of free variables, is relatively straightforward. Again, this gives additional opportunities for constant propagation and dead code elimination.
5. Array inlining. Traditional array indexing imposes unacceptable demands on the PIC architecture. Not only is array indexing awkward to code, the manipulation of interesting arrays is incompatible with the PIC's very small data store. The compiler thus eliminates all array references, replacing them as necessary with generated scalar variables.

As is typical with an optimizing compiler, the various optimizations interact strongly and multiple passes through the optimizer are required to realize all the benefits of the simplifications. It will be seen that many of the optimizations have the effect of reducing run-time dynamic state (array indexes, loop counters, subroutine stacks, at the expense of increasing the code size. This is a fortuitous combination as for both microcontrollers and model checkers, the code length (or model size) is not normally a limiting factor; the dynamic state, on the other hand, is strongly restricted. The reader should also note that occam would support even more aggressive optimizations; it is, for example possible to remove[3] almost all the parallelism and internal channels from an occam code.

3 The PIC run-time

There are a variety of obvious simplifications for the PIC. Firstly, floating-point arithmetic makes little sense on such a simple architecture and all support for floating point is dropped. Similarly, `INT`s are configured as sixteen bit. `INT32` and `INT64` are, however, supported although little used. Another occam feature which is expensive on the PIC is overflow detection for integer arithmetic. This is supported by default, but a compiler option allows the support to be removed, improving code performance. Multiplication is not supported in the PIC hardware, so a special code block, which takes its operands from fixed memory locations, is used to perform multiplies in software. This is against the inlining philosophy of the compiler, but the considerable size of the multiply code precludes its inlining. Division is currently not supported.

As noted earlier, replicated `PAR`s are unrolled by the optimizer. We still need, however, to maintain state to manage the join at the end of the PAR block; this takes up one byte in memory. The process scheduler uses one bit per deschedule point to indicate which code blocks are runnable; when the running process becomes blocked, the scheduler searches these bits for another runnable code block.

Channels are implemented as one bit per out (`!`) process; an inputting (`?`) process scans the out bits to find a runnable transfer. `ALT`s offer no special problems here although this does effectively result in a polling implementation `ALT`.

Proper support for timers and ports in the compiler makes it easy to write a demonstration code which will, say, flash several LEDs at different rates or implement full duplex serial communications.

4 Model checking

Model checking has become a preferred technique among researchers for assuring the correctness of concurrent hardware or software systems. In short, a complex system is built from a set of interacting finite state machines. An exhaustive simulation is then performed of all possible histories for the complex system. In each state of each history, the property under test is evaluated. The model checker can then report whether the property can ever be violated and, usefully, can directly emit a shortest history to a violation. It is essential in standard model checkers that the number of possible states (instantaneous configurations) of the system be finite; furthermore the complexity of the model checking run grows rapidly with the size of this state.

There are several commercial-grade model checkers available to the programming community. Well-known ones include SMV[4,5], SPIN[6] and FDR[7]. These are remarkably efficient programs but are not yet taken up widely in industry. Perhaps the main obstruction to their adoption is the substantial *semantic-gap* between programming languages and model-checker input. The issues here are by no means as dramatic as with, say, proof systems, but are nevertheless substantial for the working programmer. The program typically has to be rendered from C, Java or occam source into a wholly alien state-machine representation. Furthermore, the property under test has to be described in a form of temporal logic: a predicate calculus with a few extra quantifiers. These two foreign languages are enough to deter all but the most persistent programmers. An alternative formulation of model checking tests whether one program is a refinement of another (i.e. has a lower degree of nondeterminism); this is hardly easier to explain to the typical programmer.

Fortunately, there are ways to de-skill both the key steps in setting up a model checker. First, we can directly check the program in its high level language. There are clearly potential problems here; the variables in a typical program imply a great deal of state. Thus it is normally necessary to simplify the program under test by only tracking the variables which are essential to the property being tested; this is a form of *program slicing*. This simplification does not damage the reliability of the model check; wherever an untracked variable is used in, say, an `IF` test, its value is treated as undetermined and both branches of the test are explored. Thus the model checker is always pessimistic; if insufficient variables are tracked, then the check gives a false failure, never a false success.

We eliminate the need for temporal logic by replacing the check of our program in isolation with a check in an environment. For occam, we just join the external channels up to another occam program representing the environment. Then we typically need only two tests:

1. Safety properties are tested by checking if an error state can ever be reached.
2. Liveness properties are tested by checking if deadlock (`STOP`) can ever be reached.

We do not explicitly treat fairness. Ideas of *weak fairness* add a great deal of complexity to models of concurrency. We instead take the aggressive view that, for real-time codes, failure of any component to make progress within a fixed and finite number of *ticks* represents failure and causes a transition to the error state.

5 A model checking example: LTSA

The LTSA[8, 9] model checker is especially designed as a simple introduction to the technology. It is easy to use and can be run as an applet over the World Wide Web. A very simple example of its use gives some indication of what can easily be achieved, and where problems arise. Let us try to evaluate the correctness of the following code which purports to offer mutual exclusion between two threads. We define a turn variable:

```
int turn = 1;
```

and in each thread (me=0, 1 respectively), we guard the critical section as follows:

```
while (turn != me);
/* loop waiting for my turn */
/* CRITICAL SECTION */
turn = 1- me;
```

We model this system in LTSA's FSP language as the parallel composition of three machines, representing the turn variable and the two threads which share it:

```
TURN1 = (set0->TURN1|unset1->TURN0|isset1->TURN1),
TURN0 = (set0->TURN1|unset1->TURN0|isunset0->TURN0).
P0 = (isunset0-> claim0 -> release0-> set0->P0).
P1 = (isunset1-> claim1 -> release1-> unset1->P1).
||SYS=(TURN1||P0||P1).
```

This does not look much like the original C. Note also that we have replaced a busy polling implementation with a non-busy one. We have included claim and release transitions so that we can join up an environment which ensures that claims and releases can take place only in the permitted order. This new system, which checks the safety property of this mutex becomes:

```
TURN1 = (set0->TURN1|unset1->TURN0|isset1->TURN1),
TURN0 = (set0->TURN1|unset1->TURN0|isunset0->TURN0).
P0 = (isunset0-> claim0 -> release0-> set0->P0).
P1 = (isset1-> claim1 -> release1-> unset1->P1).
property MUTEX=(claim0->release0->MUTEX|claim1->release1->MUTEX).
||SYS=(TURN1||P0||P1||MUTEX).
```

Here the `property` keyword is a shorthand which adds extra transitions into an error state from all states on all events in the vocabulary of `mutex` which are not explicitly given transitions to other states. Thus the explicit transitions of `mutex` become the only safe transitions which the machine should accept. This system is accepted by LTSA as being without error or deadlock, so the safety property is true; the example *is* a mutex.

We can go on to check a liveness property. We require that the mutex be able to offer the (free) critical section to either thread. We check this by composing the mutex with code which makes an *internal choice* about which thread will request the mutex. Possible deadlock would imply that the mutex fails to accept this imposed choice. In occam, such internal choice is easy to express; we just use an `ALT` with two `SKIP` guards. It's a bit trickier in LTSA; we need to represent in the LTSA language an ALT with two inputs on the same channel, then hide this spurious channel. In FSP, we get:

```
TURN1 = (set0->TURN1|unset1->TURN0|isset1->TURN1),
TURN0 = (set0->TURN1|unset1->TURN0|isunset0->TURN0).
P0 = (isunset0-> claim0 -> release0-> set0->P0).
P1 = (isset1-> claim1 -> release1-> unset1->P1).
EXT = (t->claim0->release0->EXT|t->claim1->release1->EXT)\{t}.
||SYS=(TURN1||P0||P1||EXT).
```

As the reader might expect, LTSA reports a deadlock when presented with this input; our crude mutex actually forces strict alternation and to do better we need Peterson's[10] algorithm.

So what has been learnt? First, crude tests (no temporal logic) are entirely adequate to test both safety and liveness. Second, even simple tutorial model checkers have an alien input language. Finally, occam is at least as expressive as FSP both for the system under test and for its environments.

LTSA is a very attractive system for learning about model checking. Checking software is, however, an intensive task and it is not appropriate to use a visual learning tool for this purpose. Production-quality model checkers are readily available [5,6,7]; SMV makes a good choice as it has been adopted by Cadence, thus improving our chances in the future of model checking across the hardware-software boundary.

6 Model checking occam

We have chosen to implement the model check as a new back-end for SPoC. This makes heavy use of the optimiser developed for the PIC back-end to simplify the generated SMV code and to reduce the amount of program state. As explained above, it is necessary to annotate the occam program with compiler directives to indicate which variables are to be tracked by SMV. This is done by introducing directives of the form

```
#PRAGMA TRACK x 0::10
```

into the occam source. This directive not only requests the tracking of variable x, it also asserts that x has range $0 \leq x \leq 10$. A violation of this range assertion will be flagged as an error by SMV. The generated SMV code is remarkably similar to that for the PIC in overall structure. For example, the *runnable* bits associated with code fragments on the PIC are simply combined to give a program counter per concurrent thread which increments only at deschedule points. The implementation of ALT depends on the version of SMV that is in use. There is a simple version for Cadence Berkeley SMV which uses guarded set expressions. Unfortunately, this version is not open source, and a rather more complex implementation with potentially many cases has to be used with Carnegie-Mellon's open source SMV.

As well as generating the SMV code for the model check, it is important that, in the event of failure of the check, the SMV history of the failure can be rendered in language the occam programmer can understand. This back annotation of SMV state into occam state is handled by a Perl script which maps the SMV program counters back into occam source lines. This trace history can then be followed to the failure. Information to drive this script is generated by SPoC. By default the model checker tests the Temporal Logic (CTL) expression

```
AG AF termination
```

This expresses in temporal logic the assertion that the occam program will terminate. Deadlock and most forms of livelock will have the effect of making this test fail.

References

[1] M. Debbage, M. Hill, D. A. Nicole and S. M. Wykes, Southampton's Portable Occam Compiler (SPOC). In: R. Miles and A. Chalmers(ed.) *Progress in Transputer and OCCAM, Research (WoTUG-17).* ISBN: 9051991630. IOS Press, Amsterdam, 1994.

[2] Information on PIC Microcontrollers may currently be found on the World Wide Web at: http://www.microchip.com/

[3] A. W. Roscoe and C. A. R. Hoare, The Laws of Occam Programming, Programming Research Group Technical Monograph PRG-53, ISBN: 902928341, 1986.

[4] K. L. McMillan, Symbolic Model Checking, ISBN: 792393805, Kluwer Academic Publishers, Norwell, Ma., 1993.

[5] Information on SMV may currently be found on the World Wide Web at: http://www-2.cs.cmu.edu/~modelcheck/smv.html or http://www-cad.eecs.berkeley.edu/~kenmcmil/

[6] Information on SPIN may currently be found on the World Wide Web at: http://spinroot.com/spin/whatispin.html.

[7] Information on FDR may currently be found on the World Wide Web at: http://www.fsel.com/software.html.

[8] J. Magee and J. Kramer, Concurrency: State Models & Java Programs, ISBN: 471987107, Wiley, Hoboken, NJ., 1999.

[9] Information on LTSA may currently be found on the World Wide web at: http:// http://www-dse.doc.ic.ac.uk/concurrency/ltsa-v2/index.html

[10] G. L. Peterson, Myths about the mutual exclusion problem. *Information Processing Letters 12*, 3, 115—116 (1981)

[11] The source code and working examples of the software described in this paper may be found on the World Wide Web at: http://www.hpcc.ecs.soton.ac.uk/software/spoc/

Communicating Process Architectures 2003
Jan F. Broenink and Gerald H. Hilderink (Eds.)
IOS Press, 2003

The Trebuchet

John CAMPBELL
Northrop Grumman Corporation, 888 South 2000 East, Clearfield, Ut 84015-6216
john.campbell@ngc.com

G. S. STILES
Computing Laboratory, University of Kent, Canterbury CT2 7NF
dyke.stiles@ece.usu.edu

Abstract. The Trebuchet is a hardware architecture for machines implemented from conventional software source programs. It is pseudo-asynchronous in that it decouples the system clock from the computational logic, reducing electromagnetic interference, smoothing current draw, and reducing pipeline latency, key benefits of asynchronous designs. Performance of example programs is given.

1 Introduction

This paper discusses the design of the Trebuchet, a pseudo-asynchronous micropipeline, and proposes a new architecture for high performance reconfigurable computing. It grows out of an earlier Trebuchet effort we called SMAL [1].

This version of the Trebuchet consists of a linear pipeline of configurable computing elements. Each element (composed of one or more FPGAs (Field Programmable Gate arrays)). The system is composed of many interacting state machines (Fig. 1.) Each state machine, while entirely synchronous, behaves asynchronously in the sense that stage latency depends on the complexity of the computation performed rather than on the clock period. The clock period, rather than being dependent on the longest logic chain, is set by the time required to exchange handshaking signals with immediate neighbors [2].

On the global scale, the Trebuchet resembles Sutherland's asynchronous Micropipelines [3], inheriting many of their behavioral characteristics, despite being completely synchronous in the details. Hence we describe the Trebuchet as Pseudo-Asynchronous. We claim that pseudo-asynchronism allows the flexibility required to implement software as a cohesive hardware machine, and that the benefits normally ascribed to asynchronous machinery [4] may be achievable. Specifically, throughput should depend, as in conventional asynchronous circuitry, on the average stage latency rather than the longest logic chain. Likewise, current draw should be smoothed with attendant reductions in radiated EMI (Electromagnetic Interference).

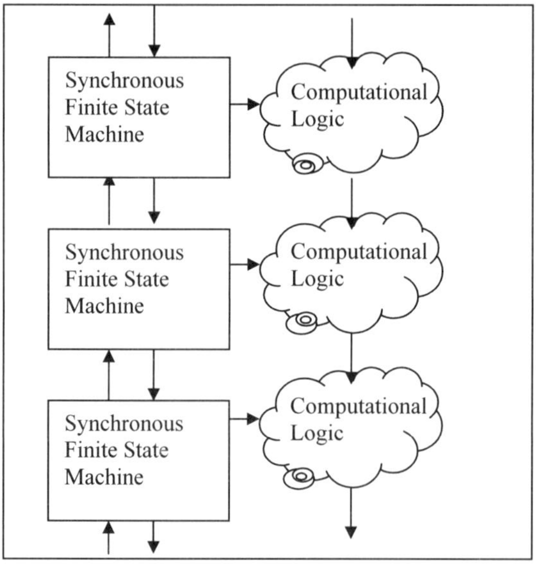

Fig. 1. A Trebuchet pipeline consisting of FSM-controlled stages.

The thrust of our research has been to develop a methodology for seamless implementation of hardware and software functionality. It has applications in both the fields of embedded systems and reconfigurable computing engines. The first (embedded systems) is concerned with dividing functionality between software and hardware (generally application specific integrated circuits (ASICs)), whereas the latter seeks to off-load time-critical functions to temporary circuits configured into FPGAs.

In contrast to conventional projects where software is prepared in an environment of existing and stable hardware, embedded computer systems typically require parallel development of hardware and software components [5]. Because the respective disciplines of hardware and software development are commonly conceived of as quite different, early decisions about task allocation are made which have profound and irreversible consequences on the ultimate cost and performance of the system. Consequently, much of the research in embedded system technology is devoted to blending the development methodologies by deriving both hardware and software from high-level descriptions so that decisions can be delayed as long as possible and are demonstrably correct when made [6].

Because conventional hardware design is tedious and foreign to most software practitioners, generating hardware from software has been an important research goal [6]. Generally speaking, these approaches target reconfigurable machinery hosted on FPGAs. However, reconfigurable co-processors tend to execute their functions rapidly and then remain idle for long periods of time. Consequently, the question of how to dynamically reconfigure the co-processor has also become important [7], [8].

Our overall goal has been to develop a methodology in which the use of CPUs, reconfigurable logic implemented in FPGAs, and permanently configured logic in an ASIC are all parameters in a scheduling problem. But because available chip real estate allows implementation of only small portions of a program in hardware, it seems desirable to develop an approach that deals with multiple chips and the consequences of chip and board boundaries.

We seek a methodology that automatically allocates portions of a program to a network of execution resources based on single-threaded software execution profiles. A methodology that constructs systems using both hardware and software can best take advantage of the available execution resources. Code written in a high level language (e.g. Java) may execute as machine instructions or directly as hardware. The decision should depend on the needs of the work to be done and the resources available to do it.

In [9] we presented a theory of hierarchical network domains and applied it to systolic networks being optimized for maximal throughput. This research expands that work to address virtual networks of reconfigurable components, allowing a uniform treatment of devices with quite different reconfiguration characteristics (i.e. dynamically and statically reconfigurable FPGAs, ASICs, and CPUs).

We chose Java as the application language because the Java Virtual Machine (JVM) has a simple and regular addressing scheme without registers [10], and because the interpreter makes it easy to gather execution statistics that may be used in mapping experiments. Our research [11] indicated that Java execution is predictable enough that transformation of portions of an application to hardware is possible.

Fleischmann and Buchenrieder [12] are also using Java to study reconfigurable hardware systems, but do not generate hardware automatically, as our system does. Hutchings et al. [13] are doing low-level hardware design with Java-based JHDL, but their tool is not aimed at high performance pipeline systems. Handel-C is a similar system based on the C language; this approach associates assignment operations with latches [14], whereas we associate assignments with wires and latch much larger aggregates.

2 The Trebuchet System

The Trebuchet runs hardware compiled from Java source code. We modified Jikes [15], an open source java compiler originally from IBM, to include extra information we needed for the conversion to hardware. The output of Jikes is a standard Java class file. We obtain profile information from a modified JVM (Java_g, part of the Sun Java JDK). The profile also includes segmentation of the Java byte-codes into basic blocks and descriptors for the structure of *for* loops, *if* statements, etc.

There is considerable opportunity for fine-grained parallelism. While parallelism is, in principal, possible to detect automatically, we added the keyword *par* to the syntax parsed by Jikes. *Par* signifies to the VHDL translator that a *for* loop is vectorizable.

The Java byte-codes are translated to VHDL by analyzing the basic block contents. Stack and Memory references become accesses to wires (thus being essentially compiled out) and successive op-codes become, for the most part, cascaded blocks of combinational logic. Array accesses become accesses to RAM. Since each array resides in its own RAM, concurrent access to different arrays is supported. Concurrent access to the same array must be arbitrated across the program.

We targeted our hardware at the Xilinx V800 FPGA. With capacity of 800,000 gates, there is room for moderate sized software experiments. In the future we intend to generalize Trebuchet to address designs involving multiple chips so that programs of arbitrary size may be executed. Hardware configuration files are generated by standard Xilinx tools [16].

2.1 Modifications to the Jikes Java Compiler

The Java code is compiled by a modified version of the Java compiler Jikes. We added keywords to the otherwise standard Java syntax recognized by Jikes. As noted above, in principle the compiler could have been modified to recognize parallelizable and systolizable loops [17]. At some point in the future, we intend to do this.

In addition to the *par*, we also included the keywords *netstart, netend,* and *expose*. *Netstart* and *netend* indicate respectively the beginning and end of the code to be analyzed for hardware mapping. *Expose* designates variables that are required as output from the execution engine. All of these are expediencies that could, in principal, be automatically recognized by a compiler.

The structures which may be inserted into the Java class files include *start* tags, *stop* tags, and *parallel loop descriptor* tags. The latter designate *for* loops for vectorization. The byte code interpreter has special code added to it to detect this extra information in the execution stream of the program. Because it is desirable that code thusly modified also be executable by conventional JVM platforms, the tags are structured so that a conventional JVM will simply jump around them.

We needed to format the tags so that they could be unambiguously recognized in the JVM execution stream. Since the compiler never generates a jump-to-self instruction, we can use this instruction as our tag: these are constructed as a jump followed by a jump-to-self, followed by tag specific information. Fig. 2 illustrates this format.

```
        Goto AfterTag
Self:
        Goto Self
        Tag ID Byte
        *
        *          \
        *           |  Tag Specific
        *           |  Information
        *          /
AfterTag:
```

Fig. 2. Format of description tags embedded in Java execution stream.

2.2 Modifications to the Java Virtual Machine

One of the purposes of the Trebuchet is to experiment with mapping of regions of software onto hardware regions. The theory of this mapping [9] depends on profile information on the number of times communication arcs are utilized, rather than the number of times nodes are visited – a standard profile. Consequently the JVM was altered to collect *transfer of control* statistics. Java_g was tailored to output a file of bytecodes segmented into basic blocks (basic blocks are sequences of code which terminate at program jumps).

2.3 Bytecode Translation to VHDL

Trebuchet, written in Common Lisp, translates the basic block and profile information provided by the modified JVM. Trebuchet symbolically traverses each basic block, generating combinational logic corresponding to the sequence of instructions. Some instructions (e.g. multiplication) are impractical to configure as purely combinational logic and necessitate further segmentation. Trebuchet also constructs hierarchical components, such as *for* loops, that consists of a controlling FSM (finite state machine) and other subcomponents.

A basic block is a sequence of code unbroken by changes in sequential flow (except at the terminus). Thus there are not multiple paths of execution within a basic block.

Trebuchet traverses a basic block, examining each instruction. Bytecodes that manipulate memory (either stack or variable store) such as IPUSH rearrange the set of working wires. Operations that produce values (e.g. IADD) take their inputs from the set of working wires (deleting them from the set) and introduce new wires with the outputs. Trebuchet translates each basic block to a combinational net of logic. Fig. 3 shows an example basic block in source code and as hardware logic.

Many compilers manipulate stacks internally and map stack locations to registers. Pushing an object on the stack, while conceptually moving the entire stack, in reality only changes the association between register names and stack offsets. Trebuchet does this with the set of wires representing program memory and the stack.

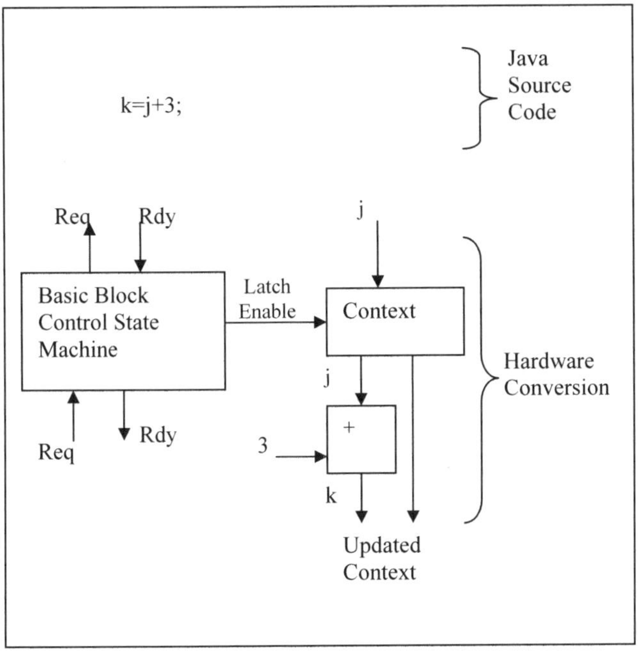

Fig. 3. Example translation of a fragment of Java code.

2.4 Vectorized Loops

Because the algorithms we want to investigate with Trebuchet manipulate programs that stream data through operators, we needed a way to generate code that could execute systolicly. Java does not have a parallel operator, so we added one.

Conceptually, our *par for* loops have four parts. These include the *initialization_clause, the end_test*, and the *step_clause* that conventional Java and C share. The *loop_body* is run overlapped. As Fig. 4 shows, a *par for* loop is organized with a tight loop that repetitively steps the loop variable, tests the termination condition, and initiates an entry into the pipeline. All three of these operations are executed in parallel to minimize the latency between subsequent pipeline entries.

Pipeline initiation may be thought of as a thread that executes a particular loop iteration. This construction is not far removed from loop vectorization, a well-studied topic in computer science [17] and could, in principle, have been accomplished automatically. But for our purposes, it is enough to use the *par* keyword to designate code that can be validly overlapped in execution.

Trebuchet generates a vectorized loop from the body of the Java *for* loop and the control clauses specified with it. It rearranges the controls and forms a pipeline from the succession of basic blocks in the body of the loop. The test condition, the stepping of loop variables, and the first stage of the pipe are all initiated in parallel.

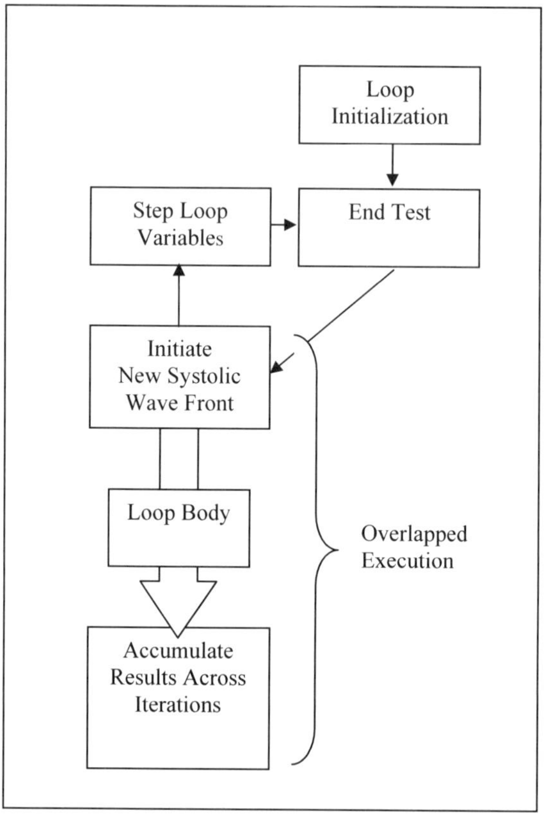

Fig. 4. Vectorized for loop.

Each iteration depends on the 'current' loop variables, as does the end test. Since these are all executed in parallel, the loop is unrolled to precompute the data. Additionally, each iteration carries forward a flag that signals the final iteration of the loop. The loop test, in Java, is intended as a condition for breaking out of the loop, and thus signals on *completion* of the final iteration.

The end condition must be propagated down the pipe because of the behavior of the last pipeline stage. It consumes each thread until signaled that the last iteration has arrived. In this special case, it handshakes its results out to whatever follows the loop.

Since each iteration propagates an end test value corresponding to the next iteration, and the test itself depends on stepped variables, each cycle must precompute index variables that are *two* iterations ahead and a test value that is *one* iteration ahead. The control loop is unrolled to obtain these phase relationships. One consequence of this is that, like old style FORTRAN *do* loops, *par for* loops must be guaranteed to execute once. Another restriction on valid *par for* loops is that step and test clauses not have side effects or access arrays.

2.5 Conditional Code

The Java compiler handles conditionally executed code by jumping around it. In vectorized code, successive executions (threads) must not be allowed to overtake and pass prior threads. Consequently, Trebuchet threads propagate from stage to stage, even where execution is suppressed. The Boolean test result propagates through the range of the conditional execution. At each stage, it suppresses update of the values passed to the next stage by switching a multiplexer. Fig. 5 illustrates this operation. Similar measures have been taken by makers of vector processing units for conventional computers [17]. Note that even though pipeline threads must traverse stages for which execution is suppressed, such traversal is rapid because the controller FSM immediately transitions to output states without waiting for the logic to propagate through the computational logic.

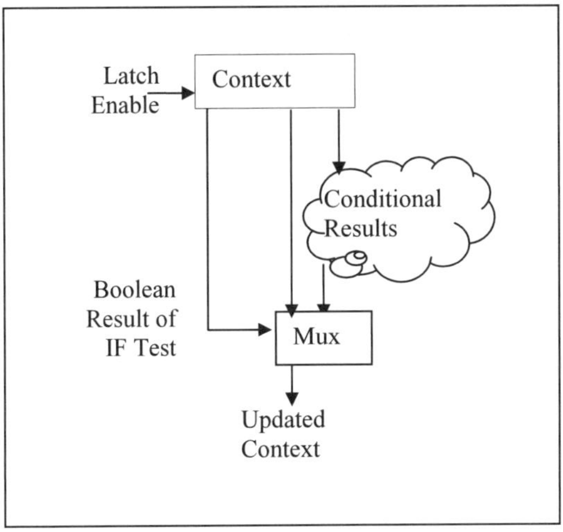

Fig. 5. Conditionally utilized results are switched with a multiplexer.

2.6 Array References

Java arrays are dynamic in the sense that they may be created at any time, moved around as needed to optimize garbage collection, and reclaimed by the garbage collector when abandoned. We did not want to subject the hardware generated by Trebuchet to the performance penalties inherent in such manipulation, so we chose to map array creations to static arrays created in the FPGA at configuration time.

If, during the course of symbolic modeling of stack and variable store, a reference is made to a location identified with a particular array, the mechanism to access that array is constructed. If the array reference cannot be ascertained, the module is marked as not being compatible with realization as hardware, and would necessarily be limited to bytecode execution by the JVM.

3 Pseudo-Asynchronous Execution

Trebuchet, when generating VHDL, calculates the length of the logic path for each pipeline stage and generates a FSM controller with enough wait states to allow signals to propagate. In order to decouple the clock period from this path, Trebuchet generates multi-cycle clock specifications for the Xilinx tools [18]. This allows the clock period to be driven by the exchange of handshaking signals rather than by the critical path through the combinational logic. This means that for lightly loaded conditions, the average stage delay dominates the pipe transit time, instead of the worst-case stage delay. This is one of the advantages touted for the asynchronous Micropipeline [4].

Another desirable trait of asynchronous circuitry is that, without a synchronizing clock, logic transitions are very well distributed in time. This minimizes current draw on the power supply and reduces the level of radiated EMI (electromagnetic interference) [4]. The tendency of synchronous logic is to have a well-defined signature of successive gates transitioning (and drawing power). It is expected that heavy usage of multi-cycle logic paths will have the effect of smearing these signatures, thus obtaining some of the advantage of purely asynchronous circuitry.

Drawbacks of asynchronous circuits include sensitivity to signal noise, performance dependent on temperature and process variations, and incompatibility with conventional FPGA tools [4]. The Trebuchet avoids these difficulties by being, at heart, synchronous. It combines the best aspects of both worlds.

4 Trebuchet Performance

The example programs [19] are derived from the well-known Hail-Stone Algorithm. The Hail-Stone applies the following algorithm:

Given N_i,
If N_i is even, $N_{i+1} = \frac{1}{2} N_i$
If N_i is odd, $N_{i+1} = 3N_i + 1$

The name "Hail Stone" derives from its characteristic of unpredictably increasing and decreasing in magnitude, the way a meteorological hailstone repeatedly rises and falls during its formation. The test program given in Fig. 6 computes the first 10 iterations when N_0 is set to 37.

The parallel bulk hailstone program is given in Fig. 7. It utilizes the nonstandard keyword *par* to direct the vectorized execution of the *for* loop, thus causing each of the computational steps to become stages in a 10-step pipeline. The loop limit was increased to 40 to keep the pipeline full longer during its execution. Test results for Simple_hail and Bulk_hail_par are shown in Table 1. There is an increase of almost 19-fold in throughput associated with the parallel version, attributable to pipelining effects.

5 Conclusions

The Trebuchet is an on-going effort. The project has already provided insight into the nature of hardware/software systems, and promises to provide some of the benefits commonly associated with asynchronous hardware design. Executing as hardware, code written in a traditional software language offers to finally join the disparate fields of hardware and software system design. The medieval Trebuchet was a marvelous example of technology used to loft heavy stones and hurl them at enemy fortresses. Our Trebuchet is, we hope, a tool for assaulting the bastions of complex system design.

```
public class simple_hail {
    public static void main(String args[]){
        netstart;
          int i,x;
          x=37;
          for(i=0;i<10;i++){
            if(((x>>1)<<1)==x) x=x>>1;
            else x+=x+x+1;
          }
          expose {x=x+1;}
          netend;
          System.out.println("\n
```

Fig. 6. Simple_hail.java source code.

Table 1. Test Results

	Simple_Hail	**Bulk_Hail_Par**
JVM Instructions	424	8,044
Equivalent Gates	18590	76,332
Fraction of XCV800 FPGA Utilized	9%	41%
Clock Frequency	36 Mhz	39 Mhz
Execution Time	11.36 μs	15.8 μs
Equivalent JVM MIPS	27	509

```
public class bulk_hail_par {
    public static void main(String args[]){
        netstart;
        int i,x;
        x=0;
        par for(i=0;i<40;i++){
            x=i;
            if(((x>>1)<<1)==x) x=x>>1;
            else x+=x+x+1;
            if(((x>>1)<<1)==x) x=x>>1;
            else x+=x+x+1;
            if(((x>>1)<<1)==x) x=x>>1;
            else x+=x+x+1;
            if(((x>>1)<<1)==x) x=x>>1;
            else x+=x+x+1;
            if(((x>>1)<<1)==x) x=x>>1;
            else x+=x+x+1;
            if(((x>>1)<<1)==x) x=x>>1;
            else x+=x+x+1;
            if(((x>>1)<<1)==x) x=x>>1;
            else x+=x+x+1;
            if(((x>>1)<<1)==x) x=x>>1;
            else x+=x+x+1;
            if(((x>>1)<<1)==x) x=x>>1;
            else x+=x+x+1;
            if(((x>>1)<<1)==x) x=x>>1;
            else x+=x+x+1;
        }
```

Fig. 7. Source code for bulk_hail_par.java.

References

[1] J. Campbell, "Experience with a Reconfigurable Java Machine," *Int'l. Conf. on Parallel and Distributed Processing Techniques and Applications*, pp. 2459-66, June 26-29, 2000.

[2] J. Campbell and G. S. Stiles, *The Trebuchet, a Pseudo-Asynchronous Micropipeline Execution Engine,* International Conference on Systems Engineering 2002, University of Nevada, Las Vegas, Aug 2002.

[3] I. Sutherland, "Micropipelines," *Turing Lecture*, *Communications of the ACM*, vol. 32, pp. 720-738, June 1989.

[4] A. Davis and S. M. Nowick, "An Introduction to Asynchronous Circuit Design," *University of Utah Technical Report UUCS-97-013*, Sep. 1997.

[5] P. M. Athanas and H. F. Silverman, "Processor Reconfiguration Through Instruction-Set Metamorphosis," *IEEE Computer*, vol. 26, no. 3, pp. 11-18, Mar. 1993.

[6] W. H. Mangione-Smith, "Seeking Solutions in Configurable Computing," *IEEE Computer*, vol. 30, no. 12, pp. 38-43, Dec. 1997.

[7] B. Hutchings and M. J. Wirthlin, "A Dynamic Instruction Set Computer," *Proc. the IEEE Symposium on FPGAs for Custom Computing Machines*, pp. 99-107, Apr. 1995.

[8] M. J. Wirthlin and B. L. Hutchings, "Improving Functional Density Through Run-Time Constant Propagation," *Field Programable Gate Array Workshop*, pp. 86-92, Feb. 1997.

[9] J. Campbell and B. Abbott, "Gear Train Theory: An Approach to the Assignment Problem Providing Tractable Solutions with Measured Optimality," *Int'l. Conf. on Parallel and Distributed Processing Techniques and Applications*, vol. II, pp. 986-95, June 30-July 3, 1997.

[10] J. Meyer and T. Downing, *Java Virtual Machine*, O'Reilly, 1997.
[11] P. Narayanaswamy, "Dynamic Arithmetic-Logic Unit Cache," *MS Thesis, Department of Electrical Engineering*, Utah State University, 1999.
[12] J. Fleischmann and K. Buchenrieder, "Prototyping Networked Embedded Systems," *Computer*, vol. 32, no. 2, pp. 116-9, Feb. 1999.
[13] http://www.jhdl.com/release-latest/docs/overview/intro.html.
[14] S. M. Loo, B. Earl Wells, N. Freije, and J. Kulick "Handel-C for Rapid Prototyping of VLSI Coprocessors for Real Time Systems *Proceedings of the Southeastern Symposium on System Theory (SSST-2002)*, pp. 6-10, Huntsville, AL, March 18-19, 2002.
[15] http://oss.software.ibm.com/developerworks/opensource/jikes/.
[16] *Development System Reference Guide*, Xilinx, Inc., 2001.
[17] J. L. Hennessy and D. A. Patterson, "Computer Architecture a Quantitative Approach," *Morgan Kaufmann Publishers, Inc.*, pp. 371-380, 1990.
[18] *Constraints Guide*, Xilinx, Inc., 2001.
[19] J. Campbell, "Efficient Automatic Mapping of Software to Hardware," Ph.D. Dissertation, Utah State University, 2002

Communicating Process Architectures 2003
Jan F. Broenink and Gerald H. Hilderink (Eds.)
IOS Press, 2003

Formal Contracts: Enabling Component Composition

Marcel BOOSTEN
Philips Medical Systems, P.O. Box 10000, 5680 DA Best, Netherlands
Marcel.Boosten@philips.com

Abstract. Traditional component interaction is based on interface calls and callbacks. Such interaction can introduce integration faults, i.e., side effects at the moment of component integration. Solutions to such problems can be hard to apply, and may require drastic changes in the design of the involved components. This paper introduces Formal Contracts, a software construct that allows side-effect free component interaction, and thereby avoids the introduction of integration faults. Furthermore, via a state machine representing the inter-component contract, Formal Contracts, in addition to the static aspects, formally specify the dynamic aspects of component interaction. Formal Contracts are a pragmatic software mechanism that supports the full development cycle: from the specification and decomposition until the debugging, composition, and test of a system.

1 Problem Statement

Research, as in Trew[2], has identified that during software Component Integration, so typically at a relatively late stage of the development cycle, the following integration issues make the development cycle costly and unpredictable:

- Race conditions in component interaction
- Re-entrant callbacks
- State-inconsistency between components and unexpected state-event combinations

The integration issues listed are only revealed at the moment components are integrated into a larger whole. These problems are largely caused by the interface definitions between components being unclear and component interaction containing side-effects that only show up during component integration. This paper claims that by using a different mechanism for component interfaces, the Formal Contracts, all these problems can be avoided by design.

Furthermore, Formal Contracts facilitate pragmatic, but also formal, specification of component interaction, and, with negligible additional effort, facilitate automated testing. Formal Contracts therefore combine "design for testability"[1] with "design for ease of testing"[1], and support the whole of the development cycle.

Formal Contracts are based on the principles of CSP: using these principles, the integration issues are avoided by design, thereby enabling component composition. A Formal Contract can be considered to be a generalised form of a Channel, or collection of Channels. A Channel is an explicitly instantiated inter-component interface in Occam - a CSP-based programming language. The basic idea behind Formal Contracts is the addition of a formal - i.e., machine readable, executable, and verifiable - specification of the interaction between components to the explicitly instantiated interface between the components. Formal Contracts specify, implement, and test the dynamics of the interaction between components, and thereby pragmatically support the whole of the software development cycle.

2 Technical Aspects of Formal Contracts

In this section, we will introduce, illustrate, and motivate the technical aspects of Formal Contracts. This section is meant to give the reader an in-depth understanding of the combination of techniques and concepts used in Formal Contracts, such that at a later stage the advantages of the use of Formal Contracts can be motivated.

2.1 Contract, Role, Component

Formal Contracts define the interaction between a number of Roles. Each Role is an abstract entity that interacts with the other Roles that are involved in the Formal Contract. When instantiating a Formal Contract, each Role of the Formal Contract has to be fulfilled by an instantiated Component. A Role is an abstract base class or interface, i.e., a collection of methods.

2.2 Formal Contracts exist as components

A fundamental idea behind Formal Contracts is their explicit existence: each Formal Contract is what one would normally consider being a, even though relatively small, component in the system. I.e., a Formal Contract is a self-contained entity that provides a number of interfaces. The interfaces it provides are the Roles of the Formal Contract. The Components use the Roles, i.e. use the interfaces provided by Formal Contracts. This is illustrated with an example in Figure 1.

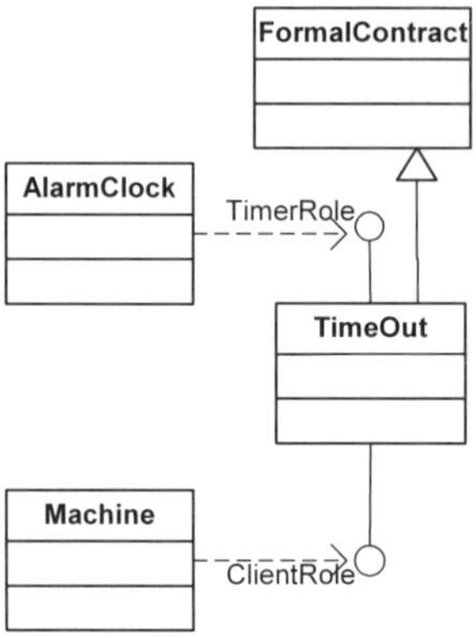

Figure 1: *The Formal Contract "TimeOut" with two Roles "ClientRole", "TimerRole", and two Components, "Machine" and "AlarmClock", that use, and thereby implement, their Role.*

Notice that this approach differs significantly from the standard approach in Object Orientation (OO) and component based infrastructures: Formal Contracts *are* the new interface between Components; Formal Contracts provide the interfaces for component interaction, whereas traditionally, Components would provide interfaces themselves, and directly use the interfaces provided by other components.

Traditional component interfaces are *asymmetrical*: a component is either a user or provider of an interface. Formal Contracts are *symmetrical*: Components have to fulfil the Formal Contract when using it.

Furthermore, notice that even though the approach conceptually differs significantly, still, standard component frameworks can be used to implement Formal Contracts.

2.3 Formal Contracts have explicit and observable state

Another fundamental idea to Formal Contracts is the fact that Formal Contracts have *state*, and that a Formal Contract *is a state machine* (or small program, if you like). Figure 2 illustrates interaction between two components in the traditional setting. Both components have an internal state, however, typically, components maintain an abstraction of each others state, the "externally observable state" to ensure that the messages send (or interface methods they call) occur as specified. This approach can easily become complicated, and is therefore error prone.

In Formal Contracts, we use a different approach: Formal Contracts have state. This state should not be considered to be the state of any of the components involved with the contract; no, it is the state of the Formal Contract itself. This way, implementers of Components need not implement abstractions of the state of other components; they can simply ask the Formal Contract for its current state, if they would want to know it. Formal Contracts export their state via read methods. These read methods -of course- do not change the Formal Contract's state.

The following code illustrates Formal Contract initialisation and the state read methods.

```
 class TimeOut : FormalContract
{
private:
  state : IDLE | COUNTINGDOWN | TIMEDOUT | CANCELLED;
public:
  TimeOut::TimeOut() { state = IDLE; }
  // Observable State
  Bool TimeOut::getState()
  {
    return state;
  }
  ...
}
```

2.4 Specifying and verifying dynamic behaviour

Another fundamental idea of Formal Contracts is that they specify the dynamic behaviour, simply by coding a small state machine that covers the full dynamic behaviour.

Formal Contracts define the relation between a number of Roles. Each Role consists of a number of state modification methods, that based on the parameters passed via the method, adjust the state of the Formal Contract. The state change is observable via the state read methods.

The implementation of the state modification methods within a given Formal Contract specifies formally the precondition and the effect on the state of the Formal Contract. Preconditions are verified via assertions, similar to Meyer's approach [8].

```
class TimeOut : FormalContract
        interface TimeOutClientRole
        interface TimeOutTimerRole
{
  ...
  int timeInSeconds;

  // Observable State
  ...
  int TimeOut::getTimeInSeconds() { return timeInSeconds; }

  // ClientRole
  void TimeOut::start(int aTimeInSeconds) {
    assert(state == IDLE);
    timeInSeconds = aTimeInSeconds;
    state = COUNTINGDOWN;
  }

  void TimeOut::cancel() {
    assert(state != IDLE);
    if (state == COUNTINGDOWN) {
      state = CANCELLED;
    }
  }

  // TimerRole
  void TimeOut::timeOut() {
    assert(state == COUNTINGDOWN);
    state = TIMEDOUT;
  }
}
```

The state diagram of this Formal Contract is shown in Figure 2.

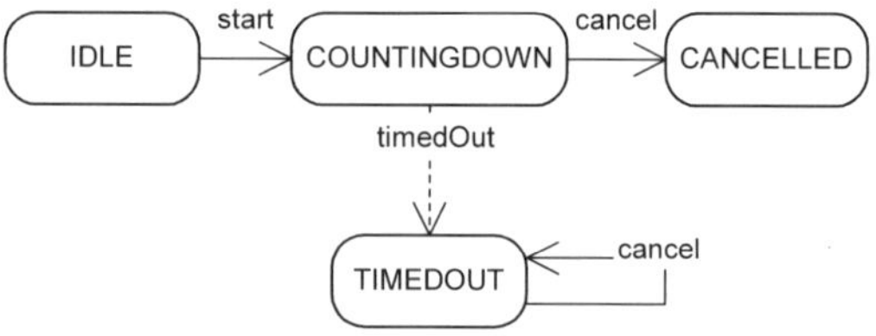

Figure 2: *State diagram of Formal Contract TimeOut. The lined arrows represent the transitions caused by the ClientRole, the dotted arrow represent the transitions caused by the TimerRole.*

In case the timeOut method is called while the state is IDLE, the assertion fails. The Component that plays the TimerRole can in that case easily be identified as the cause of the problem. Similarly, when the cancel method is called in state IDLE, the ClientRole is not well fulfilled. Typically, the Component that fails the Contract can easily be identified: either via the debugger, or via logging information of the method called and the current state of the Formal Contract. While integrating components, such assertions to pinpoint problems can help significantly.

Furthermore, while performing tests at component level, the Formal Contracts are to be present as well. Violations of Formal Contracts are to be eliminated during component-level tests.

2.5 Base Class for Formal Contracts extends component infrastructure

To make it even more explicit that Formal Contracts exist as implementation entities, a class is used as the base class for all Formal Contracts. Of course, functionality that is common by all Formal Contracts is implemented in that class. The Formal Contract class should be considered an extension of the component infrastructure.

2.6 Observer Pattern with Asynchronous Contents-Free Notification

Formal Contracts use the Observer pattern[9], but in a modified form. The Observer pattern is a very powerful decoupling and abstraction mechanism. It forms the basis for the popular Model-View-Controller pattern. The main disadvantage of the Observer pattern is the risk of introducing re-entrancy[3]. Re-entrancy problems are one of the major risks during integration testing[1]. Formal Contracts avoid the re-entrancy problem completely by design: Asynchronous Notification is used, instead of the Observer pattern's Synchronous Notification. Figure 3 illustrates Asynchronous Notification.

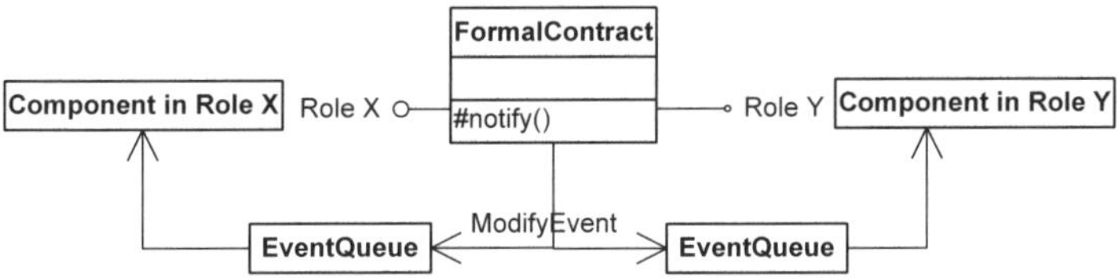

Figure 3: *Asynchronous Notification by communicating the event via a queue to the Observing Components.*

Another important aspect of the notification events used by Formal Contracts is them being contents free. The reception of a notification event indicates that the Formal Contract could have changed state (possibly it did not). It does not provide any additional information, for example not even an indication of the kind of state change that has occurred. By doing so, the Observing Components are forced to consult the Formal Contract to obtain the up-to-date status.

Due to the asynchronous (buffered) communication between Formal Contract and Observing Component, contents of a notification event would be likely not to resemble the state of the contract at that moment in time, misinterpretation of such contents could easily lead to state inconsistency problems and race conditions. These problems are avoided by using contents-free notification events.

Each Formal Contract notifies each of its state changes. This is illustrated in the following code example:

```
void TimeOut::start(int aTimeInSeconds)
{
  assert(state == IDLE);
  timeInSeconds = aTimeInSeconds;
  state = COUNTINGDOWN;
  notify();
}
```

The Formal Contract base class provides attach and detach functionality that allows Components to subscribe and unsubscribe to notification events generated by the Formal Contract. It is no problem for a Component to miss out on notification events that are generated before its first subscription to the Formal Contract: notification events are contents-free, by inspecting the state of the Formal Contract, newly subscribed Components can synchronise their internal state to that of the Formal Contract. It is the responsibility of each component to expect all specified series of state transitions (caused by other Components using the Formal Contract) to have occurred in-between inspections of the Formal Contract.

2.7 Consistency via a Reentrant Locking mechanism

In each Component infrastructure it is important to allow the use of multi threading and/or multi processing. As a consequence, Formal Contracts have to be instance MT-safe, i.e., safely usable by multiple threads or processes. Therefore, a locking mechanism is introduced: each method of each Formal Contract locks its Formal Contract during data access, and is therefore an atomic operation on the Formal Contract. Here are two examples:

```
void TimeOut::start(int aTimeInSeconds) {
  lock();
  assert(state == IDLE);
  timeInSeconds = aTimeInSeconds;
  state = COUNTINGDOWN;
  notify();
  unlock();
}

Bool TimeOut::getState() {
  enum theState;
  lock();
  theState = state;
  unlock();
  return theState;
}
```

However, to avoid race conditions, it can be important for a Component to base decisions on how to change the Formal Contract on the current state of that Formal

Contract. When the AlarmClock would use the following code, this would introduce a race condition:

```
if (contract->getState() == COUNTINGDOWN) {
   // Race condition here when contract would be cancelled,
   // since timeOut() is only allowed in the COUNTINGDOWN state.
   contract->timeOut();
}
```

The race condition occurs because the contract may change while decisions based on the contract's state are taken. We introduce a re-entrant locking mechanism to allow Components to avoid of such race conditions:

```
contract->lock();
  if (contract->getState() == COUNTINGDOWN) {
     contract->timeOut();
  }
  contract->unlock();
```

The locking mechanism is re-entrant; this means that lock/unlock pairs on the same Formal Contract can be nested without causing deadlock. Often, the use of the locking mechanism can be avoided by choosing the state modification methods --and therefore the state machine itself-- carefully. For example, by changing the `timeOut()` method from:

```
void TimeOut::timeOut() {
  lock();
  assert(state == COUNTINGDOWN);
  state = TIMEDOUT;
  unlock();
}
```

into:

```
void TimeOut::timeOut() {
  lock();
  if (state == COUNTINGDOWN) {
    state = TIMEDOUT;
  }
  unlock();
}
```

Then, to avoid the race condition in the AlarmClock, no locks are needed anymore. By introducing the design rule that a Component are allowed to have locked at most one Formal Contract at any time, deadlock problems are avoided.

Note that the example in this section is simplistic. While developing a Formal Contract, attention should be paid to make the Roles as intuitive as possible, and to avoid the need for lock/unlock outside the Formal Contract, so in the calling Component, as much as possible. From a testability point of view, one can also argue that Formal Contract should contain as little assertions as possible, and that therefore changes that replace an assertion by an if statement are always (or typically) to be preferred.

The sequence diagram in Figure 4 illustrates the interaction with the Formal Contract. The diagram shows how the Machine starts the AlarmClock via the formal contract by (1) changing the state of the contract, (2) the AlarmClock being notified of a changed contract, (3) the AlarmClock discovering that via *getState* that it has to start its internal timer. Furthermore, the diagram shows a situation in which the *cancel* of the timeout and the *timedOut* occur simultaneously: the Formal Contract decides which of the two events took place first (*cancel* in this example), and notifies both the Machine and the AlarmClock of this fact. This way, all possible inconsistencies due to differences in interpretations from either side are avoided: both Machine and AlarmClock have the same view on the situation that occurred; they both find out that the TimeOut has been cancelled.

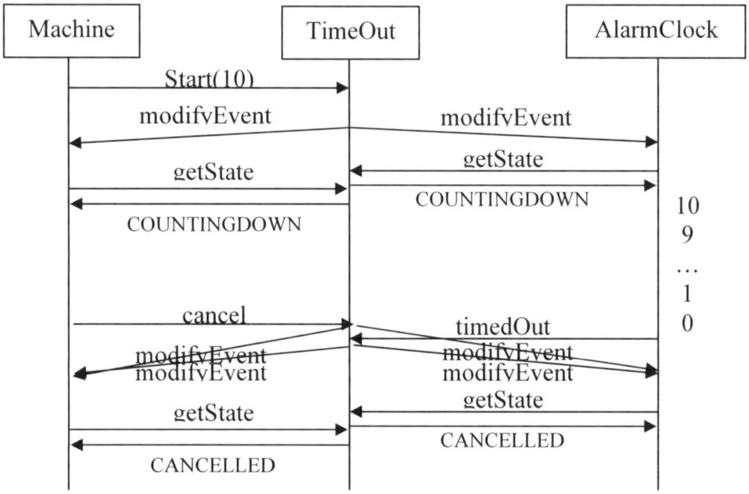

Figure 4: *Sequence diagram illustrating interaction between Machine, the Formal Contract TimeOut, and the AlarmClock.*

2.8 Shared ownership via reference counting

Formal Contracts are shared by the Components fulfilling the Roles. Life cycle management of Formal Contracts is organised via an access/de-access mechanism based on reference counting. The Component that is last to de-access the Formal Contract, automatically de-allocates it. Other solutions, for example based on garbage collection, are also fine solutions.

3 Advantages of Formal Contracts

3.1 Avoidance of State Inconsistencies and of Race Conditions

Each Formal Contract is always in a specified state, i.e., in a state that can be reached by the state machine that implements the Formal Contract. All inter component communication takes place via the Formal Contracts and their contents-less notification mechanism. Components can read and change the Formal Contract's state and thereby communicate: inconsistencies and race conditions can easily be completely avoided, while facilitating the use of multi threading and other forms of concurrency.

3.2 Avoidance of Re-entrancy Problems

Threads of different Components will *never* `pass through' a Formal Contract, and therefore never enter the execution domain of other Components. Each Formal Contract acts as a thread barrier. It allows communication and synchronisation between threads, however, it stops threads from entering each others domain. This way, inconsistency and re-entrancy problems related to multi-threading across component boundaries are avoided by design.

Often, in sequential OO programming, the use of anonymous call-backs are the cause of re-entrancy problems. However, even in situations where direct interface calls are used, re-entrancy problems[5] are likely to occur. Formal Contracts eliminate these problems by design, and are therefore also important for usage in single threaded applications.

3.3 Formal Contracts during the development lifecycle

Integration testing often has to take place relatively late in the project life cycle. It is crucial to avoid integration and integration test problems by design, since at the stage in the project that integration problems are discovered, it is typically too late to solve the problems without project slippage. Improving in the area of system integration is important for many organisations developing large and complex systems. This is only possible via an integral approach that addresses the full development cycle, and not just the last phase - at which time it is too late to solve the problems structurally:

- **Interface Specification and System Decomposition** Formal Contracts, or actually the state machine inside, is the formal specification of the interaction between Components. In addition to the definition of the static aspects of component interaction via method names and corresponding parameters, Formal Contracts also specify fully and formally the dynamic aspects of the Component interaction.
- **Component Development and Component Testing** Formal Contracts facilitate automated testing right from the start of the component development until the end of the integration phase. In principle, the "output" of each Component is automatically and always tested by the Formal Contract. Furthermore, by observing the state of the Formal Contracts that are to be fulfilled by a specific Component, a lot of insight in the behaviour of the Component becomes visible. Formal Contracts therefore are a powerful debugging tool as well. Furthermore, Formal Contracts can facilitate deadlock analysis across multiple Components. The state of each Formal Contract provides insight in "who is waiting for who". Furthermore, I believe that techniques for deadlock cycle detection in designs can exploit Formal Contracts, and thereby make automated deadlock detection possible even for large real-life component structures. Of course, significant research would be needed to proof this, and to fill in the many more details.

The use of Formal Contracts requires a new way of thinking about Component interaction. At first sight, it might look like a lot of administrative overhead. And, indeed, for very small and trivial interfaces it does introduce significant administrative overhead. However, Formal Contracts are meant as specification, communication, and verification medium between different Components, so between groups of people.

At that point, Formal Contracts become a very neat, powerful, and low-overhead tool. My own experience has shown that the approach is also applicable at a much smaller scale, so as interface between "one" or "few man" components.

4 Comparison to State-of-the-Art Techniques

4.1 Interface Definition Language (IDL)

Today, the IDL is widely used to specify component interfaces. IDL is limited to the definition of groups of methods or functions. Formal Contracts extend IDL with a means to formally specify the dynamic interface behaviour, and to verifying the Component's compliance to it. Furthermore, in contrast to IDL, Formal Contracts has been designed to avoid race conditions, re-entrance problems, and state-inconsistencies.

IDL is a language independent of any specific programming languages. It provides language de-coupling between components. This aspect of IDL can be used for Formal Contracts as well: Formal Contracts are (small) components of which the interfaces methods, the Roles, can be published in IDL or some other interface definition language used by the component infrastructure.

4.2 Model - Observer based Decoupling

Model - Observer based decoupling is sometimes used for the definition of component interfaces. Compared to Formal Contracts, these approaches typically do not eliminate race conditions and re-entrancy problems via the design of the Component interconnect. Furthermore, these approaches typically do not model the dynamic interaction at all. Typically, the Model that contains the state shared between Components is a data structure of which the fields can be changed any time into any of its values. Components fulfilling the interface should be able to deal with such state changes. The dynamic aspects of the interface are not well covered by such approach. Typically, the approach allows all state transitions on the model, and therefore the dynamic behaviour need not be specified – an approach which is practically unusable in case of non-trivial control applications.

4.3 Trew's Design for Testability approaches

In his paper[4], Trew describes design patterns that can be applied to eliminate the typical integration problems: race conditions, re-entrancy problems, state inconsistencies. Formal Contracts are a very specific combination of such design patterns. By using Formal Contracts for all interactions between Components, many of the difficult integration problems can be avoided. However, note that, even if Formal Contracts would be applied rigorously in a design, several of the design patterns identified by Trew would still be needed for more specific or exceptional situations.

Formal Contracts address the main integration problems identified by Trew, however, they also introduce a formal way of component interaction specification that decreases the chance of interpretation problems. Furthermore, Formal Contracts can very well support a new way of working: they can become the leading specifications during decomposition, and the glue during composition. Formal Contracts integrate system decomposition, interaction specification, interaction verification, and component composition into a streamlined approach.

4.4 Communicating Sequential Processes (CSP)

Within CSP research groups [5][6], there is a very clear awareness that component interaction should takes place across well-defined side-effect-free interfaces. Furthermore, they noticed that today’s Object Oriented and Component infrastructures do not provide such well-defined interfaces. Re-entrancy and concurrency problems violate the promised and promoted "Encapsulation" characteristic that should enable black box reusable components and facilitate assembly of systems by component composition.

The work presented in this paper is based on the fundamental ideas behind CSP, and has been inspired by the work of the research group to improve interaction definitions, but, at the same time, takes industrialist's experience and way-of-working into account.

Comparing the work presented in this paper with work performed on CSP is difficult due to the variety of work performed. It is important to notice that Formal Contracts use the CSP principles, therefore, it is possible to use the mathematics of CSP to specify and analyse properties of systems constructed with Formal Contracts as well.

If we compare the work performed here with work performed within the CSP-based research groups that focus on an embedding of CSP principles in programming languages, we clearly see a number of differences.

- **XToYChannels** Within JCSP and JavaPP, a relatively large number of Channels is introduced: OneToOneChannel, OneToManyChannel, ..., AnyToAnyBufferedChannel. The approach presented here generalises this to a user-definable state machine. The advantage of the state machine is that it combines logically dependent Channels into a single whole of which the dynamic behaviour is both formally and understandably specified.
- **Channels are typically only half a Contract** Channels are communication primitives between Processes. However, since channels are typically unidirectionally in Occam, JCSP and JavaPP, in order to specify the dynamic behaviour between two Processes, typically one would have to define a set of sequences of interaction across a *number of* channels between the two processes. A Contract is the complete relation between two processes, and is N-directional.
- **Event based vs. Process Based** The approach presented here allows both Process-based implementations and Event-based implementations: an Event-Based process is a main loop with a large ALT inside in which all events are handled. The decision between an Event based vs. Process based should be made on a per case bases.
- **Formal Contracts enabling combined input output ALTs** In contrast to all the CSP-based implementations that are currently available, Formal Contracts enable both input and output ALTs, and, just as natural, combinations of the two. A combined input output ALT could be the following:

```
do {
  WaitForEvent(Contract1, Contract2);
  keepwaiting = FALSE;
  if (contract1->canWrite()) {
    contract1->output(value1)
  } else if (contract1->canRead()) {
    value2 = contract2->input()
  } else keepWaiting = TRUE;
} while (keepWaiting);
```

- **Formal Contracts combine synchronous and asynchronous communication** The CSP-based approaches are mostly based on synchronous communication. Formal Contracts combines the two: communication with the contract is synchronous but basically non-blocking. The communication of the modification events is asynchronous, and 'overwriting'. Processes can decide to perform a blocking wait on modification events. The author of this article is convinced that the model presented here can easily be modelled in CSP, especially because the asynchronous communication of modification events is 'overwriting'. In Formal Contracts, Processes typically do not synchronise with each other, only with the Formal Contract. Compared most existing CSP-based approaches, this has the advantage that by default synchronisation (which can be costly, since it reduces parallelism) is avoided.

4.5 Jonkers' API Structure and Model-Based Specification

In his paper [7], Jonkers presents the API Structure and Specification method used for DVP2. There are several similarities between his approach, and the approach presented in this paper: he recognises and identifies contracts as combinations of different interfaces; he also identifies roles, and makes components fulfil those roles. He also recognises that the contract is an excellent way for people to agree on the interface.

Jonkers uses a Model-Based approach to specify the behaviour of each role: "abstract implementations" of each Role are used for specification purposes. Formal Contracts completely inverse this specification: instead of providing "abstract implementations" for each Role, the true implementation of the interaction is specified by implementing a Formal Contract. This way, the obligations and expectations of each Role and their interdependencies are made explicit. Formal Contracts are integrated as small components into the system: they are more than specification, they are the interaction medium. Formal Contracts facilitate the verification of the behaviour of each Component: the Formal Contract checks input events versus its internal state. Using Jonkers' approach, verifying that a Component fulfils its Model, its "abstract implementation", is much more complicated. The Formal Contracts approach can very well be combined with a Model-Based development approach, in that case, it can be considered an extension to the Model-Based approach. Formal Contracts avoid race conditions, re-entrancy problems, and state inconsistencies. These fundamental integration aspects have not been addressed by Jonkers' approach.

In other work, Jonkers has used Interaction Objects for communication and synchronisation. The objective of this research was different: it was meant to minimise the coupling between Processes aiming at minimisation of communication latency and throughput impact on the whole of the system.

5 Background, Context, and Experience so far

The solution presented in this paper is the result of accumulated fragments of experience in the area of component specification and integration. The first inspiration for this approach has come from the 1998-2000 Communicating Process Architectures (CPA) conferences, especially due to presentations and discussions with Peter Welch, Andrew Laurence, and Tom Locke. Thanks, it has always been very interesting!

During my working period at Philips Medical Systems, Product Management Group (PMG) Computed Topography (CT), I was faced with component integration problems, and with multi-threading aspects. I would like to thank my colleague CT architects at that time: Peter Jaspers, Phil van Liere, Jacco Wesselius, Joland Rutgers, for being as stubborn as myself, and thereby -in the end- pushing me to find the solution as described in this paper. I have implemented the approach described in this paper in the multi-threaded so-called "Multi-Scan" controller in the CT software; one of its most complex parts. The implementation demonstrated a significant reduction in complexity compared to an approach based on asynchronous message passing, in which the multitude in race conditions was hard to handle.

Furthermore, I'm very grateful for Tim Trew's work. He has very well used his architectural experience to identify the major integration problems in practical industrial situations. In my experience, it was extremely difficult to convince highly experience architects of the problems at hand, and of the need for a structural solution. However, if the consequences are in the order of many man-years of additional effort, convincing people will be easier. Tim Trew's award winning "most influential" paper[1] convinced people of the relevance of such work. Consequently, I decided to write this paper on Formal Contracts down. Some results of discussions with Tim Trew and a quick discussion with Hans Jonkers have been incorporated in this paper.

6 Conclusions

Formal Contracts are a software construct enabling side-effect-free component interaction. Thereby, Formal Contracts avoid the introduction of major integration faults. Furthermore, Formal Contracts specify in a human readable and machine executable form the dynamic aspects of component interaction. They check the state of the contract versus incoming modification requests. These characteristics make Formal Contracts a pragmatic tool that supports the full development lifecycle: from the specification and decomposition until the debugging, integration, and test of the system.

7 References

[1] Tim Trew. Demystifying Design for Testability. Philips Software Conference June 2002.
[2] Tim Trew. The aims of Integration Testing. PRL Tech. Note 3922. November 1999.
[3] Tim Trew. Software Component Composition: Still 'Plug & Pray'? Philips Software Conference 2001.
[4] Tim Trew. Testability of Component Based Software: Mastering Component Interaction. PRL Int. Note
[5] Tom Locke. The Broken Promises of OO, And How We Might Do Better. Slides presented during one of the SIG evenings at the CPA 2000 conference.
[6] Peter Welch. Java Threads in the light of occam/CSP. WoTUG-21. IOS Press, 2000. ISBN 90-5199-391-9
[7] Hans Jonkers. DVP2-API: Structure and Specification. Philips Software Conference June 2002.
[8] Bertrand Meyer. Object Oriented Software Construction. Prentice Hall, 1997. ISBN 0-13-629155-4
[9] E. Gamma et al. Design Patterns: Elements of Reusable Object-Oriented Software. Addison Wesley, 1995.

Communicating Process Architectures – 2003
Jan F. Broenink and Gerald H. Hilderink (Eds.)
IOS Press, 2003

Flexible, Transparent and Dynamic occam Networking With KRoC.net

Mario SCHWEIGLER
ms44@kent.ac.uk / research@informatico.de

Fred BARNES
frmb2@kent.ac.uk / fred@frmb.org

Peter WELCH
P.H.Welch@kent.ac.uk

Computing Laboratory, University of Kent
Canterbury, Kent, CT2 7NF, UK

Abstract. KRoC.net is an extension to KRoC supporting the distribution of occam channels over networks, including the internet. Starting in 2001, the development of KRoC.net has gone through a number of stages, each one making the system more flexible, transparent and dynamic. It now enables the occam programmer to set up and close network channels dynamically. Configuration has been simplified. All occam PROTOCOLs can now be sent over network channels, without need for conversion. Many of the new dynamic features in occam have been used to improve KRoC.net. Many of the concepts in KRoC.net are similar to those in the JCSP Network Edition (JCSP.net), KRoC.net's counterpart in the JCSP world. This paper will give an overview over KRoC.net, its usage, its design and implementation, and its future. It will also provide some benchmarks and discuss how the new occam features are being used in the latest KRoC.net version.

1 Introduction and Motivation

Originally, occam[1] [1] was targeted at multi-processor (transputer) platforms for some specific embedded system application. That platform was finite and fixed at *configuration time*. There have been projects to target multi-processor platforms other than transputers. An example is Vella's work on porting occam to Networks of Workstations (NoWs) [2, 3] which uses an interface to communicate to the network, called ocnet. This interface is a layer between the occam kernel and the operating system and the networking hardware. Like the original networks of transputers, this system was static and the layout of the network had to be configured in advance. Another example is MESH [4] which is completely hardware dependent. MESH is a messaging and scheduling system that enables accessing network hardware (e.g. ethernet cards) and provides user-level scheduling in Linux. The original version of CCSP [5], on which the current occam kernel is based, could access the messaging capabilities of MESH. But again, this was hardware dependent and static.

We want our new [6, 7, 8] occam also to target multi-processor platforms, but we want to be abe to use it for an open-ended network of machines, connected by network channels, whose size and topology are constructed dynamically at *runtime* and may be changing. One important aim is to enable occam applications to use the infrastructure of the largest network

[1] occam is a trademark of ST Microelectronics.

available — the internet. The internet plays an increasingly important role in today's society, and we want occam programs to be able to utilise this global network by simple and (in terms of CSP) safe channel communication. This requires an approach that is completely different to the transputer-style configuration and other static solutions.

This is where KRoC.net comes into play. KRoC.net is an extension to KRoC [9] which provides a framework enabling the occam programmer to set up channels between processes that are running on different machines. KRoC.net has been improved significantly since its first versions [10, 11]. In this process, many of the new features in KRoC were utilised, in particular the new Generic Protocol Converters described in section 3.

KRoC.net has now become a very dynamic tool which enables the programmer to set up network channels at runtime just when they are needed. The configuration of KRoC.net has been simplified, and a new URL-based channel setup mechanism has been introduced, allowing greater flexibility in the creation of network channels. KRoC.net's network channels have reached a high level of transparency now. They can now be plugged into any old occam process without the need to change its code. All occam `PROTOCOL`s are now supported for network channels, including components that are `MOBILE` [6].

KRoC.net is the KRoC/occam counterpart of the *JCSP Network Edition* (JCSP.net) [12, 13], the network extension of JCSP (*CSP for Java*) [14]. Many of the concepts in KRoC.net and JCSP.net are similar. Development of JCSP.net has recently been taken over by Quickstone Technologies. Some of the general mechanisms in KRoC.net are similar to the *Virtual Channel Processor* (VCP) of the T9000 transputer [15].

Sections 2 through 4 discuss the infrastructure and implementation of KRoC.net, along with the new paradigm of network channel-types and the new Generic Protocol Converters. Section 5 describes in detail how network channel-types are being set up by the occam programmer, and will give an outlook to future adaptions in the occam language which will make this setup process easier for the programmer. Section 6 describes the configuration of KRoC.net. The performance of KRoC.net and some benchmarks are discussed in section 7. Section 8 summarises the paper and gives some ideas about future work on the KRoC.net project.

2 Network Channel-types

The original idea behind KRoC.net was to distribute normal occam channels over networks, i.e. implement their behaviour, so that from the point of view of the processes which are connected by the channels, their behaviour would be identical, no matter whether the channels were local or networked.

In classical occam, there was no such notion as a channel-end, the paradigm used was always the whole channel. For a distributed system, however, it is natural for each participating processor to set up its end of a network channel separately — so we need this concept of channel-ends. Another newly introduced dynamic feature of occam greatly extends and empowers KRoC.net's model of distributed communication. We refer to *channel-types* [7, 8], which are bundles of channels, dynamically created and mobile. These also have well-defined ends, called *client-end* and *server-end* for convenience.

2.1 'Classical' Local Channel-types

To remind people about how channel-types work, here is a short example. A channel-type would be declared like this:

```
CHAN TYPE THING
  MOBILE RECORD
    CHAN INT req?:                      -- request channel
    CHAN MOBILE []BYTE reply!:          -- reply channel
:
```

THING is a channel-type which contains two channels, one request channel and one reply channel. The direction specifiers ('?' and '!') are from the point of view of the server-end of the channel-type, i.e. it would read from req and write to reply.

To create an instance of a channel-type, we have to declare two variables, one for the server-end and one for the client-end, and allocate them as in pairs:

```
THING? thing.svr:                       -- declare server-end
THING! thing.cli:                       -- declare client-end
SEQ
  thing.svr, thing.cli := MOBILE THING  -- allocation
  ...  use them
```

The server-end of a channel-type is marked by a '?', the client-end by a '!'. To use the channel-type, one would typically pass the ends of the channel-type to different processes who would then communicate over the channels inside the channel-type:

```
PROC server(THING? thing.svr)
  WHILE TRUE
    INT size:
    MOBILE []BYTE buffer:
    SEQ
      thing.svr[req] ? size             -- get size
      buffer := MOBILE [size]BYTE       -- allocate buffer
      ...  fill buffer with data
      thing.svr[reply] ! buffer         -- send buffer back
:

PROC client(THING! thing.cli)
  WHILE TRUE
    INT size:
    MOBILE []BYTE buffer:
    SEQ
      ...  set size
      thing.cli[req] ! size             -- send size wanted
      thing.cli[reply] ? buffer         -- get buffer
      ...  use buffer
:

THING? thing.svr:
THING! thing.cli:
SEQ
  thing.svr, thing.cli := MOBILE THING
  PAR
    server(thing.svr)                   -- pass server-end to server
    client(thing.cli)                   -- pass client-end to client
```

Alternatively, channel-type ends can be sent over channels:

```
PROC generator(CHAN THING? svr.out!, CHAN THING! cli.out!)
  THING? thing.svr:
  THING! thing.cli:
  SEQ
    thing.svr, thing.cli := MOBILE THING
    svr.out ! thing.svr                   -- send server-end
    cli.out ! thing.cli                   -- send client-end
:

PROC server(CHAN THING? svr.in?)
  THING? thing.svr:
  SEQ
    svr.in ? thing.svr                    -- get server-end
    ...  use thing.svr
:

PROC client(CHAN THING! cli.in?)
  THING! thing.cli:
  SEQ
    cli.in ? thing.cli                    -- get client-end
    ...  use thing.cli
:

CHAN THING? svr.chan:
CHAN THING! cli.chan:
PAR
  generator(svr.chan!, cli.chan!)
  server(svr.chan?)
  client(cli.chan?)
```

A general remark to the terms *"client-end"* and *"server-end"*: for channel-types, these are just names in order to be able to distinguish between the two ends of the channel-type. They may, of course, be used in a 'genuine' client/server relationship, but this is not a must. Channel-types may well be used for peer-to-peer communication, in which case "client" and "server" would not mean much more than, for instance, "left-hand side" and "right-hand side".

2.2 The New Paradigm of Network Channel-types

KRoC.net implements the behaviour of channel-types in a networking context. This means that internally, KRoC.net only deals with network channel-types (NCTs) that contain network channels. Using this construct, we are able to implement networked versions of both channel-types *and* classical occam channels (which are simply treated as channel-types containing just one single channel). The occam programmer creates a client-end of a channel-type on one machine, and a server-end of the same channel-type on another machine, and then connects them with KRoC.net's network infrastructure. For the application level processes dealing with the client- and the server-ends of that NCT, their behaviours are the same as if is was a local channel-type. For the occam programmer, it is also possible to connect ordinary occam channels via KRoC.net, but internally, KRoC.net treats those as NCTs containing one single network channel. Details about how NCTs are set up are described in section 5.

JCSP.net's channels are automatically *any-to-one*, which is actually a little odd and will be tidied up in a future release. KRoC.net's channels and channel-types have to be explicitly declared to be SHARED at the writing- (or client-) ends if we want that property. As with JCSP.net, KRoC net does not support networked sharing of the reading- (or server-) ends of channels/channel-types — although we are working on it!

The new occam channel-types offer a more flexible way of setting up two-way communication across a network than JCSP.net's basic channels or Connection channels. This becomes especially useful for those with SHARED client-ends. The networked "client" must CLAIM its end before it can use it — just like its non-networked version. For the duration of the CLAIM, other clients are locked out and a two-way conversation to the server can be completed without interference.

Currently, both ends of networked channel-types must be CLAIMed — even if neither is SHARED. This is for technical reasons associated with secure shut-down or movement. That does not apply, of course, to the use of non-networked channel-type ends. This is one area of non-transparency in KRoC.net that we are still considering how best to resolve. There is an argument to enforce CLAIMing on these channel-types regardless of whether they are SHARED (or networked), which would remove the anomaly. That argument concerns building in an extra level of checked security to enforce correct patterns of use of the channel components making up the type — but we leave that to a later paper.

3 New occam Features Used in KRoC.net

Our aim is to make network channels *transparent* — i.e. make their behaviour indistinguishable from that of normal channels. Two important requirements have to be met in order to achieve this aim: Firstly, network channels have to retain CSP channel semantics; and secondly, they must be able to carry any PROTOCOL — from plain data types to user-defined variant protocols.

3.1 The Extended Rendezvous

The first requirement can be met by utilising the extended rendezvous [6], that enables channel synchronisation to be 'extended' across multiple communicating processes. It operates in such a way that if a simple 'tap' process is inserted between two previously directly connected processes, those two processes will be unable to identify its presence in the network (or lack of). For example, an INT 'tap' process is simply:

```
PROC tap(CHAN INT in?, report!, out!)
  WHILE TRUE
    INT i:
    in ?? i                   -- extended input
      out ! i                 -- extended process
      report ! i              -- following process (assume always taken)
:
```

The extended process is executed whilst the outputting process (connected to in) remains blocked. This can be any process — or processes — provided that they do not attempt to communicate on in (which would immediately deadlock). The 'following' process is optional. If present, it is executed *after* the communication on in has completed (and the outputting process resumed). This 'following' process is provided mainly for use in ALTs and variant ('CASE') inputs, where expressing the same behaviour without it can be tricky.

Note that neither of the processes connected to in? and out! need any modification. The only changes are in the network setup code (the addition of an extra channel and the tap process). Processes that extend communication, such as tap, can be connected together in a pipeline, extending the communication through multiple processes.

For KRoC.net, the extended rendezvous provides the mechanism that allows communication through multiple processes (the networking infrastructure), without affecting the end-to-end synchronisation of the two communicating processes. Between two remote processes, each communication is extended whilst the networking infrastructure performs the communication.

Specifically, the OCP (Output Control Process, see section 4) performs an extended input from the local application (via DECODE.CHANNEL, explained in the following section), communicating the data and waiting for the acknowledgement inside the extended process. The acknowledgement will only be sent by the remote ICP (Input Control Process) when it has communicated successfully with the application.

3.2 Generic Protocol Converters

The second requirement, that of being able to communicate any channel PROTOCOL, is handled by Generic Protocol Converters (GPCs), two compiler built-in PROCs named "DECODE.CHANNEL" and "ENCODE.CHANNEL". Essentially, these provide a pair of processes that extend communication, of any PROTOCOL, between two processes either side, but use a well-known protocol for the channel between themselves.

The PROTOCOL used for communication between DECODE.CHANNEL and ENCODE.CHANNEL is a sequential protocol carrying two INTs. This protocol must be defined by the application. For example:

```
PROTOCOL LINK IS INT; INT:
```

The information carried by this protocol is pointer and size pairs. When the application outputs into a channel connected to DECODE.CHANNEL, the extended-process within communicates the address and size (of the application data) to its output channel. The address communicated is only valid during an extended process whose input acquired it, or where extra synchronisations are put in place to prevent the communication completing.

The reverse of this operation is implemented by ENCODE.CHANNEL. It receives two INTs on its input channel (using an extended input), which it then communicates to the process connected on its output channel — using the same application PROTOCOL as was used for DECODE.CHANNEL's input channel. The two built-in PROCs do not have fixed parameter types and are prototyped by the (illegal) occam:

```
PROC DECODE.CHANNEL(CHAN * in?, CHAN ** term?, CHAN *** out!)
PROC ENCODE.CHANNEL(CHAN *** in?, CHAN ** term?, CHAN * out!)
```

The PROTOCOL represented by *** is the link protocol — the earlier 'LINK' for example. The term channels may be typed as either INT or BOOL and are used to shut-down the GPCs (which otherwise run indefinitely). The PROTOCOL represented by * is the application protocol. These two PROCs do not check that * is the same on both sides. This is done by the KRoC.net infrastructure which ensures that the PROTOCOL.HASH value on both sides is the same. "PROTOCOL.HASH" is a simple compiler built-in that evaluates to a unique[2] constant for the given type, protocol or name (variables, abbreviations, PROCs and FUNCTIONs), that is passed as a parameter.

[2]The hashing algorithm used for PROTOCOL.HASH generates sufficiently unique values, but there exists a slim probability of the same value being generated for two completely different types/protocols.

3.2.1 Pointer Handling

When the synchronisation is extended correctly, ENCODE.CHANNEL will receive and encode a valid address. However, this is only true for same/shared memory systems. Within the KRoC.net framework, where DECODE.CHANNEL and ENCODE.CHANNEL are expected to be on physically different systems, the data output from DECODE.CHANNEL must be *copied*.

Within a networked system, the data (address/size pair) received by ENCODE.CHANNEL will be for a local memory block — allocated by the networking infrastructure. Thus, unless such data is actually *moved* into the application (by mobile communication), ENCODE.CHANNEL must ensure that the memory block is freed after the communication.

The (dynamic) memory blocks consumed by ENCODE.CHANNEL are expected to be dynamic MOBILE arrays (of BYTEs). To convert between a dynamic mobile array and address/size pairs, two additional compiler built-in PROCs are provided: "DETACH.DYNMOB" and "ATTACH.DYNMOB". These have the simple PROC signatures:

```
PROC DETACH.DYNMOB(MOBILE []BYTE var, RESULT INT addr, size)
PROC ATTACH.DYNMOB(INT addr, size, RESULT MOBILE []BYTE var)
```

These PROCs work in such a way that detaching a dynamic mobile leaves that mobile *undefined*, and attaching a dynamic mobile does the same for the INT variables. Furthermore, ATTACH.DYNMOB sets the address (in addr) to zero after attaching the mobile. These ensure that a mobile variable cannot be accessed after it has been 'detached', and that any address used to 'attach' a mobile is left undefined (and invalid).

Given the operation of ENCODE.CHANNEL, that either releases or moves the dynamic mobiles communicated into it, connecting DECODE.CHANNEL and ENCODE.CHANNEL directly is not possible. The address/size pair communicated by DECODE.CHANNEL is that of the occam data item being communicated, which is not necessarily a dynamic mobile array.

To network an encode/decode pair locally requires the use of an intermediary "link" process. Such a process also serves as a good example of using the dynamic mobile attach/detach operations:

```
PROC link(CHAN LINK in?, out!)
  WHILE TRUE
    INT addr, size:
    in ?? addr; size                   -- extended input
      MOBILE []BYTE tmp, new:
      SEQ
        ATTACH.DYNMOB(addr, size, tmp)   -- access the data as a
                                         -- dynamic mobile array
        new := CLONE tmp                 -- copy into a new array
        DETACH.DYNMOB(tmp, addr, size)   -- detach false dynamic array
        DETACH.DYNMOB(new, addr, size)   -- detach new dynamic array
        out ! addr; size                 -- communicate
:
```

When received by ENCODE.CHANNEL, the dynamic mobile created by the CLONE here is either freed, if the output PROTOCOL is *not* a dynamic mobile, or *moved* if it is.

3.2.2 Mobile Protocol Conversion

The handling of ordinary data types, such as those allocated in workspace or vectorspace, is trivial — communication always *copies*. Thus, when ENCODE.CHANNEL receives an address/size pair (using an extended input), it simply engages in communication directly with those parameters (typically the OUT Transputer instruction). Once output, it frees the memory block using MRELEASE, returning it to the free-lists for re-use.

PROTOCOLs that are implemented by the compiler as a sequential series of communications remain in that way, with the exception of some counted-array protocols. This applies to standard sequential protocol communication and variant ('CASE') communication. Counted-array protocols whose 'count' type is *not* INT64 are sent as a single communication — since the size is included. The current implementation of the GPCs only supports INT-sized counts, so any INT64 counted arrays must have their count sent separately (even there is no possibility of an application communicating more elements than can be held within an INT).

The implementation of DECODE.CHANNEL and ENCODE.CHANNEL is more complex for mobile types and mobile communication, since the application *moves* data. There are three basic forms of MOBILE types: static mobiles; dynamic mobile arrays; and mobile channel-types (that are also dynamic). Mobile channel-types are not currently supported by the GPCs, discussed in section 3.2.3.

Communication of static mobiles is done using pointer-swapping. When the application outputs into DECODE.CHANNEL, and synchronises with the extended input within, the address is extracted leaving the original pointer intact. The implementation of ENCODE.CHANNEL uses a 'temporary' static mobile, into which data is copied when the address and size are received. This is then communicated with the application using the standard mobile output instruction (MOUT).

Dynamic mobile arrays are treated slightly differently. The communication of a dynamic mobile array (using either MOUT64 or MOUTN) is performed as a single operation — the moving of dimension counts and a pointer. After a process has outputted a dynamic mobile array, the local dimension count is set to zero (such that any attempts to access it after the output result in range-check errors).

The implementation of DECODE.CHANNEL for dynamic mobile arrays outputs either one or two address/size pairs. If there is any more than one unknown dimension, the dimension counts are output first (organised sequentially in the originating process's workspace), followed by the data, both as an address/size pair. For dynamic mobiles with only one dimension (which will be the majority in most cases), only the address/size pair for the data is sent. The single dimension count can be calculated from the size of the communicated data.

Dynamic mobile handling for ENCODE.CHANNEL is particularly simple — if there is more than one unknown dimension, the first address/size pair received is a dynamic mobile that holds the dimension counts for the target mobile. ENCODE.CHANNEL uses a local (dynamic mobile) temporary into which these dimensions are copied, before freeing the communicated memory block. For mobiles with *only* one unknown dimension, the temporary is still used, but has its count field initialised from the size of the communicated data.

The next input performed by ENCODE.CHANNEL is the address/size of the data for the dynamic mobile, that is itself a dynamic mobile. If dimension counts have already been set, the communicated address is simply placed in the temporary mobile's pointer slot and the size checked against the expected size (calculated by multiplying the dimension counts with the base-type size). If the dimension counts are unset, the address is copied into the temporary mobile's pointer slot (as before), and the dimension count is calculated by dividing the size of the data by the base-type size.

This temporary is then communicated into the application, with the pointer specifically not returned to the free-lists.

3.2.3 Limitations

Mobile channel-type protocol conversion is not currently supported, but support for this is planned for the future (see section 8). Semantically, communication of channel-type ends is the *stretching* of those channel-types over the network. However, some non-trivial logic is required inside the KRoC.net infrastructure to support this. Many of these issues have already been investigated and implemented by Muller and May for the Icarus language [16]. Our approach will follow a similar path.

Support for the proposed mobile process types [17] will likely be similar in nature — requiring interaction with the KRoC.net infrastructure in order to set up links for any contained mobile channel-type ends, and to ensure that the receiving node has the code for these processes.

4 Basic Infrastructure

We want to be able to design a system in the same way regardless of the physical distribution of its processes and channels. The underlying model — channels and channel-types — ought to be the same both locally and networked. Figure 1 shows four processes running on the same machine[3]. Processes A and B are connected by a channel (with A having the channel's writing-end and B having the reading-end). Processes C and D are connected by a channel-type (with C having the client-end and D having the server-end of the channel-type). Figure 2 shows the same system of occam processes running on two different machines. The channel between A and B would now be a network channel, and the channel-type between C and D would now be an NCT containing network channels.

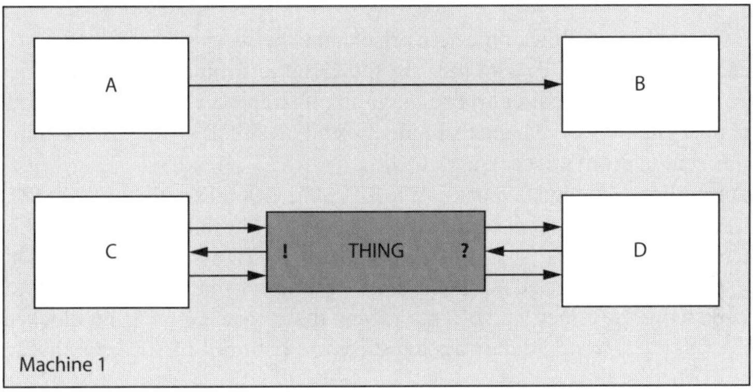

Figure 1: Local Channels and Channel-types

From the point of view of the user level processes (A, B, C and D), there is no semantical difference in the behaviour of the overall system between the two figures. Internally, however, in Figure 2, the user level processes on both machines are running in parallel with an instance of the *KRoC.net manager*, a process provided by the KRoC.net framework. The network channel and the NCT depicted in figure 2 are provided by the KRoC.net manager and multiplexed over a network link between the two machines.

The KRoC.net manager has two main tasks: setting up network channels and NCTs, and (once that is done) administrating them and managing the communication over them. The setup process is described in detail in section 5.

[3]In this paper, we define 'machine' as an occam program (i.e. the whole OS level process), that might run an instance of the KRoC.net manager described in this section. I.e. it is possible to run two occam programs on the same physical computer and to connect them with KRoC.net's infrastructure. These programs would be referred to as separate machines.

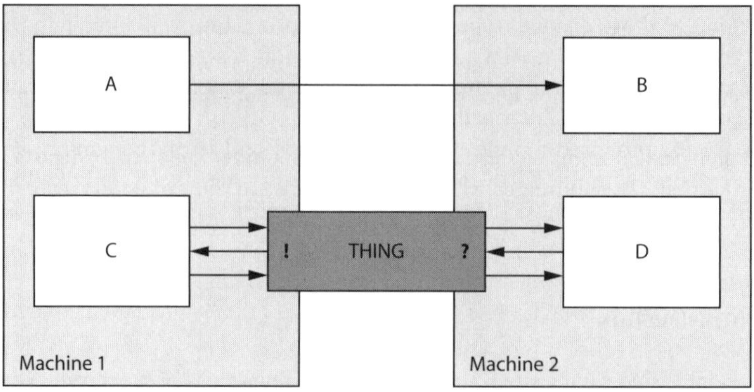

Figure 2: Network Channels and Network Channel-types

4.1 Communication Over Network Channels

As for the communication over a network channel, it makes no difference whether this network channel is a stand-alone one (cf. the one between processes A and B in Figure 2)[4] or inside an NCT (cf. the three network channels inside the NCT between processes C and D in Figure 2).

Figure 3 shows the data flow of the network channel between processes A and B in detail[5]. `DECODE.CHANNEL` and `ENCODE.CHANNEL` are the Generic Protocol Converters described in section 3. Apart from their main purpose — giving us transparency by enabling the occam programmer to set up network channels of any given `PROTOCOL` — the second purpose of the GPCs is to prevent us from unnecessary copying.

Every time process A sends something to `DECODE.CHANNEL`, `DECODE.CHANNEL` outputs the address/size pair of the data item it just received. The KRoC.net manager will then deal with the address and the size only, without copying any data around. This reduces overheads especially for larger data items. The `??` implies that `DECODE.CHANNEL` uses the extended rendezvous (see section 3). This means that process A will be blocked not only until `DECODE.CHANNEL` has read the data item from it, but until `DECODE.CHANNEL` explicitly releases it.

The KRoC.net manager maintains a network link between itself and any other machine to which a network channel has been established. All network communication between the two machines is multiplexed over this link, which saves network resources. Communication over a network link is handled by a pair of Tx/Rx processes on both of the machines that are connected by that link. The Tx process transmits network packets over the link, the Rx process receives network packets from the link.

Every writing-end of a network channel handled by the KRoC.net manager has a unique Output Control Number (OCN), every reading-end has a unique Input Control Number (ICN). These two numbers unambiguously identify each network channel.

Each writing-end of a network channel is handled by an Output Control Process (OCP). The OCP reads the address/size pair from `DECODE.CHANNEL`, also using the extended rendezvous. Each OCP knows over which link its network channel is multiplexed, and it also knows the ICN of the reading-end on the remote machine. The OCP sends the address/size

[4]which KRoC.net internally treats as an NCT containing a single network channel

[5]Channels used for setting up and administrating network channels have been omitted in order to simplify the figure.

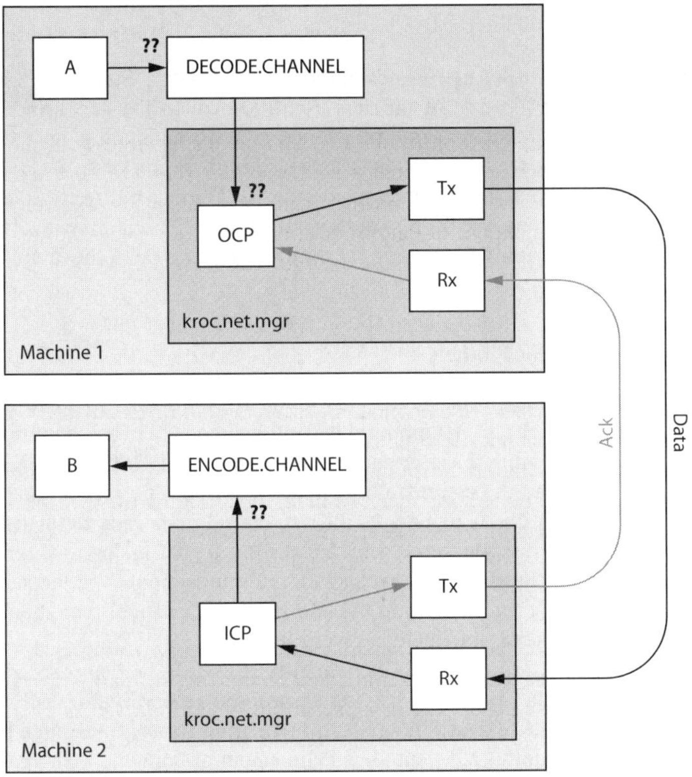

Figure 3: Network Channel Communication

pair, together with the destination ICN, to responsible Tx process. The Tx process then sends the destination ICN and the size as well as the data packet over the network link, where it is received by the responsible Rx process.

Every reading-end of a network channel is handled by an Input Control Process (ICP). When the Rx process receives a data packet, it stores it in a dynamic mobile byte array, detaches it (see section 3) and sends the address/size pair of it to the ICP. The ICP contains a buffer where the address/size pair is stored until it is requested (i.e. until the reading-end of the network channel decides to read). The buffer is necessary in order to prevent the Rx process from blocking, because the Rx process receives the data for *all* network channels that are multiplexed over this link.

The ICP sends the address/size pair to `ENCODE.CHANNEL`, which will take it using the extended rendezvous. `ENCODE.CHANNEL` will encode the address and the size into the appropriate data item and release the ICP as soon as process B has read the data item.

The ICP then sends an acknowledgement, containing the OCN of the writing-end, to the Tx process, who sends it out to the network. The remote Rx process receives it and passes it on to the OCP. As soon as the OCP has received the acknowledgement, it releases `DECODE.CHANNEL` who on his part releases process A.

This proceeding preserves the channel synchronisation semantics of CSP [18]. For process A, the whole communication procedure is an atomic action. It simply sends something to a channel, and as soon as it is released from this communication, it knows for sure that process B has received this 'something'.

4.2 Handling Network Channels-types

Setting up an NCT means to set up its ends and to connect them. Specifically, 'connecting them' means to make the location of the server-end known to the client-end. The setup procedure is described in detail in section 5. Once the two ends are connected, they can be administrated. As mentioned in section 2.2, the administration of an NCT end and the communication over the network channels inside the NCT are two disjoint states, i.e. only one of them can happen at any one time. Administrating an NCT end may involve a number of operations, including claiming/releasing it, shutting it down, and (in future versions of KRoC.net) moving it.

4.2.1 The Need For a Clear Protocol

As mentioned before, claiming/releasing is *mandatory* for both ends of an NCT, no matter whether the end is shared or not. Claiming/releasing switches between administration and communication state. Claiming a server-end means telling the KRoC.net manager that the server-end is ready to accept the claim from a client-end. Claiming a client-end means that the client-end is making a claim request to the server-end and waits until the server-end accepts this request. Thus, although the semantical meaning of a claim — 'grabbing' the channel-ends inside a channel-type for (exclusive) communication — is the same for both claiming the end of an NCT and `CLAIM`ing the end of a local channel-type, both the reason for the claim and what happens practically are very different.

A local `CLAIM` is necessary for `SHARED` channel-type ends. The claim accesses a semaphore attached to the channel-type end. This is no problem, as everything is located in the same shared memory. In the networking context, there is no 'central' resource like a semaphore, therefore a claim requires a decentralised agreement of both ends. Shared client-ends of an NCT would be located on different machines, thus they have to contact the server-end directly if they want to communicate. This means that the server-end needs to be aware of the fact that a remote client is trying to make a claim. To be so, it must be able to distinguish cleanly between when it is communicating with a client, and when it is waiting for another client to make a claim.

Without this clean distinction, there is the danger of serious race hazards that cannot happen with local channel-types. Consider the following example: The server-end is trying to send something to a shared client-end, but the client-end has already been released. Locally, this would not be a problem. The server-end would simply be blocked until another process claims the client-end and receives the data sent by the server-end. For NCTs, this is not possible. A typical raze hazard situation could arise: The server-end could send the data to the old client-end, whose ICP eventually tries to output it to the application. The application would not take it, but instead tell the KRoC.net manager to release the claim. Once we are at this stage, we are bound to have a deadlock. As there are no output guards in occam, there would be no way to stop the ICP from trying to output to the application.

In a decentralised system such as KRoC.net, an unambiguous communication protocol is therefore vital. This is the reason for the clear distinction between communication and administration. Both sides, the 'server-end side' and the 'client-end side', have to conform to an agreed pattern of communication between themselves. Claim and release become synchronisation points between the ends of an NCT. Both sides have to claim their NCT end, communicate according to their agreed protocol, and release it afterwards. Failure to do so will result in problems such as deadlock. This is, however, acceptable, as for purely local occam programs, similarly misprogrammed communication will result in deadlock, too.

The occam programmer, therefore, must be relied upon to ensure correct patterns of communication — both on local and on networked systems. [19] proposes ways to *make* processes conformant, i.e. to define the usage of the channels inside a channel-type. The communication pattern of the channel-type would have to be defined by the programmer/designer and would be enforced by the compiler. This is not yet implemented, however, and it is an issue not directly related to KRoC.net, as it would apply to channel-types in general, i.e. both local and networked.

Currently, NCT server-ends cannot be shared, and client-ends are implicitly shared, as currently it makes no difference for the implementation of KRoC.net whether there is one or more client-end connected to the server-end — they have to make a claim before communication anyway.[6]

An NCT server-end may still be `SHARED` locally, of course. Also client-ends of NCTs may be `SHARED` locally. In fact, it is strongly encouraged *not* to have more than one NCT client-end on the same machine connected to the same server-end, but to use the same NCT client-end and share it locally. In order to avoid confusion, a good design rule is to declare each client-end that is supposed to be network-shared *always* locally `SHARED` as well.[7]

4.2.2 Administrating Ends of Network Channel-types

Figure 4 shows the data flow involved in claiming the client-end of the NCT between processes C and D of Figure 2 in detail. We assume that the server-end has just been claimed, i.e. the application connected to the server-end has just told the KRoC.net manager that it is ready for communication.

Every client-end of an NCT is handled by a Client Control Process (CCP), every server-end of an NCT is handled by a Server Control Process (SCP). These are connected to the application by a channel-type called `NET.NCT.MGR`. `NET.NCT.MGR` contains a `request` and a `reply` channel. In order to claim an NCT end, the application sends a claim request down the `req` channel, and waits for a reply.

If a client is claimed, the CCP sends the claim request to the remote SCP. This is done via Tx and Rx, in the same way as the OCP/ICP handle the communication: server-ends and client-end are identified by a unique Server/Client Control Number (SCN/CCN) which is sent along with each communication over Tx/Rx in order to identify the destination SCP/CCP.

When the SCP receives the claim, it informs all the OCPs/ICPs belonging to the network channels in the NCT about it and tells them over which link the remote client-end can be accessed, as well as the respective remote ICNs/OCNs. Then it sends an acknowledgement back to the remote CCP, who on his part will inform its own OCPs/ICPs about the new connection.

If a claim arrives while the server-end is already engaged in a communication with another client, the claim request is put into a queue where it waits until the server is released from this communication. Then the server takes the next claim from the queue and proceeds as described above.

5 Setting up Network Channel-types

The goal of KRoC.net is *transparency* (i.e. application level processes to be blind as to whether their external channels are networked or local) and *simple/flexible/dynamic* setup. Transparency is nearly achieved and will eventually be 100 per cent. It is also flexible and

[6]Later, when KRoC.net will be further integrated into the occam language (see section 5), we *will* distinguish between shared and non-shared ends of NCTs! This is required especially for the differing behaviour of shared/non-shared ends when they are moved.

[7]Later, when the setup of NCTs will be further integrated into occam, the code generated by the compiler will automatically enforce this rule.

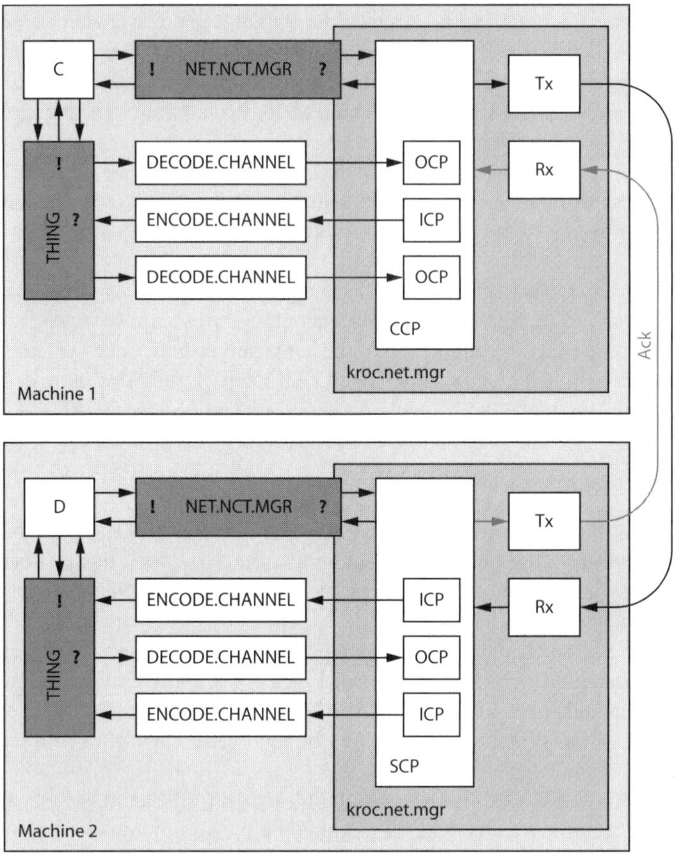

Figure 4: Claiming a Client-end

dynamic. But our current setup procedures are tedious, although straightforward. Future language enhancement will provide direct support for networked channel declaration that will make setup simple (and ensure that it is done correctly).

This section describes the current, rather tortuous but *currently* still necessary, setup mechanisms. The example used is for a networked channel-type — the `THING` defined in Section 2.1.

5.1 The KRoC.net Manager Process

The KRoC.net manager process[8] has the following header:

```
PROC kroc.net.manager.tcp(NET.MGR? net.mgr.svr)
```

`NET.MGR` is a channel-type containing two channels, a `reqest` and a `reply` channel. These are used to set up NCTs. An occam program using KRoC.net would typically create an instance of `NET.MGR` with a `SHARED` client-end and a non-shared server-end. The server-end would be passed to the KRoC.net manager, the client-end would be passed to the processes running in

[8]In this section, we refer to the KRoC.net manager for TCP networks.

parallel with the KRoC.net manager who want to set up NCTs with KRoC.net. Here is the start of the code for setting up the *server-end* of a networked channel-type (THING):

```
NET.MGR? net.mgr.svr:             -- used to communicate
SHARED NET.MGR! net.mgr.cli:      -- with the kroc.net.mgr
THING? thing.svr:                 -- application level channel-type
THING! thing.cli:                 -- (currently needed) supporting infrastructure
SEQ
  net.mgr.svr, net.mgr.cli := MOBILE NET.MGR
  thing.svr, thing.cli := MOBILE THING
  PAR
    kroc.net.mgr.tcp(net.mgr.svr)
    SEQ
      ...  use net.mgr.cli
```

The `thing.svr` end will be plugged into the application level server process. The component channel-ends of `thing.cli` will be plugged into respective `DECODE.CHANNEL` or `ENCODE.CHANNEL` processes, whose other ends are connected to KRoC.net infrastructure channels manufactured *on-the-fly* by the KRoC.net manager and supplied to us in the following conversation.

5.2 Communications to the KRoC.net Manager

Setting up an NCT means to set up either the client-end or the server-end of it, and to connect them to the remote server-end/client-end via KRoC.net. This is done by sending a request to the KRoC.net manager:

```
net.mgr.cli[req] ! setup.server.end; ANY2ONE; <url>; <prot-hash>; <end-types>
```

Note: if we were setting up the *client-end* of an NCT, the tag in the above communication would be "`setup.client.end`" rather than "`setup.server.end`".

`ANY2ONE` tells the KRoC.net manager that the client-end is shared and the server-end is not. Currently, this is a dummy tag, but in later versions it will have to match for both ends of the NCT, and it will influence the behaviour when a (shared/non-shared) end is set up or moved.

`<prot-hash>` is the `PROTOCOL.HASH` value (see section 3.2) of the channel-type we want to connect (i.e. `THING` in our case). Both the server- and the client-end have to pass the `PROTOCOL.HASH` value to the KRoC.net manager who can then ensure that matching channel-types are connected.

`<end-types>` is a counted array protocol over which the application tells the KRoC.net manager about the end-types (i.e. reading- or writing-end) of the channels in the channel-type. Again, they have to match. (For instance if we have a reading-end/writing-end/reading-end sequence in the server-end, the client-end must have a writing-end/reading-end/writing-end sequence.) These values are necessary for the KRoC.net manager to know whether to use OCPs or ICPs.

`<url>` is a string (a dynamic mobile byte array) that tells the KRoC.net manager about the location of the server-end of the NCT. We created a new URL-based mechanism for setting up NCTs. For this, we have defined a new URL called "`nct:`". The general definition of an `nct:` URL is as follows[9]:

[9] `<...>` are variable names, `[...]` means optional, | is a choice, `(...)` is for grouping, everything else is literal. Spaces are not allowed (as usual for URLs).

```
nct: ( <server-name> [ @ ( cns: <cns-name> |
                           cns. <net-type> : <location> ) ] |
       ( <server-name> | $ <scn> )
                      @direct [ . <net.type> : <location> ]
     )
```

<server-name> is a name given to the server-end. This can be any string which does not contain "@" or "$". It could be a simple name like "fred", but also a structured name like "my-application/fred" or so. This is basically up to the programmer.

5.3 The Channel Name Server

The second part of the URL communicated to the KRoC.net manager says where to register (or look up) the <server-name>. The default place for this would be the Channel Name Server (CNS). This is a central server where server-ends can be registered and where client-ends can ask for the location of a server-end according to its name. The CNS can hold a huge number of names, even of completely different distributed applications at the same time. In the latter case, structured names would therefore be sensible.

If the "@" suffix of the URL is omitted, the default CNS will be used to register and ask about the server-end. The location of the default CNS is stored in a special configuration file, described in section 6. Alternatively, the application might want to use a different CNS whose location is stored in another configuration file (cf. section 6). To do so, it would use

```
@cns:<cns-name>
```

whereby <cns-name> is the name of a non-default CNS. The third way of using a CNS is connecting directly to it using

```
@cns.<net-type>:<location>
```

<net-type> is the type of the network; it must be the same as the network type of the KRoC.net manager. <location> is the location of the CNS specific to the network type. For TCP networks, it would be "<ip>:<port>". <ip> would be the IP address or the host name (preferably including the domain) of the CNS, <port> would be the port number. A valid URL would be for instance:

```
nct:fred@cns.tcp:gaia.kent.ac.uk:4400
```

The port can be omitted, in which case the default CNS port of 4400 will be used.

If the requested name cannot be found in the CNS, it will return an error. We will soon add the possibility to specify a certain number of retries or a timeout before an error is returned.

Alternatively to using a CNS, an application may also connect a NCT directly. The server-end would be registered using

```
@direct
```

after the name. The client-end would then ask about the server-end directly at the remote machine, using

```
@direct.<net-type>:<location>
```

The variables mean the same as above for connecting directly to a CNS. If the network type is TCP, and the port is omitted, a default of 4500 will be assumed.

5.4 Replies Back From the KRoC.net Manager

When an NCT end is set up, the KRoC.net manager will establish a CCP/SCP, along with the OCPs and ICPs for all the network channels inside the NCT. If anything goes wrong, the KRoC.net manager returns an error, otherwise it returns the client-end of an NET.NCT.MGR channel-type (cf. Figure 4). The server-end of this channel-type is held by the CCP/SCP handling the client-end/the server-end of the NCT just created. As mentioned in section 4.2.2, NET.NCT.MGR is used to administrate the NCT end.

The first thing the CCP/SCP returns over NET.NCT.MGR's reply channel are client-ends of channel-types of type NET.OUT or NET.IN for each of the network channels inside the new NCT. The server-ends of these NET.OUT/NET.IN channel-types are held by the OCPs/ICPs handling the ends of the network channels in the NCT. The occam programmer would then plug each of the NET.OUT/NET.IN client-ends into a DECODE.CHANNEL or ENCODE.CHANNEL process, on whose other end the appropriate channels to the application process would be plugged in. Both NET.OUT and NET.IN contain two channels, one for the address/size pair, and one termination channel for the GPC.

5.5 Unregistered (Anonymous) NCTs

Another way of setting up an NCT is to do so anonymously. For this, there is a special request:

```
net.mgr.cli[req] ! setup.server.end.anon; ONE2ONE; <prot-hash>; <end-types>
```

Unlike the previously mentioned requests, this one does not contain a URL. The server-end is created anonymously, and upon request, it would return an 'anonymous URL' which a client-end could then use to connect to. An anonymous URL has the form

```
nct:$<scn>@direct.<net-type>:<location>
```

where <scn> would be the SCN of the server-end. The @direct<...> suffix is the same as mentioned above.

This returned string contains all the KRoC.net information normally held in the CNS and supplied to processes requesting client-end connections to this NCT. The server process, having not registered its server-end anywhere, may now pass that anonymous URL along existing channels (e.g. to a new client with whom it has established a connection over a published NCT). The receiver may use this string to set up its private client-end, without going via the CNS.

Note: There is no "setup.client.end.anon" tag for the request to the KRoC.net manager, as it is obvious from the URL that we want to set up the client-end anonymously. Instead, the normal setup.client.end tag is used. The KRoC.net manager will construct the necessary KRoC.net infrastructure to set up that client-end — just as though it was obtained normally from the CNS. In this way, a client may use a CNS-published NCT only to establish a private connection to a server, who would then FORK off a special server process to deal with that client. Otherwise, a server can deal with only one client at a time over its published channel bundle — and we may want to deal with millions concurrently!

In a later version of KRoC.net, anonymous NCTs will be blended into the existing occam semantics for moving ends of channel-types. Instead of declaring an anonymous NCT server-end and passing its URL to the client side who then connects to it anonymously, the programmer would simply allocate an instance of a channel-type in the normal way (i.e. both ends of it) and send one end over a network channel to another machine. This is the same mechanism as it works for local channel-types already. With this new level of transparency, anonymous NCTs will no longer be necessary.

5.6 Modular Design of the KRoC.net Manager

Currently, the only network type supported is TCP (i.e. the network links are TCP/IP socket connections). KRoC.net, however, is easily extensible because of the modular design of the KRoC.net manager. Figure 5 shows the component processes of the TCP KRoC.net manager. For each NCT end that the KRoC.net manager handles, there is a SCP/CCP, along with the OCPs/ICPs for the network channel-ends inside that NCT end.

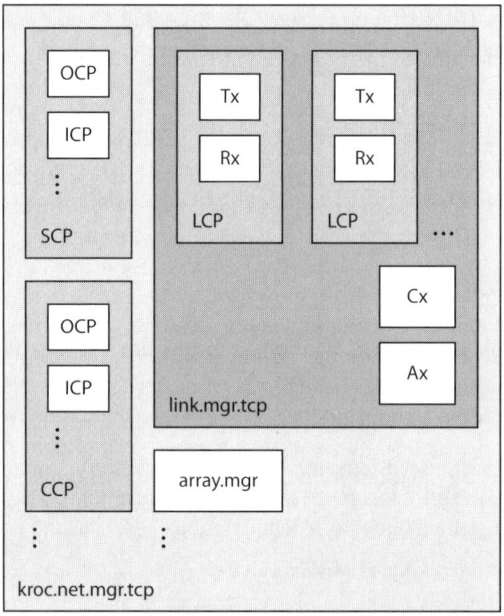

Figure 5: Components of the KRoC.net Manager

The back-end, which deals with the network communication, is inside the link manager, which forms a part of the KRoC.net manager. For each network link, the link manager contains a Link Control Process (LCP), each of which is identified by a unique Link Control Number (LCN). Figure 5 shows the link manager for TCP networks. Each LCP contains a pair of Tx/Rx processes. Additionally, the link manager contains a Cx and an Ax process, dealing with connecting to remote sockets and accepting incoming socket connections.

The TCP version of the link manager uses Barnes' occam Socket Library [20] which enables TCP socket communication under Linux/x86. The Socket Library uses the mechanism to make blocking system calls without blocking the occam kernel, described in [21].

In order to create a KRoC.net manager for a network of a type other than TCP, we would only have to replace the link manager with one for the new network type. Every other code could remain unchanged.

Inside the KRoC.net manager, the LCPs, OCPs/ICPs and SCPs/CCPs are accessed via their respective Control Numbers. For each of them, there is a channel-type called `TO.LCP`, `TO.OCP`, etc., each of which contains just one `req` channel. The server-end (holding the reading-end of `req`) is held by the respective Control Process, the client-end is inside an array at the index that corresponds with the respective Control Number. In this way, all Control Processes can be accessed by their Control Number.

The arrays are managed by special array manager processes who keep a free-list of the empty elements. When an array is full, its size will be doubled. For this, each array is protected by a CREW lock [22]. Accessing the array to communicate over one of the `TO.xCPs` requires setting a read lock on the respective array. When the array is doubled, the responsible array manager will set a write lock, create the new array and copy all elements of the old one into it.

The KRoC.net manager is completely modular in its design, i.e. processes are only `FORK`ed off when they are needed. This applies to the LCPs as well as to the 'front-end' Control Processes. Processes that are not needed anymore are terminated. The procedure follows the 'poisoning' approach described in [23].

When the application makes a request to shut down an NCT end, the SCP/CCP sends a shut-down signal to the OCPs/ICPs it has created. These on their part will terminate the GPCs. The link manager keeps a count about the number of NCTs multiplexed over each link. If this number reaches 0, the LCP will be sent a shut-down signal and take care of closing the link and shutting down the Tx/Rx processes. When the KRoC.net manager gets the final shut-down request, it passes it on to all its component processes. Only when everything has cleanly terminated, it will shut-down itself. The basic rule is graceful termination from 'inside out'.

5.7 Proposals for the Further Integration of KRoC.net Into occam

The setup process described in sections 5.1 through 5.5 is relatively easy to understand, but involves quite a number of steps. Although the communication over network channels is nearly transparent now, the setup process is not yet. In order to change this, we will extend the KRoC compiler to hide this setup and administration code behind an extended language syntax for declaring networked channels and channel-types. For this, we have proposed some ideas. These are not 'fixed' yet, however, and subject to discussion.

The idea of forcing a `CLAIM` for all use of channel-types would enable removal of the explicit network claims that an application process currently has to add to code using channel-types. A run-time test would also need to be generated for these `CLAIM`s to determine whether the associated channel-type was networked or local. The extra test would only cost a few nanoseconds and be insignificant even for `CLAIM`s on local channel-types.

For setting up an NCT end, there may be a new keyword called "`NET`". Contrary to normal channel-type ends, NCT ends would not be allocated in pairs (as the other end will be on another machine!). Here is an example that declares and dynamically constructs the server-end of an *any-to-one* NCT:

```
THING? net.thing.svr:
INT result:
...
SEQ
  net.thing.svr, result := MOBILE NET ANY2ONE TCP THING? "nct:fred"
```

Here is an example for setting up one of many possible networked client-ends for the same NCT.

```
SHARED THING! net.thing.cli:
INT result, timeout:
...
SEQ
  timeout := 5 * seconds
  net.thing.cli, result := MOBILE NET ANY2ONE TCP THING! "nct:fred" timeout
```

TCP is the network type. This is needed to ensure that the right KRoC.net manager will be used. "nct:fred" would be the URL, and result would either be an OK or an error message. timeout specifies how long (or possibly how often) to retry if the required name is not stored in the CNS yet.

This is all that will be needed. All necessary DECODE.CHANNELs/ENCODE.CHANNELs will be created automatically and wired up correctly. All the user needs to do is plug the constructed network channel-type end into the relevant server or client process (after, of course, checking the result).

When a CLAIM is made on an NCT end, KRoC would generate the necessary code to make a network claim with the KRoC.net manager, and also to make a network release when the CLAIM has been finished. Apart from that, if the NCT end is shared, the CLAIM triggers the usual action (i.e. setting the semaphore etc.).

Simple *one-to-one* network channel-ends could be declared like this:

```
INT result:
NET ONE2ONE TCP CHAN INT c? "nct:sue" result:
```

on one machine, and on the other:

```
INT result:
NET ONE2ONE TCP CHAN INT c! "nct:sue" result timeout:
```

KRoC would then create the necessary code as if c? and c! were the ends of an NCT containing a single network channel-end (the server-end containing the reading-end, and the client-end containing the writing-end). As this channel would not be shared, KRoC could do the network claim for both ends immediately after declaration, and the release before the occam program terminates or as soon as c is getting out of scope.

A simple network channel with a SHARED writing-end would be declared:

```
INT result:
SHARED NET ANY2ONE TCP CHAN INT c! "nct:sue" result timeout:
```

on one machine, and its reading-end on the other:

```
INT result:
NET ANY2ONE TCP CHAN INT c? "nct:sue" result:
```

6 Configuration

The only settings which need to be configured are the location (i.e. IP address and port number in the TCP/IP world) of both the own machine[10] (remember that 'machine' refers to an occam program running an instance of the KRoC.net manager) and the CNS. In previous versions, this information was compiled directly into the occam programs. This was OK for experimental use but rather inflexible for real life applications, therefore we have simplified KRoC.net's configuration and made it more flexible.

The locations of the CNS and the own machine are kept in special configuration files. KRoC.net is looking for these files in the program directory and in the user's home directory (in this order).

[10]The location of the own machine is used as a unique identifier for it within the KRoC.net system. This is the location that gets registered with the CNS when the machine registers an NCT server-end.

The 'own' file should be called "`.kroc.net`". Its content should be:

```
[<network-type>]
<location-info>
```

For TCP, this would be:

```
[tcp]
ip=<ip>
port=<port>
```

with `<ip>` being the IP address and `<port>` being the port number. If the port number is omitted, a default of 4500 will be assumed. If the IP address is missing, the standard outgoing IP address will be used.

The CNS uses a similar file, called "`.self.cns`". In this file, however, only the port number is needed, as the CNS does not use its IP address as part of an identifier. If the file is missing, the default CNS port of 4400 will be used.

Finally, an application using KRoC.net needs a file called "`.cns`", where the location of the default CNS is stored. Here, we can use the IP address or the host name (preferably including the domain) of the CNS as a value for `<ip>`. The port can be omitted, in which case the default port number would be 4400.

Information about a CNS other than the default CNS can be stored in a file called "`.cns.<cns-name>`", where `<cns-name>` is the name of the CNS used in the "`@cns:<cns-name>`" part of the `nct:` URL.

7 Performance of KRoC.net

We have run a number of benchmarks to measure the performance of KRoC.net. The benchmarking system consists of a sender on one machine and a receiver on another machine. The sender resides on gaia, a 1GHz Pentium III PC. The receiver is on catch22, a 2.4GHz Pentium 4 PC. Both PCs are connected to a local 100MBit/s ethernet.

7.1 Bandwidth Measurement

For measuring the bandwidth, the sender sends a sequence of equal-sized packets to the receiver over a network channel provided by KRoC.net. This is repeated for different packet sizes between 1 Byte and 64 KBytes (doubling the size each time).

Figure 6 shows the bandwidth (in Bytes/s) for the different packet sizes. For comparison, we have also included the results for sending the packets over raw sockets (without acknowledgement), as well as for sending them over raw sockets where the receiver acknowledges each packet.

Especially the comparison between the 'KRoC.net' and the 'Sockets with ack' results shows the small overhead of the KRoC.net framework. With 1 Byte sized packets, the bandwidth for KRoC.net is about half the bandwidth for 'Sockets with ack'. This is due to the fact that in the 'Sockets with ack' version, the receiver knows which packet size to expect, which is not the case for KRoC.net. Therefore, KRoC.net always needs to send the size first, and then the data[11]. With increasing packet sizes, the impact of the second communication diminishes, however.

[11] We are currently experimenting with a `WRITEV` implementation that sends size and data at once; this is not implemented yet, however.

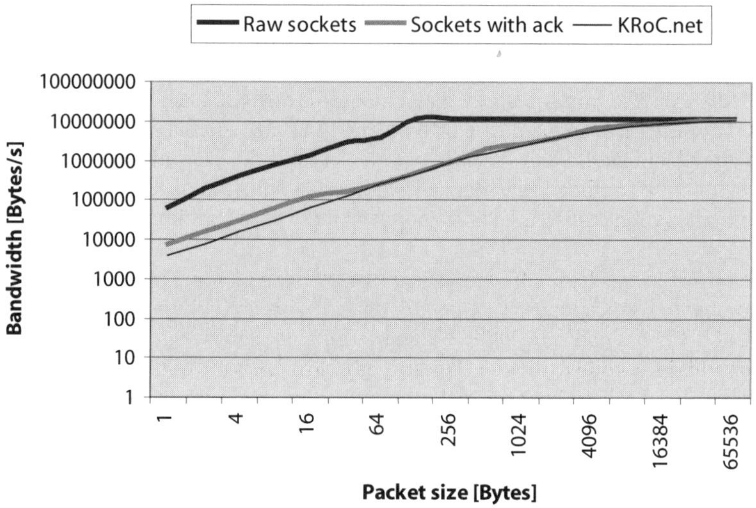

Figure 6: Bandwidth Benchmarks

It is also, perhaps, a little unfair for the 'Sockets with ack' version to know about packet sizes — for flexibility, the sizes would also need to be transmitted. Tests with a modified version of 'Sockets with ack' that also sends the size have produced practically the same results as KRoC.net, which shows the low overheads of KRoC.net.

We have also measured the overheads introduced by KRoC.net with a modified version of the benchmark program. Imagine Machine 1 of Figure 3, but instead of starting network communication when receiving data from the OCP, the Tx process sends the header of the network message over an ordinary occam channel to the Rx process. When the Rx process receives the header, it treats it like an acknowledgement from the network, i.e. it sends an acknowledgement to the OCP, who then releases the extended rendezvous as usual.

On gaia, the ping-pong time for sending a 1 Byte packet (i.e. the time between the sender sends it and receives the acknowledgement) is $1.9\mu s$ on the sending machine. This is completely negligible, as the error range for the benchmarks is about 10% (i.e. about $20\mu s$ for 1 Byte sized packets). Tests with the Distributed occam Protocol (DoP) [11], a predecessor of KRoC.net from 2001, have shown no measurable difference in the benchmark figures.

All this shows the big impact of socket communication, which involves making blocking system calls to the operating system and heavyweight OS-level IPC. At some point, the functionality of KRoC.net will be introduced into RMoX [24], the Raw Metal occam operating system. This will give us a much better performance, as all the socket communication could then be performed from within occam, i.e. without involving heavyweight OS-level context switches.

7.2 Load Measurement

Another benchmark is measuring the load of KRoC.net against other occam processes running in parallel with it. The setup of this benchmark is based on a similar one in [25]. It is shown in Figure 7.

The benchmark is a modification of the previous one. On catch22, the receiver is running

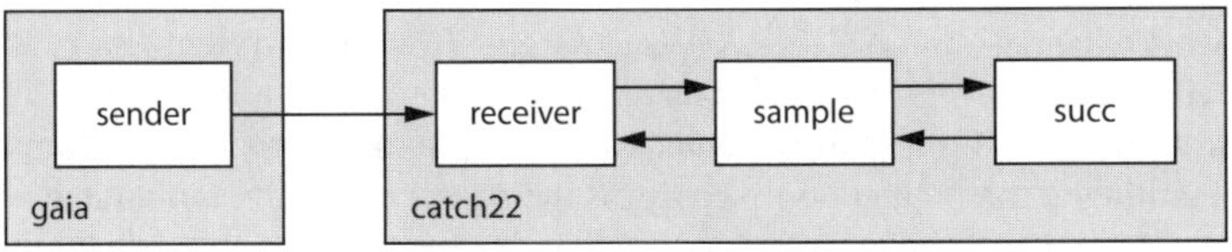

Figure 7: Load Measurement Setup

in parallel with two other processes, called `sample` and `succ`. These two processes are running at the lowest (KRoC occam) process priority [7]. They are always active and running indefinitely — but only, of course, when neither the `receiver` nor any KRoC.net infrastructure processes are active. `sample` is passing a number to `succ` which `succ` increases and sends back. The receiver can interrupt `sample` at any time, whereupon `sample` will return the current count.

The receiver measures the increase of the count against the time passed. First it does so before it starts to receive, normalising this value to 100% background processing capability. Then it measures and compares this value against the ones for the different packet sizes. The result is shown in Figure 8.

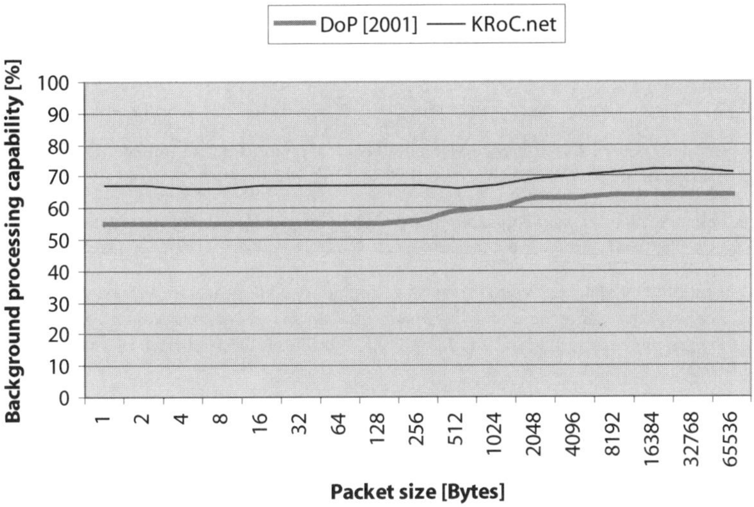

Figure 8: Load Benchmarks

This shows that KRoC.net has a relatively low impact on processes running in the background. It is even slightly better than its predecessor DoP, which is most likely due to the dynamic setup of the network infrastructure — processes are only are started when they are needed — as well as the fact that there is no copying of data involved in the communication process, as KRoC.net only deals with address/size pairs.

8 Conclusions and Future Work

This paper has presented the basic ideas behind the KRoC.net project, its infrastructure, usage and configuration. KRoC.net is now a dynamic tool that enables the occam programmer to set up network channels and NCTs easily and safely. New occam features in KRoC have been utilised to reach a high level of transparency in the use of KRoC.net. These will be extended to achieve total transparency and a much simpler and automated dynamic setup.

Error handling at the moment only concerns the setup procedure, but not the actual communication. We will add appropriate error messages for the case that something with the communication goes wrong.

We will examine ways to reduce network latency for certain types of communication. This can be achieved by reducing the number of acknowledgement packets sent over the network. There are three concrete approaches for this. Firstly, we will introduce buffered channels, which only require an acknowledgement after the buffer on the receiving side is full. Secondly, we will introduce ping/pong style NCTs (similar to the 'Connections' in JCSP.net). These will follow a strict 'request/reply' scheme, whereby each data packet also acts as an implicit acknowledgement for the previous one.

Additionally, we will improve the network latency for `PROTOCOL`s that involve more than one communication (e.g. sequential protocols or multi-dimensional dynamic mobile arrays). For this, we will introduce a third field in the link protocol of the GPCs that tells the KRoC.net infrastructure whether subsequent communications are yet to come for this `PROTOCOL`. Network acknowledgements would only be sent after the last communication of each `PROTOCOL`.

Another field that our research will look into, is to find ways of sharing server-ends of NCTs, and to implement the possibility to move ends of NCTs around networks, just as ends of local channel-types can be moved around on a local machine.

References

[1] Inmos Limited. occam 2.1 Reference Manual. Technical report, Inmos Limited, May 1995. Available at: `http://www.wotug.org/occam/`.

[2] Kevin Vella. *Seamless Parallel Computing on Heterogeneous Networks of Multiprocessor Workstations*. PhD thesis, The University of Kent at Canterbury, Canterbury, Kent. CT2 7NF, December 1998.

[3] K. Vella and P.H. Welch. CSP/occam on Shared Memory Multiprocessor Workstations. In B.M. Cook, editor, *Architectures, Languages and Techniques for Concurrent Systems*, volume 57 of *Concurrent Systems Engineering Series*, pages 87–119. WoTUG, IOS Press, the Netherlands, April 1999. ISBN: 90-5199-480-X.

[4] M. Boosten, R.W. Dobinson, and P.D.V. van der Stok. MESH: MEssaging and ScHeduling for Fine-Grain Parallel Processing on Commodity Platforms. In *Proceedings of the 1999 International Conference on Parallel and Distributed Processing Techniques and Applications (PDPTA'1999)*. CSREA press, June 1999. ISBN: 1-892512-15-7.

[5] J. Moores. CCSP – a Portable CSP-based Run-time System Supporting C and occam. In B.M. Cook, editor, *Architectures, Languages and Techniques for Concurrent Systems*, volume 57 of *Concurrent Systems Engineering series*, pages 147–168, Amsterdam, the Netherlands, April 1999. WoTUG, IOS Press. ISBN: 90-5199-480-X.

[6] F.R.M. Barnes and P.H. Welch. Mobile Data, Dynamic Allocation and Zero Aliasing: an occam Experiment. In Alan Chalmers, Majid Mirmehdi, and Henk Muller, editors, *Communicating Process Architectures 2001*, volume 59 of *Concurrent Systems Engineering*, pages 243–264, Amsterdam, The Netherlands, September 2001. WoTUG, IOS Press. ISBN: 1-58603-202-X.

[7] F.R.M. Barnes and P.H. Welch. Prioritised Dynamic Communicating Processes: Part I. In James Pascoe, Peter Welch, Roger Loader, and Vaidy Sunderam, editors, *Communicating Process Architectures 2002*, WoTUG-25, Concurrent Systems Engineering, pages 331–361, IOS Press, Amsterdam, The Netherlands, September 2002. ISBN: 1-58603-268-2.

[8] F.R.M. Barnes and P.H. Welch. Prioritised Dynamic Communicating Processes: Part II. In James Pascoe, Peter Welch, Roger Loader, and Vaidy Sunderam, editors, *Communicating Process Architectures 2002*,

WoTUG-25, Concurrent Systems Engineering, pages 363–380, IOS Press, Amsterdam, The Netherlands, September 2002. ISBN: 1-58603-268-2.

[9] P.H. Welch and D.C. Wood. The Kent Retargetable occam Compiler. In Brian O'Neill, editor, *Parallel Processing Developments, Proceedings of WoTUG 19*, volume 47 of *Concurrent Systems Engineering*, pages 143–166. World occam and Transputer User Group, IOS Press, Netherlands, March 1996. ISBN: 90-5199-261-0.

[10] I.N. Goodacre. occam NetChans, 2001. Project report.

[11] M. Schweigler. The Distributed occam Protocol - A New Layer On Top Of TCP/IP To Serve occam Channels Over The Internet. Master's thesis, Computing Laboratory, University of Kent at Canterbury, September 2001. MSc Dissertation.

[12] P.H. Welch, J.R. Aldous, and J. Foster. CSP networking for java (JCSP.net). In P.M.A. Sloot, C.J.K. Tan, J.J. Dongarra, and A.G. Hoekstra, editors, *Computational Science - ICCS 2002*, volume 2330 of *Lecture Notes in Computer Science*, pages 695–708. Springer-Verlag, April 2002. ISBN: 3-540-43593-X. See also: `http://www.cs.ukc.ac.uk/pubs/2002/1382`.

[13] P.H. Welch and B. Vinter. Cluster Computing and JCSP Networking. In James Pascoe, Peter Welch, Roger Loader, and Vaidy Sunderam, editors, *Communicating Process Architectures 2002*, WoTUG-25, Concurrent Systems Engineering, pages 213–232, IOS Press, Amsterdam, The Netherlands, September 2002. ISBN: 1-58603-268-2.

[14] P.H. Welch. CSP for Java (JCSP), 1999. Available at: `http://www.cs.ukc.ac.uk/projects/ofa/jcsp/`.

[15] M.D. May, P.W. Thompson, and P.H. Welch. *Networks, Routers and Transputers*, volume 32 of *Transputer and occam Engineering Series*. IOS Press, 1993.

[16] Henk L. Muller and David May. A simple protocol to communicate channels over channels. In *EURO-PAR '98 Parallel Processing, LNCS 1470*, pages 591–600, Southampton, UK, September 1998. Springer Verlag.

[17] F.R.M. Barnes and P.H. Welch. Prioritised dynamic communicating and mobile processes. *IEE Proceedings – Software*, 150(2), April 2003.

[18] C.A.R. Hoare. *Communicating Sequential Processes*. Prentice-Hall, 1985. ISBN: 0-13-153271-5.

[19] Frederick R.M. Barnes. *Dynamics and Pragmatics for High Performance Concurrency*. PhD thesis, University of Kent, June 2003.

[20] Fred Barnes. *Socket, File and Process Libraries for occam*. Computing Laboratory, University of Kent at Canterbury, June 2000. Available at: `http://www.cs.ukc.ac.uk/people/rpg/frmb2/documents/`.

[21] F.R.M. Barnes. Blocking System Calls in KRoC/Linux. In P.H. Welch and A.W.P. Bakkers, editors, *Communicating Process Architectures*, volume 58 of *Concurrent Systems Engineering*, pages 155–178, Amsterdam, the Netherlands, September 2000. WoTUG, IOS Press. ISBN: 1-58603-077-9.

[22] Peter H. Welch and David C. Wood. Higher Levels of Process Synchronisation. In A. Bakkers, editor, *Parallel Programming and Java, Proceedings of WoTUG 20*, volume 50 of *Concurrent Systems Engineering*, pages 104–129, Amsterdam, The Netherlands, April 1997. World occam and Transputer User Group (WoTUG), IOS Press. ISBN: 90-5199-336-6.

[23] P.H. Welch. Graceful Termination – Graceful Resetting. In *Applying Transputer-Based Parallel Machines, Proceedings of OUG 10*, pages 310–317, Enschede, Netherlands, April 1989. Occam User Group, IOS Press, Netherlands. ISBN 90 5199 007 3.

[24] F.R.M. Barnes, C.L. Jacobsen, and B. Vinter. RMoX: a Raw Metal occam Experiment. In *Communicating Process Architectures 2003*, WoTUG-26, Concurrent Systems Engineering, IOS Press, Amsterdam, The Netherlands, September 2003.

[25] Peter H. Welch and Michael D. Poole. occam for Multi-Processor DEC Alphas. In A. Bakkers, editor, *Parallel Programming and Java, Proceedings of WoTUG 20*, volume 50 of *Concurrent Systems Engineering*, pages 152–174, Amsterdam, The Netherlands, April 1997. World occam and Transputer User Group (WoTUG), IOS Press. ISBN: 90-5199-336-6.

Communicating Process Architectures – 2003
Jan F. Broenink and Gerald H. Hilderink (Eds.)
IOS Press, 2003

A CSP-based Processing Architecture for a Flexible MIMO-OFDM Testbed

H.S. CRONIE, F.W. HOEKSEMA and C.H. SLUMP
University of Twente, Department of Electrical Engineering, Mathematics and Computer Science (EEMCS), Signals and Systems Group, P.O. box 217 - 7500 AE Enschede - The Netherlands
email: h.s.cronie@student.utwente.nl, {f.w.hoeksema,c.h.slump}@utwente.nl

Abstract. Future wireless communication systems require novel techniques to increase the bitrate, coverage and mobility. One of these techniques is spatial multiplexing and we have investigated the use of a CSP-based kernel in the implementation of a spatial multiplexing testbed. It turns out that the use of the CSP-based kernel not only provides a good way of system modeling, but also provides a very scalable software architecture for the testbed. In future, we can change several system parameters without changing the software architecture. With the testbed we were able to verify the concept of spatial multiplexing in an office environment.

1 Introduction

Future wireless applications, such as wireless LANs, demand faster and more reliable communication links. Wireless LANs are typically used in an indoor environment. The indoor radio channel suffers from multipath propagation, which is caused by objects between the transmitter and receiver. Multiple components of the transmitted signal are received. The situation is illustrated in figure 1. These components interfere with each other and the in-

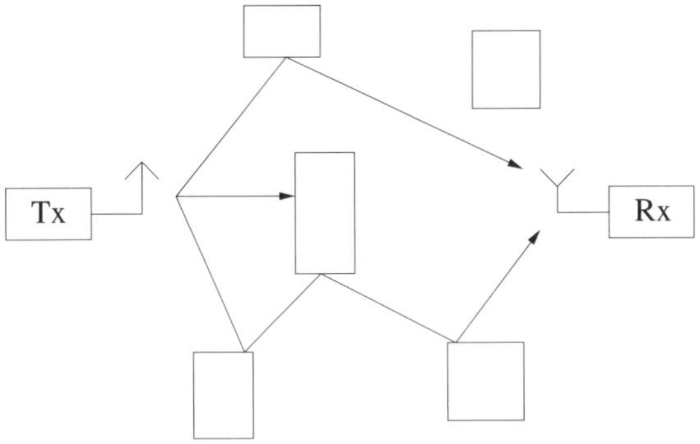

Figure 1: Multipath propagation between transmitter and receiver

terference can be constructive or destructive. The case of destructive interference is called a signal fade and its effect is that it causes a reduction in received signal power. The channel also varies with time[1] and the communication system has to be designed properly to cope

[1]The coherence time of the indoor channel at 2.4 GHz is around 20 ms.

with the negative effects of multipath fading. Recent research has shown that it is also possible to use the negative effect of signal fading in our advantage [1], [2]. For this, multiple transmit antennas are used at the transmitter and multiple receive antennas at the receiver. The idea is that fades become uncorrelated[2] for antenna separation $> \lambda/2$, in which λ is the wavelength of the radio signal. We transmit an independent datastream from each transmit antenna. At each receive antenna a linear combination of the signals transmitted is received. The input-output relation of the *Multiple Input Multiple Output* (MIMO) indoor fading channel corrupted by noise, can be written as:

$$\boldsymbol{y} = \boldsymbol{H}\boldsymbol{x} + \boldsymbol{n}, \tag{1}$$

in which $\boldsymbol{n}$ is a complex gaussian noise vector added by the channel. The complex vector $\boldsymbol{x}$ contains the symbols transmitted. These symbols are drawn from a complex constellation. At the receiver, we estimate the channel transfer matrix $\boldsymbol{H}$ and detect the data by a process called *space-time decoding*. The effect is that we are able to increase the amount of data that can be transmitted and communication becomes more reliable, while keeping the bandwidth and total transmit power constant.

The datarate achievable on a communication link is bounded by its information theoretic capacity. The use of a MIMO communication system results in an increased channel capacity. Figure 2 shows the behavior of the capacity if both the number of transmit antennas N_t and the number of receive antennas N_r is increased. As can be seen from the figure, the possible

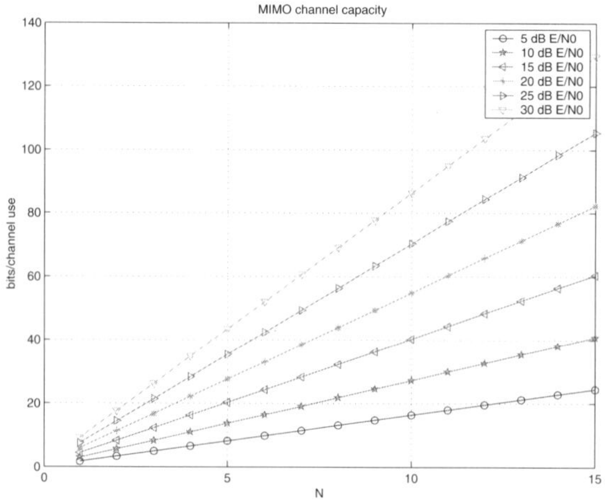

Figure 2: The capacity as function of the amount of antennas used ($N = N_t = N_r$)

gains are enormous. The capacity grows linear with the amount of transmit and receive antennas used. The received data however, has to be detected. For this, the channel transfer matrix $\boldsymbol{H}$ has to be known and a detection algorithm has to be developed which uses this matrix. To transmit information, a modulation method is needed and research has shown that *Orthogonal Frequency Division Multiplexing* (OFDM) is a suitable method in a fading environment [7]. To investigate the practical behavior of a MIMO communication system, we have developed a MIMO-OFDM testbed.

[2]Fades do not necessary have to be uncorrelated, but the possible gains are higher for uncorrelated fades.

2 Functional Testbed Architecture

In our research we address the question whether the use of a MIMO channel lives up to its promises. Claims, often based on models, have to be tested in real-world environments. We want to investigate under what conditions the channel model is valid. Moreover, other points of interest are the required processing capability for the different algorithms and the design of the computational platform for this type of systems (including HW/SW partitioning). We have build a testbed to answer these questions. Figure 3 shows an overview of the testbed. Both the transmitter and receiver use multiple antennas. The interesting signal processing

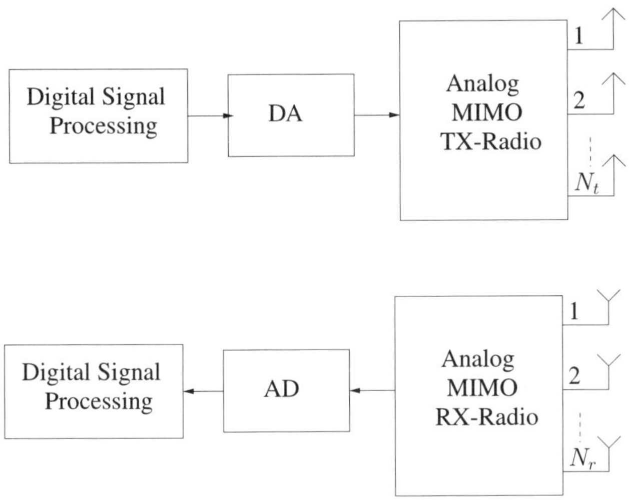

Figure 3: The MIMO-OFDM testbed

takes place at the receiver. First, analog processing converts the RF signal to a baseband signal. The baseband signal is sampled and the subsequent digital signal processing functions have to be implemented at the receiver:

- *OFDM synchronization:* The received signal has to be symbol-synchronized and some other OFDM signal processing has to be performed.
- *OFDM demodulation:* The synchronized data has to be demodulated.
- *Frame decoding:* The demodulated datastream contains symbols for the channel estimator, so called *pilot symbols* and symbols for the space-time decoder, so called *payload symbols*. These two have to be separated.
- *Channel estimation:* The MIMO detection algorithms require channel knowledge and the channel matrix $\boldsymbol{H}$ has to be estimated using the pilot symbols.
- *Space-Time decoding:* The received payload symbols have to be decoded.

The focus of our research is in developing MIMO algorithms and to investigate the feasibility of MIMO in a real-world office environment. So, we want to implement the algorithms into a real-time real-world system while not being fully knowledgeable of the computational requirements and hardware-software partitioning of the system. On one hand we want to investigate new functions and algorithms and on the other hand, we have to be prepared to change the computational platform depending on the computational requirements of these algorithms, as we constrain ourselves in that the system has to operate in real-time.

Our MIMO system consists of different functions (tasks) that exchange data with one another. A function is implemented by some algorithm. However, we want to map these functions on a processing platform (hardware and/or software) in order to operate in real-time. We expect the functions and their interconnection structure (dataflow between the functions) to be more or less constant, while the algorithms and their computational complexity may vary. We want a way of system modeling which captures these communication aspects of the system and that allows mapping of an individual function to a particular processing platform independently of the other functions.

The theory of *Communicating Sequential Processes* (CSP) can be used to model our MIMO system. CSP is a notation for describing concurrent systems [3] and these systems interact with each other by means of communications. By using CSP, we can model all relevant aspects of our system. Moreover, we are able to focus on the functions and algorithms, real-time constraints and change computational platforms *while using a single system model only*.

3 CSP System Modeling

The required MIMO functions which are identified in the previous section, map directly to a CSP system model. The model for the receiver is given in figure 4. The CSP model consists

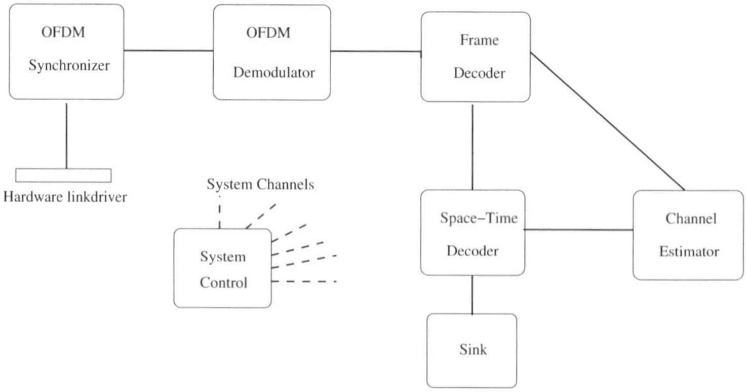

Figure 4: The CTCPP receiver architecture.

of five key processes:

- *OFDM Synchronizer:* the process acquires symbol synchronization and it communicates with the `OFDM demodulator` process. Synchronized symbols are written to the `OFDM demodulator` process.
- *OFDM Demodulator:* the process demodulates the OFDM symbols. Demodulated symbols are written to the `Frame Decoder` process.
- *Frame Decoder:* the `Frame Decoder` process extracts pilot symbols from the demodulated datastream. Pilot symbols are written to the `Channel Estimator` process and payload symbols are written to the `Space-Time decoder` process.

- *Channel Estimator:* the process performs channel estimation and channel estimates are written to the Space-Time Decoder.
- *Space-Time Decoder:* the process implements a MIMO detection algorithm.

The `System Control` process controls the other processes and status information is exchanged between these processes. Detected data is sent to a `Sink` process. The hardware linkdriver provides the sampled data from the analog radio. The communication in our system is not sample-based. The `OFDM Synchronizer` process delivers a group of OFDM symbols. A typical OFDM symbol consists of 32 samples and around 20 symbols are sent to the `OFDM Demodulator` process each time a message exchange takes place.

4 CTC++ Real-time kernel

To implement the CSP based architecture, we have used a real-time CSP kernel developed at the Control Engineering Group of our department [5], [6]. The C++ version of the kernel is used which provides us with `Process` classes and `Channel` classes. The kernel includes a scheduler and different priorities can be assigned to each process. The CTC++ kernel provides a novel hardware abstraction layer as is shown schematically in figure 5. Communication takes place over channels and each channel can have a linkdriver. The

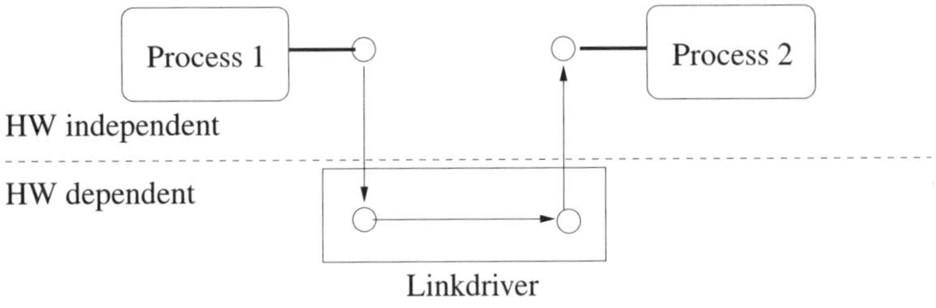

Figure 5: The hardware abstraction provided by CTC++.

linkdriver contains all the hardware dependent code. If the processing platform is changed, the linkdrivers are the only elements which require modifications. The different processes are implemented independently of the hardware they run on. Our processing platform is likely to change, however the receiver software itself does not have to be altered. Figure 6 shows the deployment diagram of our current testbed. The figure also shows a possible deployment diagram for a multi-processor platform.

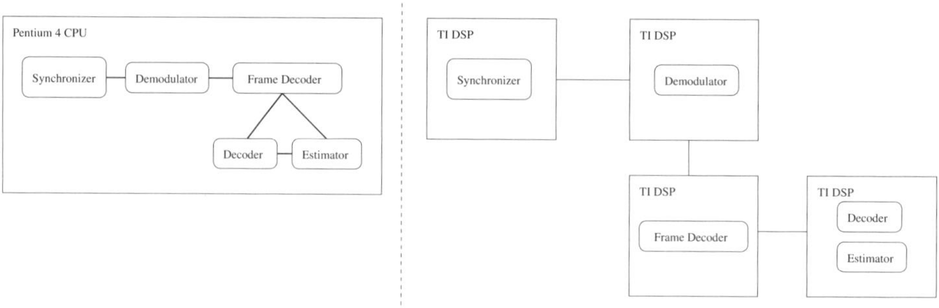

Figure 6: Two possible deployment diagrams, the first one is implemented.

4.1 Process communication

Whenever communication between two processes takes place, the kernel will provide the synchronization. The current implementation of the CTC++ kernel is more or less sample-based. Each communication instance consists of e.g. writing an integer on a channel. The channel itself makes a copy and a reading process is provided with this copy. In our system we would like to communicate on another scale. A buffer with OFDM symbols is written on the channel. This whole buffer would be copied by the channel and this introduces a performance loss. A solution is to write a reference to the buffer on the channel. This is illustrated in figure 7 for the `OFDM Synchronizer` process and the `OFDM Demodulator` process. The figure shows the two processes and the data structure which has to be communicated.

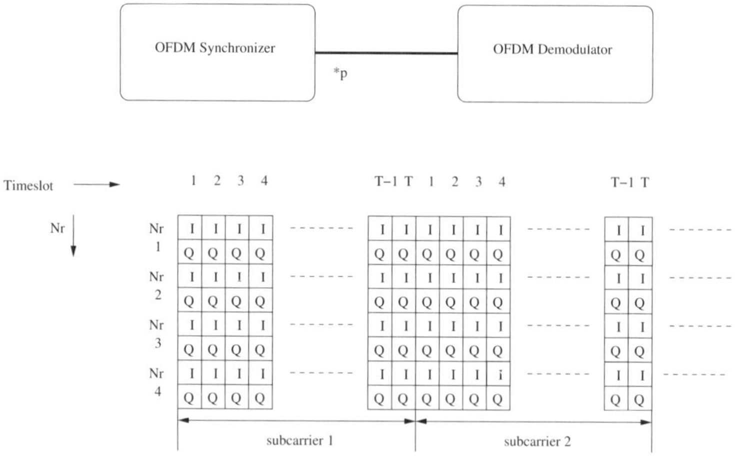

Figure 7: The communication on a channel.

The structure contains around 1 kB of data. A reference to this structure (*p) is written on the channel instead of the entire structure itself. However, this introduces some other difficulties with ownership. Only one process can make use of a data structure at a time. We have not solved these issues, but e.g. semaphores can be used to lock and unlock data structures.

4.2 Process priorities

In CTC++, we can also assign different priorities to the different receiver processes. The CTC++ does not require to assign priorities in the form of numbers. Priorities are assigned on the base of inter-process relations, so only a priority ranking is given. Some processes run on a higher priority as other processes. The priorities of the receiver processes are ranked as follows:

1. The highest priority is assigned to the `OFDM Synchronizer` process. If the synchronization is lost, synchronization has to be acquired again.
2. The second highest priority is reserved for the `OFDM Demodulator` process.
3. The third highest priority is reserved for the `Frame Decoder` process.
4. The fourth highest priority is reserved for the `Space-Time Decoder` process.
5. The fifth highest priority is reserved for the `Channel Estimator` process. Some channel estimates can be missed, because the Space-Time Decoder can also decode with older channel estimates. However, the system performance will degrade.

5 Testbed as an Experimental Platform

The testbed has been implemented and some experiments were made [4]. The receiver software runs on a Pentium 4 2.4 GHz machine and initial tests indicate that the system is able to run in real-time. Figure 8 shows a figure of the experimental setup in our laboratory. The

Figure 8: The experimental setup.

transmitter and receiver are separated a few meters and there is no line of sight component between the transmitter and receiver. Initial experiments have been performed and verify the concept of spatial multiplexing. Two independent datastreams have been transmitted from two antennas and these two have been successfully separated at the receiver. For the experiment, the `Sink` process compares the received data with the data transmitted. The resulting bit error rate performance of the system is shown in figure 9.

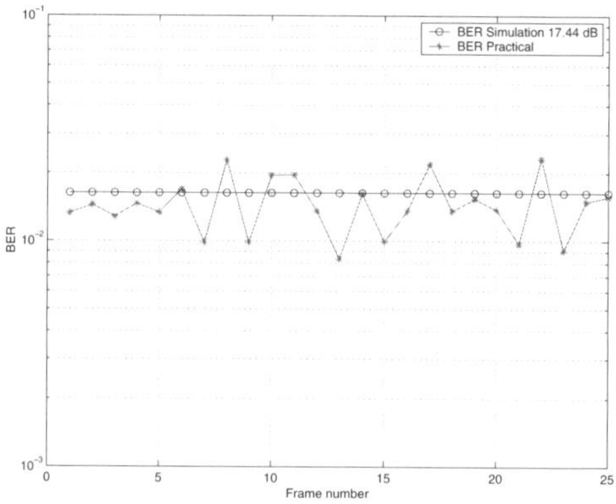

Figure 9: The raw bit error rate performance of the testbed.

In future research we plan to enlarge the channel bandwidth used, so the required amount of processing power increases. A multi-processor platform will be required to satisfy the need for computational power. The inter-process communication has to change and for this linkdrivers can be written.

6 Conclusions and Recommendations

In developing the testbed, the transition from traditional functional modeling to a CSP model is not large. The required functions map easily to CSP processes and the interaction between the processes is easily identified. As an example, we have described the receiver architecture for our testbed. The receiver software has been implemented with the CTC++ kernel and it seems to work functionally correct and testbed has been used to verify the concept of spatial multiplexing.

The use of the CSP and the CTC++ kernel is not only a nice way of system modeling, but also provides us with a scalable system (scalable with respect to the amount of antennas used and the transmission bandwidth). The software written has a proper hardware abstraction layer, which make it highly portable to other platforms. In future, a multi-processor system can be used and only the channel linkdrivers have to be rewritten.

For our type of application a design pattern could be developed and incorporated as an extension to the CTC++ kernel. The design pattern should include facilities for communicating references over channels and ownership management. This functionality could be incorporated into a special kind of channel. Also some additional questions have to be answered. For multi-processor platform e.g., the use of communicating references over channels implies a shared memory architecture. The alternatives to implement such a system, which is a combination of hardware and software design, can be investigated.

The CSP model also gives us the opportunity to mathematically verify the functional correctness of different receiver architectures and this can be done in future.

References

[1] I.E. Telatar, “Capacity of Multi-antenna Gaussian Channels”, IEEE Trans. Commun., vol. 42, No. 2/3/4, pp. 1617-1627, November-December 1999.

[2] D. Gesbert, H. Bölcskei, D. A. Gore, and A. J. Paulraj, “MIMO wireless channels: Capacity and performance prediction”, IEEE Globecom 2000, San Francisco, CA, pp. 1083-1088, Nov. 2000.

[3] A.W.Roscoe, “The Theory and Practice of Concurrency”, Prenctice Hall, 1998, ISBN 0-13-674409-5

[4] H.S.Cronie, “The Design and Implementation of an OFDM Spatial Multiplexing Testbed”, Ms.S. assignment, Signals and Systems, July 2003

[5] G.Hilderink, J.Broenink, W.Vervoort and A.Bakkers, “Communicating Java Threads”, www.ce.utwente.nl/javappp

[6] G.Hilderink, J.Broenink and A.Bakkers, “Communicating threads for Java”, www.ce.utwente.nl/javappp

[7] O.Evers, M.Sandell, J.Van de Beek “An introduction to orthogonal frequency-division multiplexing”, September 1996.

Communicating Process Architectures 2003
Jan F. Broenink and Gerald H. Hilderink (Eds.)
IOS Press, 2003

Real-time and Fault Tolerance in Distributed Control Software†

Bojan ORLIC and Jan F. BROENINK
Twente Embedded Systems Initiative,
Drebbel Institute for Mechatronics and Control Engineering,
Faculty of EE-Math-CS, University of Twente,
P.O.Box 217, NL-7500 AE Enschede, The Netherlands
E-mail: B.Orlic@utwente.nl, J.F.Broenink@utwente.nl

Abstract. Closed loop control systems typically contain multitude of spatially distributed sensors and actuators operated simultaneously. So those systems are parallel and distributed in their essence. But mapping this parallelism onto the given distributed hardware architecture, brings in some additional requirements: safe multithreading, optimal process allocation, real-time scheduling of bus and network resources. Nowadays, fault tolerance methods and fast even online reconfiguration are becoming increasingly important. All those often conflicting requirements, make design and implementation of real-time distributed control systems an extremely difficult task, that requires substantial knowledge in several areas of control and computer science. Although many design methods have been proposed so far, none of them had succeeded to cover all important aspects of the problem at hand. [1] Continuous increase of production in embedded market, makes a simple and natural design methodology for real-time systems needed more then ever.

1 Introduction

Being inherently parallel, control systems need transparent and structured way of using multithreading in a distributed environment. Formal checking based on CSP algebra has already been successfully applied to solve those problems in the *occam* programming language. This programming language was used for design and implementation of the scalable distributed systems constructed from *transputer* nodes and links. After the transputer disappeared from the market, several occam-like libraries [2],[3] were implemented in the popular programming languages. Section 2 starts by emphasizing the need to modify a way in which occam constructs are implemented in a distributed environment. Application model independent of target hardware architecture model is key to efficient and flexible design. However, in occam those two are not totally independent since not all constructs can be distributed. Constructs can actually easily be adapted to distributed form, if they are treated in the same way as all other process. They should communicate with their subprocesses using only channel communication and not knowing their identities. Afterwards, applicability of the CSP based framework in control is explored. In Section 3, various scheduling techniques for distributed occam-like programs are analysed. In Section

† This research is supported by PROGRESS, the embedded system research program of the Dutch organization for scientific research, NWO, the Dutch Ministry of Economic affairs and the Technology Foundation STW.

4, the proposed approach to scheduling is extended to additionally lead to graceful degradation in case of processor overloads and network congestions. Fault tolerance issues and possibilities of system reconfiguration through mode changes are also discussed.

2 CSP application model

Hard real-time control data is usually exchanged over serial fieldbus protocols. The proposed distributed framework is an attempt to conceptually extend the occam-like library made by Hilderink [3], to achieve fault tolerance and real-time guarantees in fieldbus-based distributed control systems. Occam programs are organized as a nested hierarchical composition of constructs and processes. Nesting is allowed by the fact that constructs themselves are also processes.

2.1 The application model should be independent of the hardware architecture model

Separating application and target hardware architecture models of embedded system plays a key role in establishing a high degree of modelling and exploration flexibility [4]. An application model is independent of the target architecture if an application designer can make it exclusively based on the required functionality. Later in the design phase, this application model can be mapped onto the range of hardware architectures. Process allocation can be proposed in order to achieve real-time guarantees, best or fair utilization of all processors and communications.

CSP-based models are claimed to be architecture independent, but this is not quite true because only the parallel construct can be distributed (PLACED PAR). The actual design consists of several hierarchies of CSP constructs and processes that will run in parallel on different nodes.

The most desirable application model is the one where any process (keep in mind that constructs are processes too) can be deployed on any node. Thus all constructs should be able to be distributed in same way as a PLACED PAR from occam. Actually, this is true for almost all constructs. The PRIPAR construct was derived to assign priorities in using a basic shared resource - microprocessor. In a single processor system, PRIPAR priorities will indeed determine the order of execution. However, in a distributed system the role of PRIPAR construct is questionable and it does not make much sense to use it in the usual way. Besides easier readability, maintainability, and reconfigurability, purpose of using the SEQ construct, instead of putting whole sequential behavior inside one process, is so far an unused opportunity for distribution. Different parts of sequential behavior can be executed on nodes where the needed resources are. Sequential behavior can be implemented by using parallel processes, or in this case processes on different nodes, where the end of the first process in sequence triggers the next one and so on. Distributed SEQ organized in this way should not be confused with a *pipeline*, because the first process in sequence can be triggered again only after the last process in the sequence finishes. The ALT construct seems to be the most problematic for distribution. Guards can be attached to remote channels in same way they are attached to internal channels. There is just the additional communication overhead when remote guard becomes ready.

2.2 Constructs are processes too

In practical implementations of occam-like libraries, constructs are often not realized in the same way as other processes and instead of communicating only through channels, function calls are used. This creates a situation in which a parent construct holds pointers to its subprocesses, and each of them in turn has a pointer to its parent construct. When constructs are distributed over several nodes, this hardcoding is the source of many difficulties and artificial solutions must be applied to solve them. Furthermore, the essential reconfiguration power of CSP lies in the fact that processes do not know about identity of processes they are communicating to. Consequently, parent constructs should be protected from knowing the identity of its subprocesses.

A solution is to view the constructs only as processes that take care of the execution of a group of processes. Being a process itself, construct should communicate only through channels. Instead of holding pointers to each other, process and its parent construct need just to be correctly connected through a pair of channels. Thus, instead of using dedicated function calls, every process execution could be triggered over a channel. Similarly, the end of process execution is an event important for its parent construct. Instead of calling the parent construct's function, a process should use a dedicated channel to notify its parent about that event. Since each channel communication is also a potential scheduling point, this solution will be more preemptive in a natural way, eliminating the need for time-slicing algorithms. Not only is this approach more in the CSP spirit, but it will also allow easier distribution of processes over remote nodes. For instance a SEQ construct can trigger its subprocesses in sequence without knowing their location. Less flexible, but more efficient solution is to let each subprocess from a sequential construct, directly triggers next subprocess after it has finished. A distributed PAR construct can trigger all of its subprocesses on remote nodes using some kind of broadcast message and wait till all of its subprocesses finish their executions.

Also to dynamically update the list of nested processes, channels for adding and removing processes from/to a construct can be added to its interface. Those channels would be checked for changes on beginning of every execution cycle.

2.3 *Applicability of CSP application model in control*

In existing control systems, there is often a gap between timing constraints of the control theory model and its practical software implementations. This time inaccuracy can lead to uncontrollable performance degradations [5]. Control system can encompass many control and supervisory loops. Each loop can be further decomposed into the set of mutually dependent cooperating tasks, described using *precedence constraints* (see Figure 1).

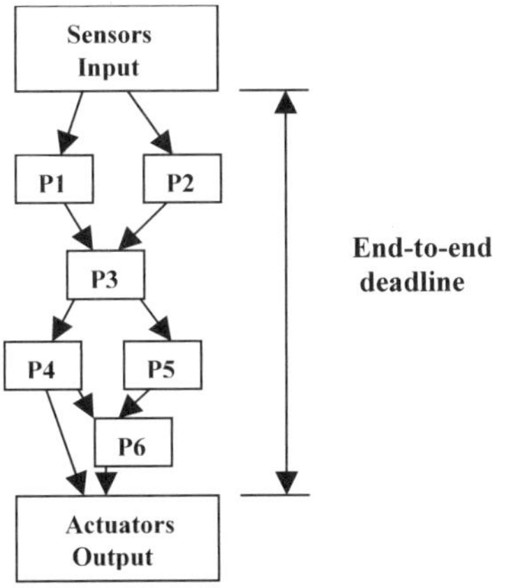

Figure 1: Possible precedence graph for one of control loop chains

Tasks are possibly distributed over different processors, near data sources they use or because some of them need special processors or hardware. Control loop chain starts with sensor data measurements and finishes with delivering command data to actuators. The order of task execution is on each processor determined by the use of an appropriate scheduling strategy. Scheduling requirements are determined by classifying tasks and messages, according to changes in their *utility functions* over time, in several well-known categories. *Hard real-time* tasks require *guarantee* to be given that it will finish its execution before *deadline*. In case of *firm real-time* tasks, there is no point to continue the task execution after missing a deadline. However their execution can occasionally be skipped without disastrous consequences. For *soft real-time tasks* reaching deadlines do not have to be guaranteed and applying only *best-effort* strategy is sufficient. *Non-real-time* tasks have no deadline. This classification sorts tasks based on the importance of achieving deadline. For control systems, except mentioned deadline-oriented tasks, especially important are *precise* [6] or *time-bounded* tasks [7]. *Precisely periodic* task [5] must execute, rather than just to be released, exactly one period apart in predefined time moments. Due to finite clock precision and the time needed for a processor to perform other activities, a small amount of jitter is unavoidable. Larger the jitter less is the utility obtained by task execution (see Figure 2). To reduce influence of other tasks on the size of this jitter, genuine precise tasks should not synchronize on events other then time events.

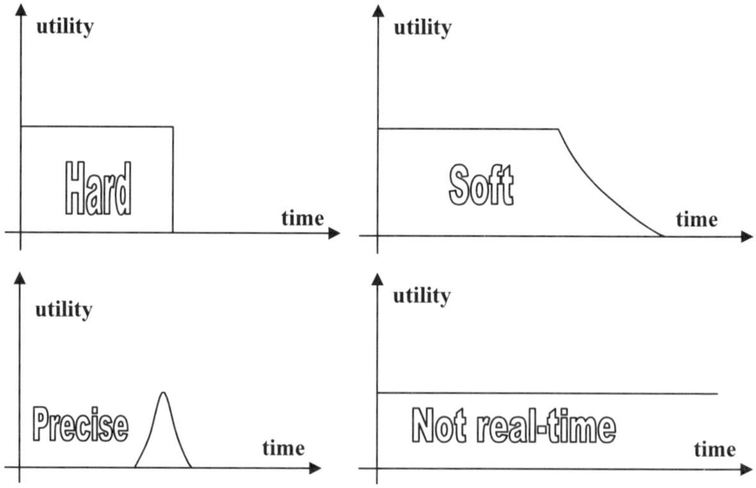

Figure 2: Utility functions for different tasks and message types (adapted from [6])

Although modern control theory is making some advances towards control algorithms that handle time varying delays, most of the contemporary control systems are based on control theory that assumes constant control delays and equidistant sampling [5]. Translated to the task domain, this means that sampling and actuation tasks should be *precisely periodic*. Note that precise and synchronized execution is more important in case of sampling tasks then in case of actuation tasks. The system state variables change value over time subject to physical laws. Jitter in sampling time will thus introduce uncertainty into the measured value. Because an uncertainty is introduced even before control computations are done, jitter in sampling time is more dangerous then jitter in actuating time. Since only a finite amount of energy can be delivered to an actuator in one sampling period, in many cases a little late actuation will still be better then no actuation. This leads to the implementation of actuation tasks as processes which will be released after both input data from preceding computational processes has arrived and a predefined actuation delay time is about to expire.

Often, instead of specifying explicit deadlines to every intermediate task forming a precedence graph of computational algorithm, this behavior is defined only through *end-to-end deadlines* [8]. End-to-end deadline is the only hard deadline for whole chain of tasks inside control loop.

Precedence constraints (see Figure 1) of tasks from each control loop can easily be translated into a hierarchy of CSP constructs and processes. Instead of specifying whole construct hierarchies, the application designer can only specify relationships between processes for which relationships are known. A design tool can then derive relationships that stayed unspecified, like proposed in [9]. Non-uniqueness of this mapping leaves additional space for various optimizations during design process.

Achieving real-time guarantees is most important in implementation of control systems. Especially appealing sub area of research in CSP theory, is attempt to give CSP support for real-time [10]. This real-time CSP theory is still not sufficiently mature for application design purpose. Actually real-time CSP theory can only make time specifications. Execution times of processes will always depend on properties of employed processors and communication links and those are described in the hardware architecture model. It is hard to expect that CSP theory will be extended in such an extent to behave like a simulator of hardware properties and behavior. It seems wise to view real-time scheduling and formal checking for multithreading hazards as totally orthogonal problems. Real-time behavior of systems is target-architecture dependent and rely on appropriate priority assignment scheduling, while occam-like process/channels application models enable the CSP formal check. Applying formal checks can guarantee freedom of *deadlocks*, *livelocks* and *race conditions* during normal system operation. But after component failure, components waiting on communication with failed components will be deadlocked. This situation must be handled by using timeouts and other error detection mechanisms, and by invoking recovery mechanisms possibly based on redundancy.

In order to really simplify system testing and validation by using CSP formal checks, the use of asynchronous communication must be maximally restricted. Furthermore, all parts of CSP based library including scheduler and various communication link receivers & transmitters should be built using construct / process / channel model. Only in this way, it is possible to be sure that overall system will not deadlock.

3 The target-architecture model: Can scheduling achieve real-time guarantees?

Instead of just gluing a communication layer to existing single system architecture, design of distributed real-time systems should be treated in the holistic way. Scheduling in distributed systems is more complicated then scheduling in its single processor counterpart. Reason lies in the fact that processor time scheduling and network resources scheduling are not totally independent of each other.

3.1 End-to-end deadline scheduling in distributed environment

In a distributed system, scheduling decisions are made locally for each node and our hierarchy of dependent occam-like processes spans over many nodes. If processes communicate reasonably often, natural scheduling/descheduling on channel communication will implement approximately fair distribution of processor time. But in real-time systems, fair distribution of processor time would conflict the fact that not all processes have same importance of execution. Some processes contribute more to the system behaviour then others. Some have more stringent timing requirements. The hierarchy of constructs and process encompasses part of the knowledge on process ordering. More precisely it defines which processes are done in a particular order and which are executed in parallel. Creating hierarchy of dispatchers by attaching a dispatcher to each construct works fine in case of a single processor system. Applying the same concept in a distributed system is not so good idea, since each dispatcher would under its control have subprocesses that might be on different nodes. Hierarchy of constructs and processes cannot directly be used to guarantee real-time scheduling. Still, this information reduces non-determinism considering the order of execution on each processor by imposing constraints on valid schedules. Those constraints can be used to transform the end-to-end deadline of each control loop into conservative deadlines attached to each process. Let us designate *worst-case execution time* of any process as WCET and *minimal case execution time* as MinCET. One of possible mappings for a set of processes from Figure 1 to CSP hierarchy of channels and processes is: (P1||P2);P3;(P4||P5);P6. WCET, MinCET and utility functions derived for the application model of that control loop are shown in Figure 3.

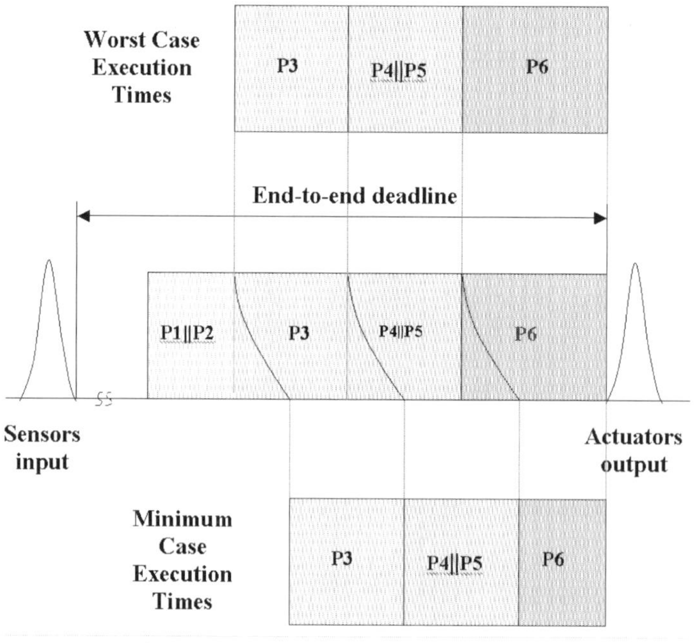

Figure 3: Deriving utility functions and deadlines for one control loop chain

According to [8] in case of a hierarchical approach to end-to-end scheduling, the best strategy is to use distributed scheduling based on both global and local information. In this case *global* information is obtained by offline use of application-specific constructs and process hierarchy to derive intermediate deadlines for each process. Those deadlines could then be applied *locally* for run-time scheduling. The *absolute deadline* is determined by subtracting _WCET of all following processes and remote communications from the time when a control loop chain execution cycle ends. If there are multiple paths through the chain for which real-time guarantee is necessary, the deadline should be calculated considering each path and the worst-case value should be chosen. The *relative deadline* is obtained by subtracting the task release time from the absolute deadline. The simplest and most often applied priority assignment method is *rate monotonic* (RM) approach.

3.2 RM approach

Complex control systems encompass several control loops with possibly different sampling periods. Those control loops might *not* be totally independent. For instance, supervisory loop can occasionally update a parameter inside some control loop. The rate monotonic approach would assign priorities to be proportional to the respective sampling frequencies. This gives the scheduler freedom to execute task anywhere between two consecutive periodic releases of the task. Note that this is somewhat different from *precise in time* execution. To guarantee real-time execution of all tasks, this approach assumes that scheduled processes are mutually independent and that deadlines are equal to the period, which does not hold in this case. Even on a single processor, RM approach can be strictly applied only if each control loop chain is implemented as one process with a deadline equal to its period. Applying rate monotonic scheduling locally on each node and on the network is obviously not identical to the case of applying it to the same software system on a single processor. Priorities are properties defined for each process on the system level, but applied on the local level. If we consider two active processes with different priorities on different nodes, process with lower priority on the less loaded node can be executed before process with higher priority on the overloaded node. Thus, in distributed systems global priority does not determine order of execution.

Even if we could prove schedulability of system designed in this way, this approach would still have a major disadvantage that it does not lend itself naturally to applying fault tolerance methods. Reason lies in the fact that in RM scheduling there is no distinction of importance priorities and deadline priorities. If we try to achieve graceful degradation by organizing a control loop process further as concurrent execution of several process hierarchies with priorities based on importance, priority inversion situation can happen. For instance let us consider two control loop chains with different sampling frequencies running in a RM-based PRIPAR. If the RM approach is used, RM priorities must have priority over importance priorities. Not important lowest priority optional process belonging to more frequently executed control loop, will have a higher priority then essential hard real-time process belonging to control loop executed less frequently. This example of priority inversion, resulting from mixing RM priority assignment and importance priorities, is illustrated in following occam-like pseudo code:

```
PRIPAR
 Control loop 1
  WHILE TRUE
   SEQ
    TIME? samplingTime1
    ...Sensor measurments
    ...Control computations
    PRIPAR
     ...Essential (hard r.t.)
     ...Optional 1
     ...Optional 2
    ...Actuation
 Control loop 2
  WHILE TRUE
   SEQ
    TIME? samplingTime2
    ...Sensor measurements
    ...Control computations
```

PRIPAR
...Essential (hard r.t.)
...Optional 1
...Optional 2
...Actuation

3.3 *Deadline based scheduling*

Whenever the deadline is not equal to the period, modification of rate-monotonic approach known as *deadline-monotonic* (DM) priority assignment can be applied. Instead of getting priority proportional to the sampling frequency, each process will get priority inversely proportional to its relative deadline. The scheduler has now freedom to execute tasks in any time between task release and its deadline. In our case, conservative absolute deadlines can be derived for each process by subtracting the sum of WCET of all following processes from the end-to-end deadline. However, relative deadlines will depend on process release moments, which are dependent on the execution state of all processes in system and cannot be determined in practice. Strict application of deadline monotonic priority assignment is therefore not possible.

With absolute deadlines derived from hierarchy of constructs and processes, applying *earliest deadline first* (EDF) scheduling is possible. This approach is usually avoided because a large amount of overhead is introduced by recalculating priorities, especially in systems with several sampling frequencies. Introducing acceptable constraints can in some cases though lead to more efficient deadline and priority recalculation. Choosing sampling periods in such a way that there is not a to small *biggest common divisor* is usually not too much to ask. This value should not be small, because it will be the *basic period* designated by time reference messages, announced periodically over the fieldbus by a *time master*. Time reference messages must have highest priority on the fieldbus. If the maximum latency time for the highest priority message is not small enough, other ways (like interrupts) must be used to synchronise clocks. Local schedulers will use these messages to perform sampling and to synchronize their timers.

To efficiently use an EDF approach a *least common multiple* of applied sampling periods should be calculated as well. This value will designate the *large cycle* used as time line for expressing deadlines. Deadlines have not to be recalculated all the time. The scheduler knows the current time and the deadlines of processes expressed as offsets from start of the large cycle. A process only needs to recalculate its deadline at the very end of its execution by increasing the last absolute deadline for one period of its control loop. When the deadline value exceeds the large cycle, it is given its first value again. The scheduler always chooses to execute process with deadline nearest to current time. Thus, due to the special properties of control systems, recalculation of absolute deadlines for EDF scheduling can be done in very efficient way.

Advantage of using fieldbus time reference message compared to using local timers is that all sampling will be triggered simultaneously and with a time precision dictated by the most precise processor in system. As explained in section 2.1., synchronized and precise execution is more important for sampling then for actuating tasks. Therefore, while time reference message should be used to determine sampling points, time points of actuating can be determined using local timers. To ensure that sampling is done precisely in time and that sensor measurements can efficiently be distributed to all processes that need it, sampling processes should not synchronize with computational processes directly. Same sensor data, needed by several processes on different nodes, can be broadcasted. Synchronization and triggering control loops can be done in separate *trigger process*. This process execute triggering scheme describing the time execution pattern of the current operation mode. The trigger process is executed after all required sampling processes have finished or after the timeout in case of sensor failure. Thus it can also handle cases when various sensors data is unavailable. To minimize communication overhead trigger process is distributed. On each node part of triggering scheme, concerning only the control loops initiated there, is placed.

To illustrate some of architecture decisions more clearly simple control system will be considered as an example. Let system consists of Control loop 1 and Control loop 2 with sampling periods T and 2T respectively, and Supervisory loop executed every 3T, where T is length of the basic cycle. Required length of one large execution cycle is obviously equal to 6T. More insight into the time patterns of loop executions can be given using some kind of sketch as in Figure 4. The vertical axis shows the flow of time inside one large cycle and the horizontal arrows present points in time when precise sampling and actuation tasks should be executed.

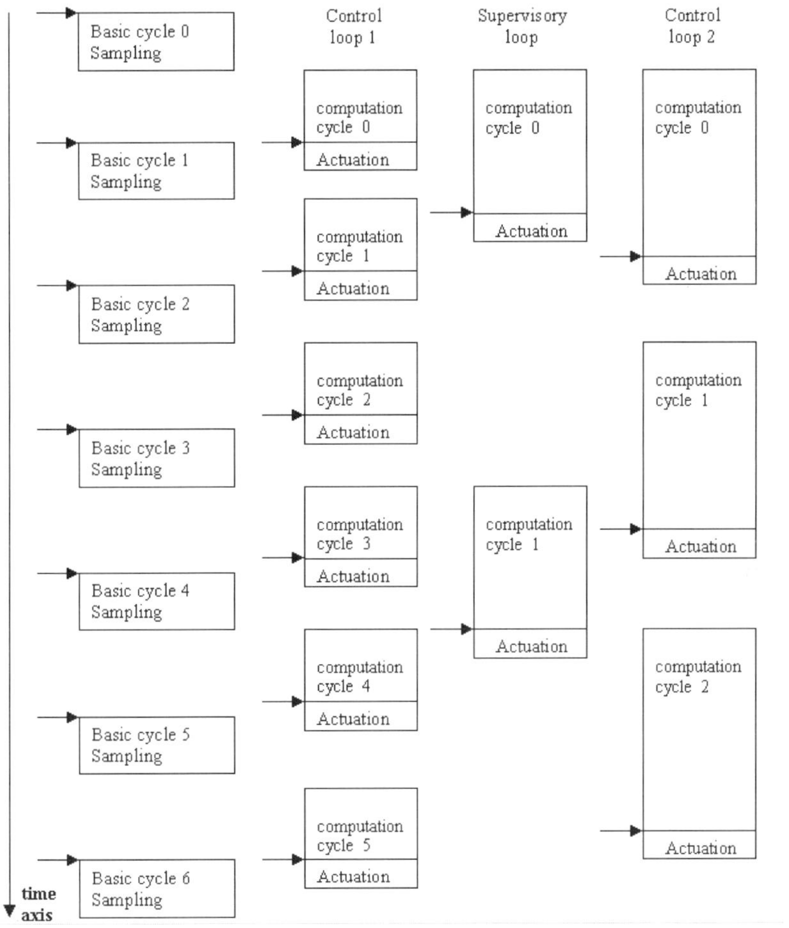

Figure 4: System consisting of three control loops executed on different sampling frequencies

A readable way to describe the functionality of the highest level of the system is using some kind of occam-like pseudo code:

```
largeCycleSize=6
Interrupt Service Routine SAMPLING_ISR[i] {Signal data[i] ready}

PAR
  SEQ
    WHILE(true)
      if((basicCycleNum++)==largeCycleSize) basicCycleNum=0
      FIELDBUS_Broadcast? time ref. msg. {largeCycleNum, basicCycleNum}

      -- after receipt of this message on each node START_CONVERSION(basicCycleNum)
      -- process will trigger sampling of appropriate A/D converters based on
      -- basicCycleNumber; after conversion completion interrupts are triggered

      START_CONVERSION(basicCycleNum)
        CLOCK? SamplingTime
        PAR
          SamplingProcess[0]
            -- wait data[0] ready
            A/D_Conv_register[0]?data[0]
            OVERWRITE_BUFFER[0]!data[0]
          SamplingProcess[1]
            -- wait data[1] ready
            A/D_Conv_register[1]?data[1]
            OVERWRITE_BUFFER[1]!data[1]
           ...
      -- This PAR construct on exit broadcast SamplingOver message

      Trigger process
FIELDBUS_Broadcast? SamplingOver{list Of unavailable Sensor data}
        -- here unique triggering scheme is shown. When not all control loops
-- start on same node, triggering is done only on node where loop starts
  SWITCH (basicCycleNum)
           CASE 0:
             Control_loop1.START!SamplingTime
             Control_loop2.START!SamplingTime
             Supervisory_loop.START!SamplingTime
           CASE 1:
             Control_loop1.START!SamplingTime
           CASE 2:
             Control_loop1.START!SamplingTime
             Control_loop2.START!SamplingTime
           CASE 3:
             Control_loop1.START!SamplingTime
             Supervisory_loop.START!SamplingTime
           CASE 4:
             Control_loop1.START!SamplingTime
             Control_loop2.START!SamplingTime
           CASE 5:
             Control_loop1.START!SamplingTime

  Control_loop 1
     SEQ
       WHILE(true)
         START? SamplingTime
 PRIALT
           PARAMETERS1_UPDATE?par
             parameterUpdate()
           SKIP
          ...Get sampled data from overwrite buffers
          ...CALCULATIONS (hierarchy of distributed CSP processes and constructs)
  CLOCK? AFTER SamplingTime+ControlDelay[1]
          ...ACTUATING
  Control_loop 2
     SEQ
       WHILE(true)
          START? SamplingTime
          ...Get sampled data from overwrite buffers
          ...CALCULATIONS (hierarchy of distributed CSP processes and constructs)
  CLOCK? AFTER SamplingTime+ControlDelay[1]
          ...ACTUATING
  Supervisory_loop
     SEQ
       WHILE(true)
          START? SamplingTime
          ...Get sampled data from overwrite buffers
          ...CALCULATIONS (hierarchy of distributed CSP processes and constructs)
```

```
...PARAMETERS1_UPDATE!par
```

`START_CONVERSION` is dependent of `basicCycleNum` value because not all sensor measurements are performed every basic cycle. This variety of sampling frequencies reduces the possibility to trigger all A/D conversions using a single broadcast signal. A satisfying alternative is to implement this part of code as a short sequence of assembly lines, triggering conversions always in the same order and with the same offset within sampling period. Once the measured data is collected in the registers of A/D converters, the execution order of sampling processes fetching that data is not so important. All those processes can be executed in parallel. Sampling processes that do not require data every basic cycle will have an internal counter to effectively reduce the sampling frequency. Same sensor data can be used by several processes from different control chains. Some of those control loops might have an execution frequency less then the sampling frequency of the used sensor. The sampling process might in that case write data to an overwrite channel or buffer in case all samples should be collected.

Each calculation block represents a hierarchy of distributed mutually dependent processes. Each of those processes has an initial deadline value calculated offline based on WCET times and the end-to-end deadline. On the very end of each process execution, its deadline is recalculated by adding the period. Obtained deadlines are used in local processor and network schedulers to implement EDF policy.

For fault tolerance reasons, several nodes equipped with high precision oscillators are capable to take over the role of the time master in case it fails. If some process misses its relative deadline, this still does not imply failure of the system. Reason is that extra conservatism is added while transforming end-to-end deadline into deadlines of intermediate processes. It is important to perceive that only the end-to-end deadline of the whole control loop chain is a hard deadline. If any subprocess misses its deadline, derived under worst-case assumptions, that still does not imply that end-to-end deadline will also be missed. Deadlines derived for intermediate processes can thus be classified as soft real-time, as it was depicted in Figure 3 using concept of utility functions. From Figure 3 it is possible to conclude that missing end-to-end deadline is unavoidable only after the current process is not completed and time reaches value determined by subtracting _MinCET of all following processes from the end of the loop chain cycle time. At this point, an exception is raised and the system performs what ever fault-tolerance measures are needed to prevent disaster [11].

4 Graceful degradation and reconfigurability

Being mission-critical by definition, hard real-time control systems usually have to encompass some support for fault tolerance. Sometimes, perhaps due to failed components, there is not enough processor time available to perform all computations or not enough network bandwidth to send/receive all messages in time. In same time it is important to observe that not all tasks and messages contribute equally to the overall system behavior. Instead of complete failure, reaction to processor overloads or network congestions must be omitting execution of less important processes resulting in somewhat lower, but still acceptable performance of delivered service. Reconfiguration is introduced due to hybrid nature of the designed system or due to the need for software restructuring as respond to, planned or failure caused, changes in hardware architecture or environment.

Several approaches to graceful degradation can be used:

1) Increasing sampling periods of control or supervisory loops can reduce time needed for application execution. This is possible because the sampling period is usually not determined based only on Shannon's theorem, but also to satisfy some

performance requirements [5]. However, changing a sampling period implies that prefilters must be changed too [12]. Possibly, this can be implemented. Anyway, filter dynamics and sampling periods are parts of system model and thus the whole model must be recalculated. [12] A solution could be to predefine alternative computational algorithms for several values of sampling periods.

2) Simplified alternative processes that need fewer resources can be employed, resulting in degraded quality of service.

4.1 How can graceful degradation be naturally achieved?

The methods described in section 3.4 attach priorities directly to each process and leaves the PRIPAR construct unused. One of the reasons is that PRIPAR priorities can be defined only during design time, and deadline based priorities are changing dynamically. More structured approach would be to derive PRIPAR priorities based on the importance level of process (essential, optional level one, optional level two …). If priorities were derived hierarchically, their values would be dependent of application structure. Although nested hierarchies of PRIPAR processes exist, *importance priorities* should not be dependent of the position in the hierarchy. This means that process importance priority defined inside PRIPAR is also its global importance priority. Essential processes would have the highest importance priority and several additional levels of optional processes can exist in the system. Thus, the lowest importance priority that can be executed in time, will define the level of system performance. Assigning this kind of priorities can help to divide system execution on several layers and accomplish graceful degradation in case of overload. If any essential process needs to be executed, it will always have priority compared to other optional processes. If the system is additionally made as totally pre-emptive, processes from a lower priority level can never jeopardize real-time execution of higher priority layers. Thus, the essential layer can be seen as separate hard real-time system. Several optional levels with high priority would form the soft real-time layer of the application and the rest of the levels would not have deadlines and will form the non real-time layer of the system. Execution of firm processes can be omitted occasionally, but not every time. To prevent starvation, firm process can, in addition to its basic importance priority, experience temporary priority boost each time it is omitted. This boost is cancelled after such a firm process is executed. To keep guarantees for the hard real-time layer unaffected by this, importance priority of firm process after applying priority boost must always be less then the essential priority.

The deadline value and importance priority are encapsulated in each processes, and made viewable to the scheduler. The schedulability is tested offline during the process allocation phase. Some processes will have a fixed allocation due to their connection to specific hardware. Allocation of the other processes can be determined in a way that will guarantee real-time behaviour of overall system. The allocation is the result of a complex optimization process aiming at fair distribution of the load to all nodes and minimizing network communication, especially for high priority processes, while satisfying time constraints. Local schedulers are executing EDF policy with several priority groups according to graceful degradation levels.

4.2 Graceful degradation in the communication layer

Absolute deadlines can also be derived for each remote communication. Priority of messages on the fieldbus will be determined based on the importance priority and its deadline. Dominant part of message priority is the importance priority, which is set to the value of priority of the sending process. In this way, more essential layers of execution will always be given priority over more optional layers. This means that in case of *network congestion*, the ability to transfer messages in time will gracefully degrade. The second part of the message priority will be derived based on its deadline if there is one. For instance, in case of CAN, the message ID is used both for arbitration and addressing purposes. Such a CAN message ID can, therefore, be divided into several fields sorted by relevance: importance priority field, deadline priority field, one bit for specifying whether data or protocol control information is transmitted, one bit for specifying if content is rendezvous message acknowledgment or plain message, one bit for broadcast and two fields needed to uniquely identify each channel: node ID field and channel ID field. By using node identifiers Channel IDs can be assigned locally on each node.

Priority(importance|deadline) (PRIORITYFIELDSIZE)	Protocol/data (1B)	/ack (1B)	broadcast (1B)	nodeID (NODEIDFIELDSIZE)	channelID (CHANNELIDFLDSIZE)

Main principles of communication subsystem are to encapsulate package protocol details in a passive link driver[13] object and all hardware specific functionality in a passive device driver object. Link drivers are plugged into the remote channels transparently of user process. In this way, the process using a channel does not know whether that channel is remote or local. Each remote channel has a link driver object and there is one unique device driver object per node. Any two link drivers speaking the same language (having same package protocol) can communicate. Interrupt service routines initiated by the hardware side of communication subsystem will only release blocked processes: producers or consumers to allow them to continue executing write () or read () function of link driver. Synchronization to access the communication resource (like CAN transmit mailbox) as well as all other device specific functionalities and register configurations are hidden in the device driver object. The prioritization in accessing communication resources is done in a natural way because semaphores sort waiting-queues based on priorities of blocked processes.

4.3 Reconfigurability

Many control systems are *hybrid* and encompass several *modes of operation* depending on current place in the system's state-space. For instance, in airplanes separate modes can be defined for: take-off, normal flight, landing and various emergency situations. In systems with several modes of operations, each process can be used in one, several or all modes. In each mode, the same process can be employed inside different configuration and with different priority level. For every mode, separate hierarchy of constructs is predefined. Only one of several alternative hierarchies can be active at the same time. Thus, mode change will trigger fast system reconfiguration by choosing an alternative hierarchy of constructs to be executed in the next cycle. Information on the current mode or information on the mode change must still be passed to a process whenever it is triggered. This is another situation in which the described solution using start channels to trigger process execution from its parent construct appears to be useful. In each mode, different levels of graceful degradation are defined by designating different priorities to system parts. During the design phase, each mode must pass a schedulability test, a CSP formal check and a control system performance simulation. Once the system is designed in this way, a central reconfiguration management process (possibly being implemented as human operator) can only trigger mode changes, which will be performed automatically.

More flexible online reconfiguration can be achieved by implementing a reconfiguration tool on a remote node. This tool would monitor program execution and perform optimization techniques for load balancing and it would execute schedulability tests and formal check all potential alternative configurations. Due to the design principle, described in section 3.1, that processes are not hard-coded into the parent constructs, hierarchies of constructs and processes could be modified easily for inactive modes. New modes could also be defined, tested and loaded online. Such software system would be able to respond without restarting to major planned changes in software as well as in its environment structure. New nodes could be added and employed easily; nodes that need maintenance would be removed from system through planned reconfiguration and possibly transient period of system performance in appropriate level of graceful degradation.

4.4 Fault tolerance issues

Fault tolerance is important aspect of mission critical control systems, and it must be considered during design of such a system in as much as possible natural way. Reliability of control systems is usually increased by choosing robust control laws, non-sensitive on order of operations or finite precision of calculations, or by calculating and reducing the sensitivity of control loop on time delay, inserting redundant sensors or actuators into system, using sensors to check if actuators had performed requested action [12]. But besides this design time approach, additional software methods are needed to enable run-time error detection and determine how the system should respond in case of error. Therefore, the next research area would be to investigate possibility and adjustments needed to apply well-known fault tolerance techniques in the context of this framework. Refining every PAR construct to PRIPAR, by defining importance priority of each process, will divide the system into several layers of graceful degradation. But this is not enough. Fault tolerance must be taken care at all levels in the system architecture: on level of signals, data structures, processes, channels, constructs, nodes, network, cluster of nodes and on global system level. On each level, appropriate error detection and recovery mechanism can be applied and a minimum level of service is defined to distinguish operational and failed components.

Every primitive process is sequential and every sequential execution can easily be divided into one or several *recovery blocks*. Recovery block [1] including several concurrently executing processes is much harder to implement. After error is propagated through inter-process communication, all processes participating in that communication can be affected. Rigorous application of recovery blocks in primitive processes and sequential constructs will minimize the probability of error propagating through channels. However, this might *not* be enough, since acceptance test might need to check invariants including global state variables defined out of the scope of such recovery block. In a software system organized as hierarchical layers of communicated processes, the global state is distributed locally in every process. While sequential execution can be considered to change the state in deterministic way, concurrent execution yields different patterns of state traces, depending on actual order of execution. The correct point for executing acceptance test in PAR constructs is immediately after all inner processes in that parallel construct have finished execution and the final state of the current execution cycle is reached. One of possible implementations is that each process makes part of its internal state viewable via an asynchronous channel. If the acceptance test fails, all processes from the recovery block reload previous state. Acceptance tests must contain knowledge of state variables relations as well as which processes are maintaining needed state variables. Thus, those tests must be defined separately for each mode and redefined whenever some operation mode is redesigned. If processes from a PAR construct can be partitioned into groups with no communication between groups, separate recovery block can be defined for each group. Only for those groups that did not pass the acceptance test, alternative execution is started.

Processes or constructs must report their failures to their parent constructs using status messages in their end channels. This is another important factor in deciding to use end channels instead function calls to notify parent processes of finished execution. Service status messages can deliver combination of error flags: *value error* in case of failed acceptance test or *timeout error* in case of missed deadlines. If the detected error cannot be handled locally, it is propagated to higher levels in hierarchy. When the error reaches the highest level and cannot be handled, some emergency mechanism is used to get the system to a *safe state*. If the part (node, sensor, actuator, network) that needs maintenance is identified, the problem is solved by excluding that part from system operation and replacing its functionality by activating passive replicas on other nodes.

5 Design tool requirements

A designer must be supplied with a tool that offers a simple transformation path from complex requirements to a simple, safe and testable implementation. A control algorithm and its implementation represent two views of the same system model. Therefore, the strength of the used design methodology would especially be improved if the design tool is made in combination with simulation and control design tools like Matlab, Simulink or 20SIM. In those tools, system models and control algorithms are traditionally designed using a functional block diagrams. When translating this to CSP, each block could be turned into a process, and passing data between blocks would become communication over CSP channels. In the implementation phase, the tool would take as inputs this occam like structure derived from the control algorithms. Real-time requirements of model would be refined by specifying sampling periods and end-to-end deadlines for each control loop chain. To ensure that automatically generated code is safe, design tool must be equipped with a CSP-based formal checker. Combining this with some graphical modelling language, like GML proposed by Hilderink [9], would make the design process even more natural and easy. CSP models should include this library's underlying multithreading kernel model and network protocol models. The application level design methodology could be further improved by translating existing design patterns to occam-like application models and constructing new CSP design patterns as guide to solving typical problems in the real-time control application area.

The design tool should be able to recalculate reliability and other dependability features of the system. Refining every PAR construct to PRIPAR, by defining importance priority of each process, will divide system into several layers of graceful degradation. For any signal, the allowable range and maximum rate of change and additional ways of protection could be specified. A library of fault-tolerance enabled components and design patterns can be made to facilitate automatic code generation. For instance, replicated PAR components can be made to ensure replica's synchronization and consistency. Also based on given invariant constraints, a tool could propose how to construct recovery blocks and generate code automatically. Influence of various faults and component failures should be easily simulated, leading to discovery of weak points in system.

Architecture model should describe used hardware components. Programmable processing node models (memory model, instruction execution times, cache…) and detailed fieldbus models should be made. Using those models, time and reliability behavior of single application model could be simulated for range of target architectures. Based on those simulations and analysis, appropriate architecture and optimal process allocation would be chosen. It would be possible to validate complete system by refining application level simulation considering influence of chosen architecture. For instance, special libraries like TrueTime and Jitterbug [14] can be used to explore influence of determined worst case time-delays on control algorithm, giving us opportunity to adapt control algorithms.

6 Conclusions

In this paper, several design principles were proposed leading to consistent, designer friendly, CSP-based architecture for design of fault tolerant, distributed, hard real-time control systems. Avoiding hard coding of processes inside parent constructs leads to enhanced flexibility and easier distribution. The proposed priority assignment method integrates graceful degradation and real-time scheduling. This approach tries to make a unified view on several distinct important areas in real-time control software design. It expresses need to treat design of distributed real-time systems as whole and not as gluing real-time communication layer to existing systems. Properties of design tool needed to achieve this unified design view are explored.

References:

[1] Burns A., Wellings A., *Real-Time Systems and Programming Languages Ada95, Real-time Java and Real-Time POSIX.* Third ed. INTERNATIONAL COMPUTER SCIENCE SERIES, ed. M. A.D. 2001, Essex, UK: Pearson Education Limited.

[2] Welch, Peter H. *Java Threads in the Light of occam/CSP.* in *Architectures, Languages and Patterns for Parallel and Distributed Applications, WoTUG-21.* 1998. Amsterdam: IOS Press.

[3] Hilderink, G.H., Broenink, J.F., and Bakkers, A.W.P. *Communicating threads for Java.* in *Proc. 22nd World Occam and Transputer User Group Technical Meeting.* 1999. Keele, UK.

[4] A.D. Pimentel, P. van der Wolf, E.F. Deprettere, L.O. Hertzberger, J. T. J. van Eijndhoven, S. Vassiliadis. *The Artemis Architecture Workbench.* in *Progress workshop on Embedded Systems.* 2000. Utrecht, the Netherlands.

[5] Wittenmark B., Nilsson J., Torngren M. *Timing problems in Real-time control systems.* in *American control conference.* 1995. Seattle.

[6] Buttazzo, Giorgio C., *Hard real-time computing systems: Predictable Scheduling Algorithms and Applications.* The Kluwer international series in engineering and computer science. Real-time systems. 2002, Pisa, Italy: Kluwer Academic Publishers.

[7] Bakkers A.W.P., van Rooij R.M.A., James L . *Design of a real-time operating system (RTOS) for robot control.* in *OUG-7 Technical Meeting.* 1987. Grenoble, France.

[8] Liu, Jane W. S. *Issues in distributed real-time systems.* in *IDA Workshop on Large, Distributed, Parallel Architecture of Real-Time Systems.* 1993. Fairfax, Virginia.

[9] Hilderink, Gerald H. *A graphical Specification Language for Modeling Concurrency based on CSP.* in *Proc. Communicating Process Architectures 2002.* 2002. Reading, UK.

[10] Schneider, S., *Concurrent and Real-Time Systems: The CSP approach.* 2000: Wiley.

[11] Orlic B., Broenink, J.F. *CSP channels for CAN-bus connected embedded control systems.* in *Progress 2002 Workshop.* 2002. Utrecht.

[12] Astrom, K.J., Wittenmark B., *Computer-Controlled systems : theory and design.* 3rd ed. 1997: Prentice Hall, Inc.

[13] Hilderink, G.H., Bakkers, A.W.P., and Broenink, J.F. *A Distributed Real-Time Java System Based on CSP.* in *The third IEEE International Symposium on Object-Oriented Real-Time Distributed Computing ISORC 2000.* 2000. Newport Beach, CA: IEEE.

[14] Cervin A., Henriksson D., Lincoln B., Eker J., and Arzen K., *How Does Control Timing Affect Performance?* IEEE Control Systems Magazine, 2003. **23**(3): p. 16--30.

Communicating Process Architectures – 2003
Jan F. Broenink and Gerald H. Hilderink (Eds.)
IOS Press, 2003

occwserv: An occam Web-Server

Fred BARNES
Computing Laboratory, University of Kent, Canterbury, Kent. CT2 7NF
frmb2@kent.ac.uk

Abstract. This paper presents 'occwserv', the occam web-server. This is a highly concurrent web-server, written in the occam multi-processing language, that supports the majority of the HTTP/1.1 protocol. Dynamic process and channel creation mechanisms are used to create scalable 'server-farms', each responsible for a particular web-server function — for example, reading client requests or running CGI processes. The design of the web-server is presented, along with some early performance benchmark results. Although performance may appear a limiting factor (when compared to other web-servers such as Apache), much is gained from the simplicity and security of occam. Extending the web-server with new functionality, for example, is intuitive and largely trivial — with the guarantees that code is free from race-hazard and aliasing errors. An experimental non-standard addition, the OGI (occam Gateway Interface), is also presented. This provides a mechanism for dynamically loading and attaching pre-compiled occam processes to the running web-server, that can then handle one or multiple client connections. A text-based style adventure game is examined briefly, that allows multiple clients to interact within a "multi-user dungeon" (MUD) style environment.

1 Introduction

Web-servers are naturally concurrent applications — they must be capable of processing more than one client connection/request simultaneously, if to be of any practical use. Theoretically, from the web-server viewpoint, serial processing of clients would provide the greatest degree of efficiency. However, this requires that clients (and the network) are capable of absorbing data at the maximum rate generated by the web-server — rarely the case in practice. A more typical situation is one in which the web-server writes data simultaneously to multiple 'slow' clients, and this model reflects the typical 'real world' situation, where the bandwidth available to end-users (using web-clients) is generally significantly less than the bandwidth available to web-servers (often housed in dedicated 'server warehouses').

Thus, concurrency is a requirement, but may be provided for in several different ways. For web-servers written in C, such as Apache [1], handling of multiple clients is performed using a mixture of *threads* and the '`select()`' system-call (that performs I/O multiplexing). The cost of '`select()`' is generally $O(n)$, where n is the number of descriptors involved. At some point, this overhead becomes significant, hence the use of multiple threads, each performing '`select()`' on a manageable number of descriptors. An extensive overview of the 'event-dispatch' mechanisms of Linux (as of 2001) can be found in [2].

However, programming concurrency this way is neither intuitive nor easy, and has a huge potential for error. The occam multi-processing language [3], based on the CSP process algebra [4, 5], provides the ideal environment for developing concurrent applications — particularly in that it safeguards against common aliasing and race-hazard errors.

The KRoC/Linux [6] system, on which this work is based, was initially unable to provide general socket I/O — due to a lack of support for handling multiple blocking system-calls (because KRoC runs as a single UNIX process, any blocking call performed by an occam process would suspend the entire program). The addition of blocking system-call support [7] has removed this limitation and provides a general mechanism for executing C code in another thread, thereby allowing the scheduling of occam processes to continue normally.

An initial version of the occam web-server was presented during a fringe session at the CPA-2001 conference in Bristol [8]. This version of the web-server was entirely functional, but somewhat limited. In particular, it lacked the ability to scale dynamically, using fixed-size process farms. Additionally, the sharing of channels was handled using a user-defined 'SEMAPHORE' type [9] combined with explicit suppression of aliasing and race-hazard checks for the shared channel.

This paper presents version 2 of the occam web-server, that makes extensive use of new language features that provide the missing functionality. Particular use is made of mobile *channel-types* and dynamic process creation using the 'FORK' mechanism [10, 11]. The new version is a significant step forward from the original, now providing a highly concurrent *dynamic* web-server, that scales automatically with demand (and whose implementation remains secure against aliasing and race-hazard errors).

The web-server supports the HTTP/1.0 and HTTP/1.1 protocols [12, 13], but lacks support for HTTP over SSL — due largely to the lack of an (Open)-SSL [14] (secure sockets layer) binding for occam.

1.1 Related Work

The quality of a web-server is typically measured in terms of its performance at processing client requests. This represents a very real-world requirement — web-servers are expected to be able to handle the demand placed on them. The effect of this is that producing web-servers that are able to compete in the same league as Apache (the web-server which most will be benchmarked against), is *hard*. In particular, implementations in interpreted languages suffer enormously when placed under high-load.

Two web-servers that are of interest are the Haskell Web-Server [15], and YAWS [16], a concurrent web-server written in Erlang. Both of these present novel approaches to web-server implementation, in the typically functional languages Haskell and Erlang. A significant gain comes from the functional abstraction of these languages — code is automatically safe from race-hazard and aliasing errors, and the necessary concurrency mechanisms are provided by the run-time systems [17, 18].

However, for both these web-servers, there is quite some distance between design and implementation — particularly in concurrent Haskell. This is largely due to the fact that these languages were originally designed to be functional, not concurrent languages. occam, on the other hand, was designed to be concurrent (using the CSP model), and thus makes a significantly more natural language for expressing concurrent systems such as web-servers (provided they can be modelled in CSP).

From another viewpoint, functional languages are generally highly dynamic, utilising mechanisms such as lazy evaluation and garbage-collection [19]. A comparable level of dynamic behaviour has only recently been added to occam, largely in the form of dynamic process and network creation/re-configuration. Digressing slightly, this leads to an interesting question: is a dynamic occam useful for implementing run-time systems for functional languages ? — with extensive use being made of dynamic processes and mobile channel-ends.

Another type of web-server, that is becoming increasingly common (largely through its inclusion in Linux), is the TUX Linux-kernel embedded web-server [20, 21]. This uses optimised (and direct) interaction with the Linux file-system and networking code, resulting in significant performance gains. An analysis of TUX (1.0) is given in [22].

1.2 Paper Overview

Section 2 describes the design (and parts of the implementation) of 'occwserv', with each core component of the web-server examined separately. The performance of the web-server is examined in section 3. It should be noted, however, that performance is not the only objective of the occam web-server — clarity of design and simplicity of implementation are of equal importance, since these affect the long-term maintainability of the software (scalability in particular).

Section 4 presents some initial conclusions and discusses future avenues of research. Of particular interest is improving the performance of the web-server, that is bottle-necked in several critical places.

2 Design and Implementation

Figure 1 shows the top-level design of the occam web-server — a network of concurrent components that communicate client 'connections'. Also shown are the socket interactions performed on connected clients.

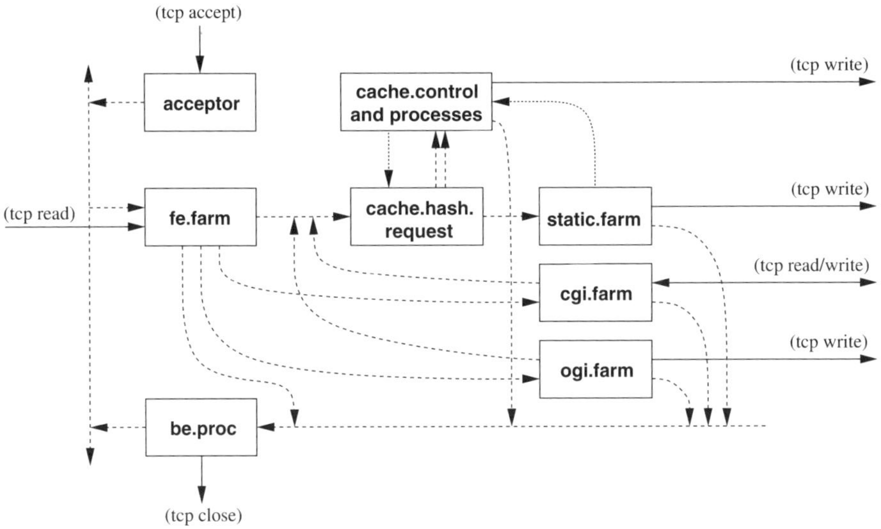

Figure 1: occam web-server main top-level processes

New connections originate in the 'acceptor' process and are communicated to the front-end process farm, 'fe.farm'. The function of the front-end is to simply read the client request, then pass the connection on to another part of the web-server based on the request.

Requests for static content are passed through the 'cache.hash.request' process (that may divert them to a cache-handling process), into the 'static.farm'. Processes within the 'static' process farm simply copy the contents of a file to the client, before passing the connection to the back-end ('be.proc').

The back-end process either closes the connection or, if marked as 'keep-alive', sends the connection back into the front-end farm for another request. Statistics are also collected at this point, and communicated to a separate process (not shown).

Requests for CGI scripts (the execution of a process whose output is returned to the client), are sent to the 'cgi.farm' process. The operation of this is similar to the 'static' farm, where a separate 'worker' process exists for each active client connection.

The 'ogi.farm' process network is used to handle requests for OGI ('occam gateway interface') modules — deliberately similar in name to 'CGI' (common gateway interface). OGIs are an experimental addition that enable pre-compiled occam processes to be dynamically loaded and attached to the running web-server. Furthermore, an OGI module can specify that *all* requests for a particular URL should be directed to that instance of itself. This has several significant advantages over the CGI and similar mechanisms, discussed in section 2.4.

2.1 Front-End Farm

The front-end farm is responsible for reading client requests, that determines where the connection will be sent next. The process 'pool' is managed by the 'fe.farm' process, whose purpose is to ensure a readily available supply of 'fe.process' worker processes, as shown in figure 2. Note that the incoming connection channel (from the 'acceptor') is given to the worker processes directly — i.e. connections are not farmed out by 'fe.farm', it simply ensures that enough free worker processes are available.

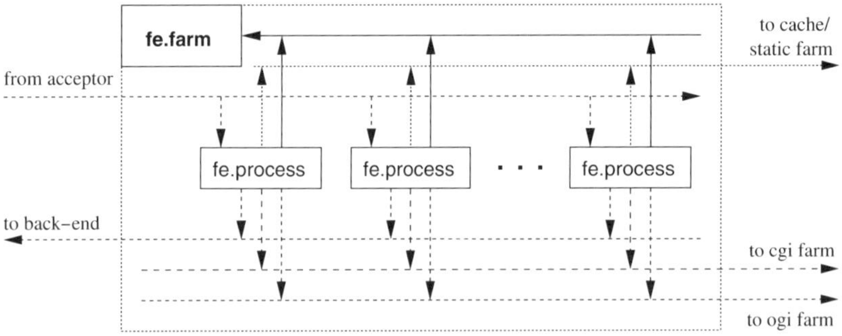

Figure 2: Front-end process farm

The 'fe.farm' process maintains a count of the number of idle worker processes, initially zero. Following an idea suggested by Welch and presented in [23], the body of 'fe.farm' is simply a loop — that first makes sure that there are *n* idle workers (FORKing more if necessary), then performs an input from the shared channel connecting it to the workers. When a worker process starts processing a request, it communicates the value -1 on the shared channel. When it finishes processing a request, $+1$ is communicated. The value input inside 'fe.farm' is added to the idle-worker count. Thus, whenever a worker process starts, this count is decremented, and incremented again when the worker finishes, with 'fe.farm' constantly ensuring the availability of *n* workers.

Internally, the 'fe.process' worker process invokes 'line.reader' — a PROC whose purpose is to read lines of data (the request) from the client socket. Parts of the request that are of interest (specifically the 'action') are stored within the mobile connection structure, passed as a parameter.

Once the request has been read, the 'fe.process' passes the connection to another part of the web-server, based on the file (name and path) requested. Requests that contain the sub-string '/../' are dropped, as are those that time-out (if read timeouts are enabled).

Requests whose path starts with the configured 'CGI' path (typically '`/cgi-bin/`') are passed to the '`cgi.farm`'. Similarly, requests that start with the configured 'OGI' path (currently just '`/ogi/`') are passed to the '`ogi.farm`' process for handling. All other requests are passed towards the '`static.farm`', via '`cache.hash.request`'.

2.2 Serving and Caching Static Pages

The original version of the occam web-server had no support for page caching. For frequently accessed files, caching can be an advantage, up to the point where the frequently requested set of files exceeds the cache size (where cache effectiveness starts to degrade, and may even become an overhead — if pages are repeatedly cached and un-cached).

Before sending a connection to the '`cache.hash.request`' process, the front-end farm generates a hash-code for the request URI, that it stores inside the connection structure. The '`cache.hash.request`' process maintains an array of hash-codes for currently cached pages, and is connected to a fixed number of '`cache.process`'s. When a connection is received, the hash-code array is scanned to find a match. If found, the connection is sent to the appropriate '`cache.process`', otherwise the connection is forwarded to the '`static.farm`' for ordinary processing.

The '`cache.process`' processes and '`cache.hash.request`' process are managed by the '`cache.control`' process, as shown in figure 3. The '`cache.control`' process controls what particular pages get cached, based on information sent from the '`static.farm`' (requests for pages that were not cached) and the cache processes themselves (requests for pages that were cached).

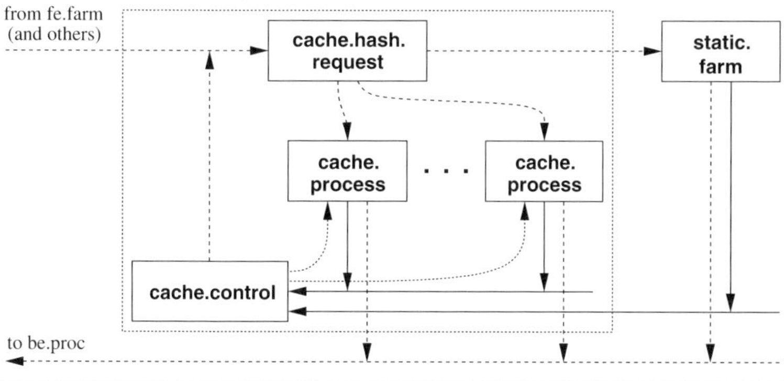

Figure 3: Cache control and processes

Internally, '`cache.control`' maintains an array of cache-entry information, twice as big as the number of actual '`cache.process`' processes. The entries are sorted (via a separate 'lookup' array) based on a 'count'. This value is incremented by page hits and reduced over time, but is always kept between 1 and the number of cache processes — that prevents eviction through aging alone. The top-half of the sorted array maintained by '`cache.control`' contains pages that are actually cached. The bottom-half of the array contains pages that are not currently cached, and acts as a 'buffer' area where pages compete before being loaded into the cache proper. Whenever a page joins or leaves the 'top-half' (cached pages), appropriate messages are sent to the cache processes and '`cache.hash.request`' process. The implementation of this is done particularly carefully, in order to avoid deadlock and to prevent connections being sent to cache processes that are unable to handle them. Some use of the extended-rendezvous [10] is made by various components to achieve this.

Pages that are not cached are simply passed to the 'static.farm'. This is a process farm that operates in very much the same way as the 'fe.farm', but with incoming connections being shared amongst 'get.page' processes. The job of 'get.page' is to simply open the requested file and send the contents back to the client (prefixed with standard headers), before closing the file and passing the connection to the back-end ('be.proc'). The Linux-specific [1] 'sendfile()' system-call is used to transfer file-contents to the client, that performs the transfer in the OS (Linux) kernel to avoid the unnecessary copying of data to and from user-space.

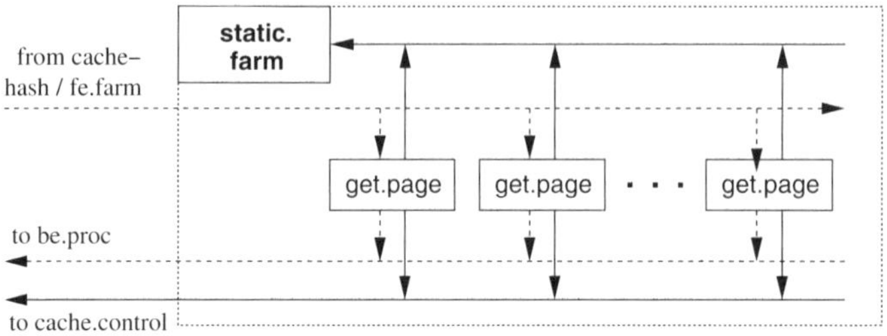

Figure 4: Static-page process farm

The 'get.page' processes are also connected back to the 'cache.control' processes, to which they send information about missed cache requests, as shown in figure 4. If the requested file cannot be opened (in 'get.page'), the request is re-written to '/error....html', with the dots replaced by the appropriate error-code (404 for 'file not found', for example).

2.3 Running CGI Scripts

CGI (Common Gateway Interface) [24] provides a mechanism for running programs, scripts in particular, from within a web-server whose output is returned to the (remote) client. This can be used from anything such as interfacing legacy applications via shell-scripts or C programs, to generating dynamic content from databases using PHP [25]. Interaction is still strictly request-response, but the response may contain substantial amounts of embedded code (such as Java-Script [26]).

The 'cgi.farm' process-farm is constructed in the same way as the 'fe.farm' and 'static.farm' networks, using 'cgi.page' processes to handle individual requests. The occam 'process' library [27] is used to execute CGI processes, that have their 'standard-output' connected directly to the client socket.

If the requested program/script cannot be run (because it does not exist or cannot be executed), the connection is passed to the static handlers (via the cache), with the request modified to return an error-file.

A reasonable amount of effort is involved in setting up environment for a CGI script (as per the CGI specification), but is relatively straight-forward. The occam process library handles the execution and termination of other programs transparently. All the application ('occwserv' in this case) need do is call 'proc.run', giving the path to the executable, an environment array and a set of file-descriptors for communication (one for each of the streams 'stdin', 'stdout' and 'stderr'). A more occam-friendly interface is available through 'proc.wrapper', that uses BYTE channels for communication instead of UNIX file-descriptors.

[1] The 'sendfile' system-call, that copies data between file-descriptors within the kernel, exists on various UNIX/POSIX systems, but often with different semantics and restrictions.

2.4 OGI 'plug-in' Modules

Although CGI provides a versatile mechanism for interfacing with arbitrary programs via a HTTP interface, it does have limitations. The most restrictive of these is performance — creating and executing a new OS process is expensive. Since this overhead is generally constant, the impact on small (quick running) CGI scripts is large, particularly noticeable when benchmarking (section 3). Long-running CGI processes, for example, performing complex remote database queries that may take several seconds to execute, tolerate the process startup/shutdown overhead more readily. For such cases, performance in benchmarking should depend on the efficiency of the remote server(s) — if the web-server is simply acting as a 'gateway'.

For Apache, some of the CGI performance issues are solved by building script interpreters into Apache run-time loadable modules. Two examples are the Perl and PHP modules, that interpret those scripting languages from within the web-server, without requiring a separate OS process. This results in a significant performance improvement when 'executing' CGI scripts of these types. There is, however, no protection between Apache and its modules — an erroneous module could easily take down the web-server (or more seriously, compromise the security of the system).

The other significant limitation of CGI scripts is the handling of persistent (client) state. Many applications interfaced through the CGI mechanism require client state to be maintained in some way. For an on-line ordering system, this might be the contents of the remote client's 'shopping-basket', or the current 'state' in an on-line banking transaction. Current implementations range from storing the whole state in the client (by ping-pong-ing the state with each request-response), to maintaining the state in the server and providing clients with a *session* identifier, that is used to access the server-stored state for subsequent requests. This state can either be stored on disk between requests, or use can be made of a dedicated back-end server that retains the state in memory, with CGI scripts that simply communicate data to and from this server (assuming it generates HTML or other suitable output).

Although these approaches to managing client state do work, they have their own limitations. In a transaction based system, for example, a 'false' client could send a number of requests concurrently, that each modify the server state in a different way. The outcome in these cases should ideally be predictable — an error message, for instance, or the successful completion of one and appropriate 'state consistency' errors generated for the others. In general, interaction with transaction based systems are (per-client) sequential, and this must be enforced somehow — by locking the server-held client 'state' during the transaction, for example.

Ultimately, CGI can be complex to program, particularly if separate requests must 'interact' with each other (in a real-time on-line transaction, for example). The general non-concurrent approach to CGI programming introduces its own problems, limited scalability in particular.

2.4.1 Introduction

For the occam web-server, the OGI (occam Gateway Interface) has been introduced. Instead of executing 'common' code behind the web-server, pre-compiled occam code is dynamically loaded into the web-server — as a concurrent process network in its own right. In a similar way to Apache modules, 'occwserv's OGI modules become part of the web-server handling connections internally. Furthermore, an OGI can specify (when loading) that it handles *all* requests for that OGI — i.e. the OGI can, if it wishes, handle multiple concurrent connections simultaneously — or it can serialise them (simply by not accepting a new connection into the OGI until an earlier one has left). Within a concurrent system, however, concurrent client interactions are much easier to manage. Furthermore, the OGI can remain loaded (when it specifies that *all* requests should be directed to it). This almost instantly solves much of the state-retaining problem, the OGI simply stores the client state in variables/processes and clients use a form of '`session`' identifier to access that state (the occam web-server currently does not support 'cookies' — state that is stored in the client but transfered in HTTP headers rather than HTML code).

2.4.2 OGI Process Handling

Figure 5 shows the structure of the '`ogi.farm`' process network in the web-server. Currently, a fixed-size pool of '`ogi.handler`' processes is used, but this could be made dynamic in the future (the required support for some nested `MOBILE`s has only recently been added to occam).

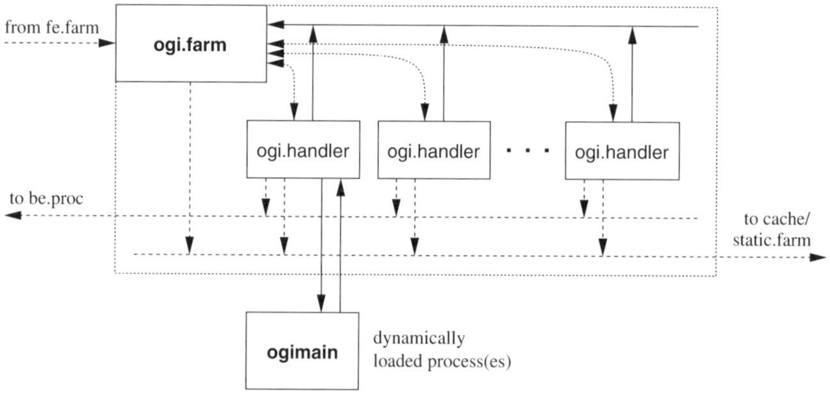

Figure 5: OGI process network

Unlike other process farms, connections are input by the '`ogi.farm`' process, then passed to the worker processes. The main body of the '`ogi.farm`' process, after creating the worker processes, forever `ALT`s between incoming connections (from the front-end farm) and 'signals' from the worker processes (using a shared local channel).

The '`ogi.handler`' worker processes exist in one of three states, that is maintained in the '`ogi.farm`' process: idle; busy and not accepting connections; or busy *and* accepting connections for the same OGI (matched using the connection's hash-code).

When a new connection is received by the '`ogi.farm`' process, the internal state is scanned to see if any OGI handler is currently accepting all requests for that OGI. If so, the connection is simply forwarded to that worker process. Otherwise, the internal state is re-scanned to find an idle worker. If none can be found, the connection is passed to the static farm (via the cache) to return an error to the client[2].

If a free worker is found, the connection is forwarded to that worker, but instead of looping back to the `ALT`, the '`ogi.farm`' process waits for a response from the worker to which the connection was sent. The response is only sent by a worker after it has started the actual OGI process, and indicates whether the OGI will handle the single connection given to it *only*, or will handle *all* requests for that OGI. This information is used to update the internal state of '`ogi.farm`', that then loops back to the `ALT`. If the OGI failed to load (signalled

[2]With an unbounded dynamic OGI process farm, the error of 'no free slots' will not occur — until memory or other system resources are exhausted.

in this response), the connection is passed to the static farm (to return an error) and the 'ogi.handler' process returns to its idle state.

The shared 'signal' channel is used by the 'ogi.handler' processes to notify 'ogi.farm' when they have finished/terminated, sending their 'ID' (allocated from zero upwards). When communicated, the 'ogi.farm' process simply changes its internal state for that worker to 'idle'. Before it returns to the idle state internally, the 'ogi.handler' process communicates the connection to the back-end of the web-server (section 2.5).

2.4.3 OGI Modules

OGI modules are pre-compiled occam processes (or networks of processes), accessed using a PROC named "ogimain", with the following signature:

```
PROC ogimain (VAL DPROCESS me, []CHAN ANY x.in?, x.out!)
```

This is the generic interface used by the "dynamic loadable processes" extension to KRoC/-Linux [28]. The actual interface used by the web-server for OGIs is recovered using channel RETYPEs:

```
CHAN OGI.LINK.IN in? RETYPES x.in[0]?:
CHAN OGI.LINK.OUT out! RETYPES x.out[0]!:
CHAN SHARED LOG! log.in? RETYPES x.in[1]?:
```

Where the 'OGI.LINK.IN' and 'OGI.LINK.OUT' protocols are defined by:

```
PROTOCOL OGI.LINK.IN IS D.CONN; MOBILE []BYTE:

PROTOCOL OGI.LINK.OUT
  CASE
    persist; BOOL
    conn; D.CONN
:
```

The 'D.CONN' structure is the *mobile* 'connection' record, passed into the OGI module through 'OGI.LINK.IN' and retrieved from the 'conn' variant of the 'OGI.LINK.OUT' protocol. The following 'MOBILE []BYTE' (dynamic BYTE array) on 'OGI.LINK.IN' contains the 'query-string' from the request, with appropriate de-mangling of encoded characters, "%20" meaning 'space' for example[3]. The third channel, 'log.in', is used to communicate a shared (client) channel-end to the OGI, that is remotely connected to the system 'log' (described in section 2.6).

When an OGI starts up (run in parallel within an 'ogi.handler' process), it inputs its first connection, then responds with a 'persist' message. Ideally, this should be done as quickly as possible, since the 'ogi.farm' will not accept any more new connections until it has received this response. If 'TRUE', the 'ogi.farm' will re-direct all requests for that OGI into the newly activated handing process. If 'FALSE', other requests for the same OGI will result in new handlers being activated.

The code within 'ogi.handler' communicates on all three channels in parallel, so an OGI module can send the 'persist' message first (if no interaction with the remote client is required first). As soon as this is received, the response is relayed to the 'ogi.farm' process so it can continue accepting connections.

[3]The first 128 characters are typically ASCII, the rest depend on the character-set used by the server (and are interpreted appropriately) — generally one of the extended ASCII character sets.

As far as error-handling is concerned, the dynamic processes extension 'catches' errors from loaded process networks. This stops an erroneous module from bringing down the web-server, but at the expense of possibly 'losing' system resources (for the web-server, dynamic memory and file-descriptors). The connection itself can be recovered (by copying, with `CLONE`, the connection structure before it is passed to '`ogimain`'). This is indicated to the '`ogi.handler`' locally, when the call that 'runs' the dynamic process returns:

```
ccsp.run ("ogimain", libhandle, [xx.in,xx.log], [xx.out], res)
```

The '`libhandle`' is a reference to the dynamically loaded library that contains an '`ogimain`' process. '`xx.in`' and '`xx.out`' are the reverse `RETYPE`s for the channel connections. The '`xx.log`' channel is used to communicate the *shared* mobile channel-end (connected to the system 'log') to the OGI process. The result is left in the `INT` parameter '`res`'. On successful termination, this will hold the constant '`DPROCESS.FINISHED`', or '`DPROCESS.FAULTED`' if a run-time error occurred[4].

Five OGI modules have currently been implemented — mostly for demonstration purposes. Two of the modules are simple tests, one persistent (with a 'hit-counter') and one non-persistent (handling only one connection then terminating). The traditional 'finger' script has also been OGI'd. This uses the occam socket library to open a connection to the requested machine (given as the query-string), and then returns the information sent back — or an error message if the connection times out (after 5 seconds) or is refused.

Another OGI module is one simply designed to crash in various ways — to test the dynamic process crash-handling mechanism as well as the web-server. Supported errors range from simple `STOP`s to array-bounds and floating-point errors. The final OGI is the "occam adventure game", described in the following section.

2.4.4 The occam Adventure Game

The occam adventure game is a basic "multi-user dungeon" (MUD), that allows (multiple) users to login and interact via the web. Rather than being a pure client to web-connections, the adventure game is an autonomous process network, with built-in 'bots' that randomly wander around the game.

The current design is fairly static, using a centralised 'game' process that manages locations, objects and users. Processes representing objects and users (either internal 'bots' or remote web-clients) are connected to the game process using a shared channel-type. The client used for interacting with such environments is typically 'telnet', or a similar text-based TCP/IP clients. The client/server interaction is largely asynchronous — the server may send data to the client whenever it wishes; the client to server data is provided by the user.

This presents a slight problem when interfacing with remote web-clients, whose interaction with the web-server is strictly request-response. The solution is to introduce of buffering and polling — the returned web-page instructs the client to auto-refresh after 30 seconds, that will then collect up any game 'events' received since the last request-response.

In addition to the web-interface, the game also offers an interface through the IRC (Internet Relay Chat) protocol [29, 30]. With this interface, the game connects to an IRC server and registers itself as a user (currently 'occgam'). Users (on the same IRC *network*) interact with the game using IRC 'private messages' (to and from 'occgam'). IRC supports the asynchronous behaviour of the game more naturally than HTTP, with 'events' sent directly to remote clients (instead of being queued, as they are for web-clients).

[4]There is a third possibility for dynamically loaded processes, '`DPROCESS.SUSPENDED`'. This is not currently supported by the web-server, however.

Figure 6 shows the game process network. Web-clients are handled within the 'ogimain' process only. The main 'game' process maintains internal arrays of users 'logged in' to the game, that are either web-clients or IRC clients. Each interacting IRC client is handled by an individual 'irc.client' process, controlled through the 'irc.interface' process. The 'game' process generates a potentially infinite stream of 'events' (for example, "person X entered room Y"). The 'event.filter' process receives these events and selects only those which are relevant to connected IRC clients. These events are passed through an overwriting-buffer (within the 'irc.interface' process) to the 'irc.client' processes, that handle them — either by discarding (if the user *just* left a room for which an event was received) or reporting back to the remote IRC user (again via the 'irc.interface' process).

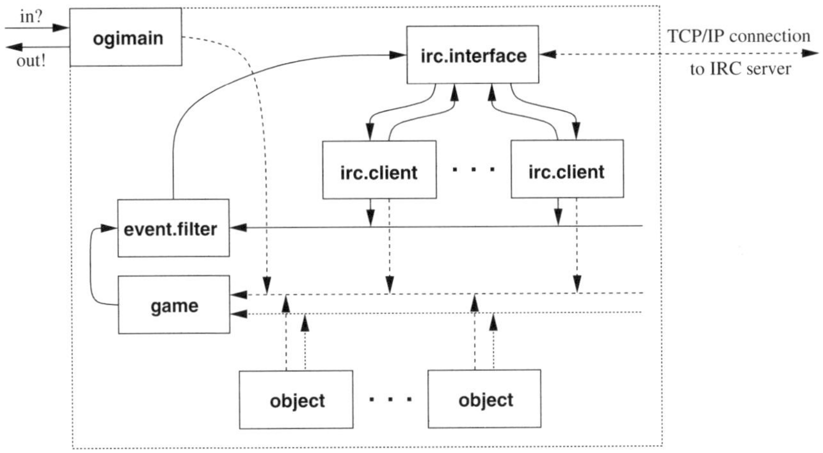

Figure 6: occam 'adventure game' process network

The other entities within the game are 'object's. These can either be 'held' by a user, or 'free' in a room. The particular state of objects is maintained by the 'game' process, that controls user interactions with those objects.

The current implementation, of 'object's in particular, is somewhat unsatisfactory — particularly in scalability. With the recently added support for basic nested MOBILE types (particularly dynamic arrays of mobile channel-ends), the opportunity for a more natural implementation arises: with rooms, users and objects as individual processes connected by mobile channel-types. The existing implementation converges in the 'game' process, that maintains the global state. In a future implementation (already begun), this state is recorded by the run-time connectivity of the network, and no longer centralised. Figure 7 shows an example of the proposed process network.

Such a network, besides being easier to implement, also provides the opportunity for dynamic re-configuration. This is already used in some respect, to *move* objects and users between rooms. But it also allows the layout of rooms to be changed at run-time — the inserting of new rooms or removal of existing rooms, or even more exotic, the migration of rooms (that can be used to implement fun features such as 'lifts').

The proposed implementation, unlike the existing network, uses 'web.client' processes to handle individual web-clients. Previously, the functionality of these was handled within the 'game' process (for instance, the queueing of messages for clients). Internally, the rooms in the game see only 'users' and 'objects', in addition to their immediate neighbours. They are not aware of whether a particular user is interacting through the web-server, the IRC

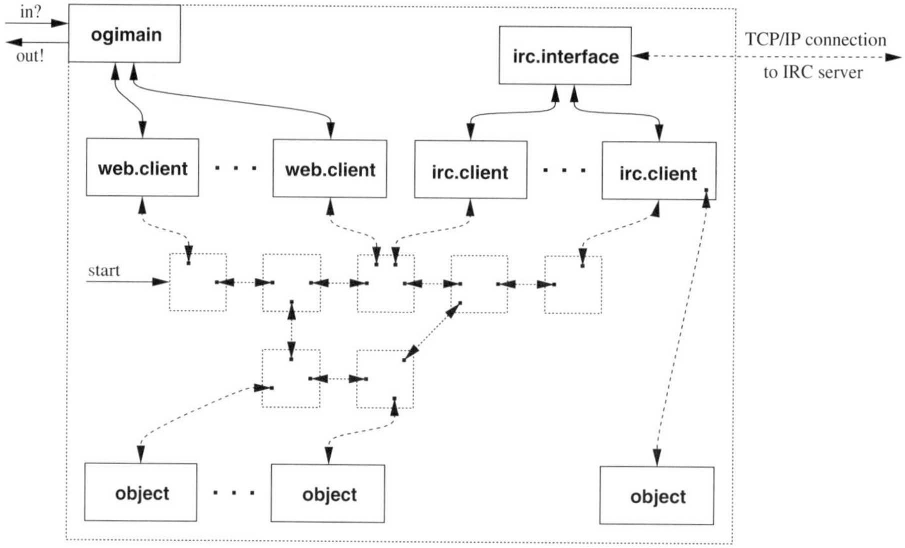

Figure 7: New 'adventure game' process network

interface, or through any other interface that may be introduced at a later date.

The 'start' client-shared channel is connected to the 'ogimain' and 'irc.interface' processes (not shown). This is used to feed 'user' channel-ends into the network when they first connect to the game (when the appropriate client process is FORKed to handle them).

2.5 Back-end Processing

The back-end of the web-server consists of the single process 'be.proc'. This acts as a drain for 'used' connections from various processes within the network. If the client request resulted in an error being returned, or the connection was not marked as 'keep-alive', then the connection to the client is terminated. Connections marked as 'keep-alive' (and that did not cause an error), have their internal state cleared and are output into the front-end farm to wait for another request.

In addition to recycling connections, 'be.proc' uses the information gathered within the connection structure to update a global 'stats' process. This includes the number of bytes received from and sent to the client, and the time spent in various parts of the web-server. These run-time statistics can be accessed through the special URL "/stats.html", that the 'cache.hash.request' process intercepts (and re-directs to the 'stats' process).

This information is also used to report a 'hit' in the web-server log-file, including the source IP address, client request string and server response code.

2.6 Auxiliary Processes

In addition to the main components shown in figure 1 (page 253) are a number of auxiliary processes: the 'stats' run-time statistics process; the 'log' process; and the 'config' process.

The 'stats' process collects various run-time statistics, reported by the web-server components using a shared channel. The information collected includes general connection statistics, the sizes of the various process farms and the contents of the page cache (including non-cached pages). The statistics can be extracted by requesting the page '/stats.html', that 'cache.hash.request' re-directs into 'stats'. The contents of the page-cache are retrieved by querying the 'cache.control' process, rather than having 'cache.control' report the cache state whenever it changes.

The 'log' process simply acts a multiplexer between two channels in a channel-type and the top-level 'screen' channel. The input to the 'log' process is a server-end of the 'LOG' channel-type, defined as:

```
CHAN TYPE LOG
  MOBILE RECORD
    CHAN P.LOG log?:
    CHAN BYTE c?:
:
```

The 'P.LOG' protocol defines three variants for BYTEs, INT counted BYTE arrays, and mobile BYTE arrays. Ideally, processes should try to use the mobile variant, since communication is unit-time. The separate BYTE channel 'c' is provided so that the standard (KRoC) library functions, 'out.string' for example, can be used easily.

The 'config' process provides the web-server components with their configuration. This includes, for example, the file-system path for static-pages, prefix for CGI scripts, and any limits on the sizes of the various dynamic process farms. Currently, this information is provided statically at compile time (to the 'config' process), and typically accessed only once by the various components as they initialise. Although there is currently no mechanism for adjusting the configuration (within the 'config' process) at run-time, other web-server components can be forced to reload their configuration by means of the 'acceptor' sending out a specially marked connection.

3 Performance

This section presents some early performance benchmark results for the occam web-server. The 'httperf' [31] benchmark program is used for the majority of benchmarks. This tries to 'hit' a web-server at a specified rate, by maintaining multiple connections. At a rate of 100 hits/second, for example, 'httperf' will attempt a new request every 10 ms. New requests may create new connections, or re-use old connections that have been marked as 'keep-alive'. This reduces some of the TCP overheads (disconnect and reconnect between requests), but most HTTP clients (web-browsers) do not make sensible use of the functionality (although this is gradually improving).

Figure 8 shows the request and response rates for Apache and the occam web-server, with increasing 'attempted' requests-per-second. As can be clearly seen in the results, the occam web-server starts to lose performance at 700 requests/second (175 connections/second). Apache only starts to lose performance at around 1300 requests/second (325 connections/-second), at which point the network throughput reaches its limit (approximately 10 megabytes per second).

Figure 9 shows the corresponding network bandwidth and maximum number of concurrent TCP connections. As soon as the web-server being benchmarked becomes unable to satisfy requests, the number of concurrent connections increases sharply — 'httperf' opens more connections and sends more requests in an attempt to maintain the requested rate. The result of this is that performance is degraded even further, as significantly more connections must be managed concurrently.

The benchmark results shown here are for a standard Apache installation (without any specific performance tweaks), and the occam web-server with both statistics-collection and

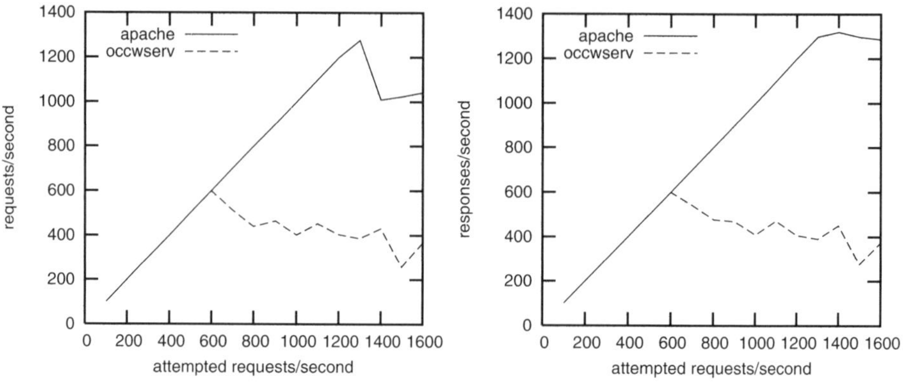

Figure 8: Request and response benchmark results (20, 000 hits, 4 requests/connection, 8k static page)

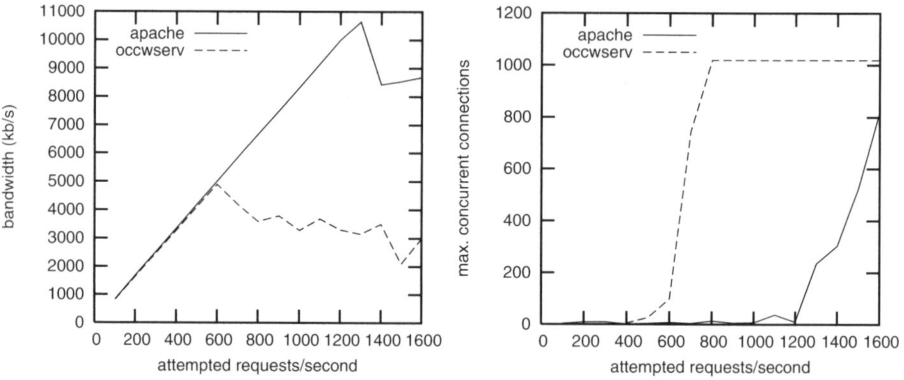

Figure 9: Network bandwidth and max. concurrent connections benchmark results (20, 000 hits, 4 requests/connection, 8k static page)

post-mortem debugging [32] enabled. This gives a moderate disadvantage to the occam web-server, but not enough to account for the difference in performance. Furthermore, the occam web-server was compiled without any inlining optimisations. When enabled, these improve the performance of communication considerably.

A certain amount of overhead is due to the blocking system-calls mechanism [7], that has a *minimum* overhead of 9 us for the dispatch-collect cycle (measured on the same P3-800). Each connection, with four requests, requires 13 blocking calls — one for the initial 'accept' and three for each request (read request, write headers, send data).

The blocking system-calls mechanism is currently undergoing some optimisations, that should dramatically improve their performance, particularly for the web-server. It is hoped that these optimisations will be in place before the conference, where new results can be presented.

4 Conclusions and Future Work

Although the occam web-server may not perform as well as Apache, there is much value in the security gained by using occam. In particular, there is no opportunity for 'buffer-overflow' exploits, which have previously been found in Apache and Microsoft's IIS. Much effort is often spent in the formal verification of C (and C++) programs, particularly as regards the use of pointers. The purpose of such verification is largely to determine the absence of buffer-overflow opportunities and ensure the correct use of pointers (that they are initialised, for example). For occam, verification (of this type) is performed automatically by the compiler, and is hugely aided by the lack of explicit pointer types. Automatic verification for C code is difficult, but has been the subject of much research. With particular regard to internet servers and buffer-overrun possibilities is [33], that advocates the use of static checking — techniques that the occam compiler already employs, although at a less complex level. occam, however, has only recently become suitable for constructing such server applications.

There is clearly no functional deficiency from using occam to implement a web-server — this one works. Performance, however, is a major issue which must be addressed if the occam web-server is become a viable alternative to servers such as Apache and IIS. Fortunately, work is underway which should significantly improve the performance of the web-server (that has, until recently, been constrained by a lack of time).

In providing mechanisms that allow multiple clients to interact, the occam web-server does particularly well. This is largely the result of utilising concurrency, where creating a process that manages multiple client connections is simple — the server can control the clients (or process representations of those clients). In a non-concurrent web-server, interaction of this nature will typically be client-driven — driven by the sequential *flow-of-control.* Concurrent implementations, as demonstrated by the 'adventure game' OGI, are largely intuitive — and are much easier to program.

The use of occam for the construction of a web-server is just one example from a wider field of application — all 'internet' servers. One server type that is planned for implementation is a mail server, at least RFC-821 [34] (SMTP) compliant and with some local delivery capability. Mail servers have a less urgent requirement for real-time responses (most connections are automated), but have an equal, if not greater, need for security and reliability. This can certainly be seen from past failings, particularly in the widely-used '`sendmail`' mail-server (some of which are examined in [35]).

4.1 Handling Large Numbers of Simultaneous Clients

One area in which the occam web-server suffers slightly is the handling of large numbers of simultaneous clients. Because each client connection is handled individually, the overheads increase linearly with the number of simultaneous connections. This is particularly relevant to the blocking call dispatch/collect mechanism, whose overhead is significant (around 9 us). A more significant issue, however, is that the number of concurrent blocking-calls supported is limited (in order to avoid excessive numbers of 'clone' processes being created).

Non-concurrent web-servers typically use '`select()`' to 'wait' for activity from a number of clients (either for reading or writing), an operation that only involves a single blocking call — and is reasonably efficient for small numbers of 'clients' (socket file-descriptors). Placing the 'multiple wait and read' functionality within a single blocking-call offers two advantages for the occam web-server: reducing the number of simultaneous blocking calls; and following from that, reducing the impact of the blocking-call overhead itself.

Figure 10 shows an example of how this network might be constructed (for reading client requests). In principal, the total number of '`n.select.read`' blocking-call 'processes' is limited (regardless of the number of connections), but the frequency of their use varies. The network is constructed using a pipeline of 'timed.buffer' processes, with the most-active process at the left and least-active at the right. Each 'timed.buffer' process acts as a 'collector' for connections, until a pre-determined timeout is reached. It then makes the blocking call to wait and read from all connections, until that same timeout is reached again (if desired). When the blocking-call returns, connections that have remained inactive are forwarded to the next 'timed.buffer' process (with a longer timeout). Connections on which the entire request has been received are forwarded to a modified front-end farm — for distribution to the rest of the web-server (this distribution could be done locally, however, to be efficient). Connections that have only sent a partial request stay put.

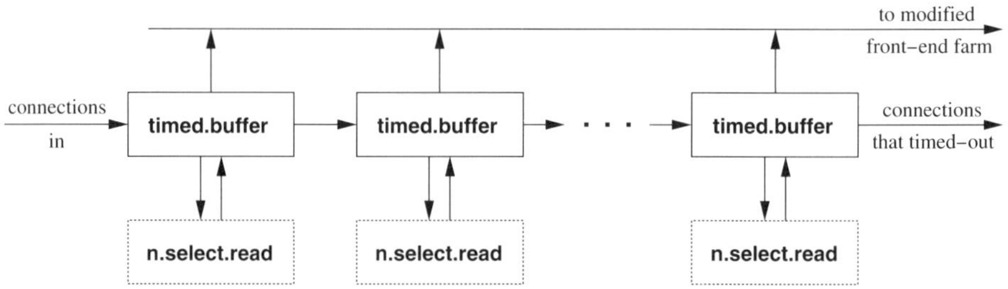

Figure 10: Example client request-reading process pipeline

In theory, the use of such a connection-handling process pipeline should increase performance dramatically. Client connections that are slow (perhaps deliberately — as is the case in 'denial of service' attacks), migrate towards the right-hand side of the process pipeline, where they will not interfere with other clients. Ultimately, inactive connections simply time-out, at which point they discarded (to avoid wasting resources associated with them).

A similar approach can also be taken for handling multiple client writes. However, the current implementation makes use of the 'sendfile()' system-call, that simply copies data between descriptors — completely within the OS kernel, making it a very suitable operation for returning static files to client connections. Whether using a 'select' based approach to writing data to clients instead remains to be investigated.

4.2 Raw-Metal Web-Serving

Currently under development is a CSP based operating-system environment, RMoX [36]. This is programmed in occam and makes extensive use of the same dynamic facilities (particularly dynamic process creation and channel-end mobility).

Within this environment, blocking system-calls (and their associated overheads) simply do not exist — any interaction with the operating-system is via a (low-cost) channel communication. Thus, the full parallelism can remain, resulting in a simpler implementation.

Integrating the web-server into RMoX is expected to be relatively straight-forward, once the necessary functionality (TCP socket communication) has been implemented.

References

[1] Apache Software Foundation. Apache HTTP Server Project, June 2003. Available at: http://httpd.apache.org/.

[2] Abhishek Chandra and David Mosberger. Scalability of Linux event-dispatch mechanisms. In *Proceedings of the 2002 USENIX Annual Technical Conference*, Boston, Massachusetts, USA,

June 2001. Available from: `http://www.usenix.org/publications/library/proceedings/usenix01/chandra%.html`.

[3] Inmos Limited. occam 2.1 Reference Manual. Technical report, Inmos Limited, May 1995. Available at: `http://www.wotug.org/occam/`.

[4] C.A.R. Hoare. *Communicating Sequential Processes*. Prentice-Hall, 1985. ISBN: 0-13-153271-5.

[5] A.W. Roscoe. *The Theory and Practice of Concurrency*. Prentice Hall, 1997. ISBN: 0-13-674409-5.

[6] P.H. Welch, J. Moores, F.R.M. Barnes, and D.C. Wood. The KRoC Home Page, 2000. Available at: `http://www.cs.ukc.ac.uk/projects/ofa/kroc/`.

[7] F.R.M. Barnes. Blocking System Calls in KRoC/Linux. In P.H. Welch and A.W.P. Bakkers, editors, *Communicating Process Architectures*, volume 58 of *Concurrent Systems Engineering*, pages 155–178, Amsterdam, the Netherlands, September 2000. WoTUG, IOS Press. ISBN: 1-58603-077-9.

[8] F.R.M. Barnes. occwserv – an occam Web-Server, September 2001. CPA-2001 SIG presentation, University of Bristol, UK. Available at: `http://frmb.org/publications.html`.

[9] Peter H. Welch and David C. Wood. Higher Levels of Process Synchronisation. In A. Bakkers, editor, *Parallel Programming and Java, Proceedings of WoTUG 20*, volume 50 of *Concurrent Systems Engineering*, pages 104–129, Amsterdam, The Netherlands, April 1997. World occam and Transputer User Group (WoTUG), IOS Press. ISBN: 90-5199-336-6.

[10] F.R.M. Barnes and P.H. Welch. Prioritised Dynamic Communicating Processes: Part I. In James Pascoe, Peter Welch, Roger Loader, and Vaidy Sunderam, editors, *Communicating Process Architectures 2002*, WoTUG-25, Concurrent Systems Engineering, pages 331–361, IOS Press, Amsterdam, The Netherlands, September 2002. ISBN: 1-58603-268-2.

[11] F.R.M. Barnes and P.H. Welch. Prioritised Dynamic Communicating Processes: Part II. In James Pascoe, Peter Welch, Roger Loader, and Vaidy Sunderam, editors, *Communicating Process Architectures 2002*, WoTUG-25, Concurrent Systems Engineering, pages 363–380, IOS Press, Amsterdam, The Netherlands, September 2002. ISBN: 1-58603-268-2.

[12] T. Berners-Lee, R. Fielding, and H. Frystyk. Hypertext transfer protocol – HTTP/1.0. RFC 1945, Internet Engineering Task Force, May 1996.

[13] R. Fielding, J. Gettys, J. C. Mogul, H. Frystyk, and T. Berners-Lee. Hypertext transfer protocol – HTTP/1.1. RFC 2068, Internet Engineering Task Force, January 1997.

[14] Mark J. Cox, Ralf S. Engelschall, Stephen Henson, and Ben Laurie et al. OpenSSL (Secure Sockets Layer), April 2003. Available at: `http://www.openssl.org/`.

[15] Simon Marlow. Writing High-Performance Server Applications in Haskell Case Study: A Haskell Web Server. Technical report, Microsoft Research Ltd., Cambridge, UK., July 2000. Available at: `http://citeseer.nj.nec.com/marlow00writing.html`.

[16] Claes Wikstrom et al. YAWS: yet another web-server, June 2003. Available at: `http://yaws.hyber.org/` or `http://sourceforge.net/projects/erlyaws/`.

[17] Simon Peyton Jones, Andrew Gordon, and Sigbjorn Finne. Concurrent Haskell. In *Conference Record of POPL '96: The* 23rd *ACM SIGPLAN-SIGACT Symposium on Principles of Programming Languages*, pages 295–308, St. Petersburg Beach, Florida, January 1996. ACM Press, New York. ISBN: 0-89791-769-3. Available at: `http://citeseer.nj.nec.com/jones96concurrent.html`.

[18] Joe Armstrong. Concurrency Orientated Programming in Erlang, November 2002. Invited talk at the Lightweight Languages Workshop 2002, MIT. Available at `http://www.sics.se/~joe/talks/ll2_2002.pdf`.

[19] Richard Jones and Rafael Lins. *Garbage Collection: Algorithms for Automatic Dynamic Memory Management*. Wiley, New York, 1996, reprint 1997.

[20] Ingo Molnar et al. Answers from Planet TUX: Ingo Molnar Responds, July 2000. Interview, available at: `http://slashdot.org/interviews/00/07/20/1440204.shtml`. Page last visited June 2003.

[21] Red Hat, Inc. *Red Hat Content Accelerator Manuals*, 2003. Available at: `http://www.redhat.com/docs/manuals/tux/`.

[22] Chuck Lever, Marius Eriksen, and Stephen Molloy. An analysis of the TUX web server, November 2000. CITI Technical Report 00-8. Available from: `http://citeseer.nj.nec.com/lever00analysis.html`.

[23] F.R.M. Barnes and P.H. Welch. Prioritised dynamic communicating and mobile processes. *IEE Proceedings – Software*, 150(2), April 2003.

[24] NCSA Software Development Group. *CGI/1.1: Common Gateway Interface Version 1.1*. W3C: World Wide Web Consortium, 1999. Available from: `http://www.w3.org/CGI/`. Last accessed June 2003.

[25] PHP Group. PHP: Hypertext Preprocessor. Available at: `http://www.php.net/`.

[26] Netscape. *JavaScript Reference*, 2001. Available from: `http://devedge.netscape.com/central/javascript/`. Last visited June 2003.

[27] Fred Barnes. *Socket, File and Process Libraries for occam*. Computing Laboratory, University of Kent at Canterbury, June 2000. Available at: `http://www.cs.ukc.ac.uk/people/rpg/frmb2/documents/`.

[28] Frederick R.M. Barnes. *Dynamics and Pragmatics for High Performance Concurrency*. PhD thesis, University of Kent at Canterbury, June 2003.

[29] J. Oikarinen and D. P. Reed. Internet relay chat protocol. RFC 1459, Internet Engineering Task Force, May 1993.

[30] C. Kalt. Internet relay chat: Client protocol. RFC 2812, Internet Engineering Task Force, April 2000.

[31] David Mosberger and Tai Jin. httperf: A Tool for Measuring Web Server Performance. In *First Workshop on Internet Server Performance*, pages 59—67, Madison, WI, June 1998. ACM. Available at: `http://citeseer.nj.nec.com/mosberger98httperf.html`.

[32] D.C. Wood and F.R.M. Barnes. Post-Mortem Debugging in KRoC. In P.H. Welch and A.W.P. Bakkers, editors, *Communicating Process Architectures, Proceedings of WoTUG 23*, volume 58 of *Concurrent Systems Engineering*, pages 179–192, Amsterdam, the Netherlands, September 2000. WoTUG, IOS Press. ISBN: 1-58603-077-9.

[33] David Wagner, Jeffrey S. Foster, Eric A. Brewer, and Alexander Aiken. A First Step towards Automated Detection of Buffer Overrun Vulnerabilities. In *Network and Distributed System Security Symposium*, pages 3–17, San Diego, CA, February 2000. Available at: `http://citeseer.nj.nec.com/wagner00first.html`.

[34] J. B. Postel. Simple mail transfer protocol. RFC 821, Internet Engineering Task Force, August 1982.

[35] Ian Goldberg, David Wagner, Randi Thomas, and Eric A. Brewer. A Secure Environment for Untrusted Helper Applications. In *Proceedings of the 6th USENIX Security Symposium*, San Jose, CA, USA, 1996. Available at: `http://citeseer.nj.nec.com/goldberg96secure.html`.

[36] F.R.M. Barnes, C.L. Jacobsen, and B. Vinter. RMoX: a Raw Metal occam Experiment. In *Communicating Process Architectures 2003*, WoTUG-26, Concurrent Systems Engineering, IOS Press, Amsterdam, The Netherlands, September 2003.

Communicating Process Architectures – 2003
Jan F. Broenink and Gerald H. Hilderink (Eds.)
IOS Press, 2003

RMoX: A Raw-Metal occam Experiment

Fred BARNES[†], Christian JACOBSEN[†] and Brian VINTER[‡]

[†] *Computing Laboratory, University of Kent,*
Canterbury, Kent, CT2 7NF, England.
{frmb2,clj3}@kent.ac.uk

[‡] *Department of Mathematics and Computer Science,*
University of Southern Denmark, Odense, Denmark.
vinter@imada.sdu.dk

Abstract. Operating-systems are the core software component of many modern computer systems, ranging from small specialised embedded systems through to large distributed operating-systems. This paper presents RMoX: a highly concurrent CSP-based operating-system written in occam. The motivation for this stems from the overwhelming need for reliable, secure and scalable operating-systems. The majority of operating-systems are written in C, a language that easily offers the level of flexibility required (for example, interfacing with assembly routines). C compilers, however, provide little or no mechanism to guard against race-hazard and aliasing errors, that can lead to catastrophic run-time failure (as well as to more subtle errors, such as security loop-holes). The RMoX operating-system presents a novel approach to operating-system design (although this is not the first CSP-based operating-system). Concurrency is utilised at all levels, resulting in a system design that is well defined, easily understood and scalable. The implementation, using the KRoC extended occam, provides guarantees of freedom from race-hazard and aliasing errors, and makes extensive use of the recently added support for dynamic process creation and channel mobility. Whilst targeted at mainstream computing, the ideas and methods presented are equally applicable for small-scale embedded systems — where advantage can be made of the lightweight nature of RMoX (providing fast interrupt responses, for example).

1 Introduction and Motivation

This paper presents RMoX, a highly concurrent CSP [1, 2, 3] based operating-system written in an extended variant of the occam [4] multiprocessing language.

On commodity hardware, operating-systems are the software component that control access to the physical hardware, manage applications, and provide services to applications (such as networking and structured file storage — where these are required). Clearly, it is essential that operating-systems function correctly (as their specification intended), including graceful error handling. Unfortunately, many operating-systems do not behave this way.

In our view, mainstream commodity operating-systems (such as Windows^{tm} and Linux^{tm}) suffer from three major problems. Firstly there is the issue of implementation errors (bugs) — often resulting in undesirable (and sometimes dramatic) system failure. The second is the problem of scalability: most operating-systems do not scale well over large numbers of processors (in either SMP or distributed environments). Finally is the issue of performance. Many of the mechanisms which provide security (at the hardware level), do so at great cost (such as memory-management and operating-system call gates). Much of this is in response to the first problem (of software error), requiring hardware support to ensure that software components do not interfere with each other destructively.

Bugs within the operating-system itself present a significant problem: at best they are an inconvenience; at worst they cost: in lost work and time (whilst the system is fixed).

The majority of software errors within an operating-system are due either to programmer error or incorrect specification. As the size and complexity of the system increases, the number of errors introduced also increases. The counter-acting force is time — the longer a system remains in operation, the more bugs that will be discovered and 'fixed'. Of course, without some level of formal verification (either manual or automated), there can be *no guarantee* of software correctness (particularly when the specification is deficient).

The C programming language [5], as used to implement the majority of operating-systems — and in some ways was designed for that purpose — provides no mechanisms for guarding against data aliasing and race-hazard errors. Within an operating-system such errors can be catastrophic, are easily introduced, and can be extremely hard to locate.

Operating systems written in C are not at a complete loss however; language research elsewhere has produced 'safer' dialects of C, such as Cyclone [6], that provide greater protection against aliasing (and some race-hazard) errors.

1.1 Concurrency in Operating Systems

The C language provides no natural support for concurrency, which presents a significant challenge to the operating-system implementor. Concurrency is required for some reasons and desirable for others.

One of the many tasks performed by an operating-system is the scheduling of application processes. Although processes might not interact between themselves, they will all interact with the operating-system. Many interactions between applications and the operating-system are *blocking* (socket reads, for example). Clearly, having one process blocked in the kernel should not prevent other processes from being scheduled. Furthermore, a process that is busy performing a computational task should not be allowed to 'hog' the processor.

Linux, for example, handles most concurrent interaction using 'wait-queues', structures where processes may wait inside the kernel, to be later rescheduled by another process (for instance, on completion of an I/O event, or a message from another process). At this level (and considering only single-processor systems), race-hazard errors are not too common and could only be caused through mis-handling of asynchronous events (typically hardware interrupts). A good description of the Linux kernel (and the various wait-queues, including the run-queue) can be found in [7].

The step in complexity from a uni-processor to a multi-processor kernel (for Linux) is significant, particularly for shared-memory multi-processing. Protection from race-hazards is largely handled using spin-locks to protect critical regions of the kernel. As Linux has evolved, this locking has become increasingly more fine-grained — now allowing multiple processes to be executing concurrently inside the kernel (across multiple processors). This is complex, and frequently a source of error (particularly in device-drivers). The slightest mis-placement of a 'lock' or 'unlock' call can have grave consequences. Errors of this type can remain undetected for a long time, and might never be discovered at all.

Unfortunately, programmers who work on the Linux kernel are often left with the common view that "concurrent programming is hard". This very generalised statement is wrong — CSP and its implementations have proved that already. The problem here is that concurrent programming in C is hard — and there is no easy or clean solution to that, particularly where operating-systems are concerned.

Concurrency (with CSP semantics) *is* desirable however, particularly for operating systems. Good concurrent design can produce systems that are easily scalable — this is simply not the case for many non-concurrent languages, C in particular.

1.2 The RMoX Operating System

The problems of existing operating-systems, particularly their lack of concurrency, has led us to RMoX: an experimental, highly concurrent, CSP-based dynamic operating system written in occam. Traditional occam is not particularly suitable for modern-day operating-system implementation. Its lack of *any* dynamic capability presents a significant problem, in addition to a lack of multi-level priority support.

Fortunately, previous and on-going work in KRoC/Linux [8] has added the dynamic capabilities required. Mobile data-types [9] are used to provide unit-time data *movement* operations (such as returning data blocks from a block device driver to a file-system driver). Much of the network connectivity (within RMoX itself) is handled using *shared* channel-types. Two significantly useful features are provided by these: the ability to share the resources of a server with multiple clients (in such a way that security and fairness are enforced by the compiler); and the ability to *move* channel-ends around a process network. These, combined with the `FORK` dynamic process creation mechanism, provide an ideal way to build dynamically scalable process networks. One type of network built in this way — the process farm — is discussed in detail in [10, 11] and [12]. RMoX, however, is a somewhat more general-purpose process network.

The underlying concept of RMoX is simple: a scalable concurrent system that provides 'operating system' functionality, including device-drivers, file-systems and networking. As a general guide, we aim towards providing a POSIX [13] compatible interface. POSIX clearly defines, for example, what operations must be supported on files, or what operations must be supported on sockets. It does not define how such functionality should be implemented, however.

One particularly significant gain from using a concurrent language is the removal of explicit scheduling control from operating system components. In Linux, for example, when a process wishes to block inside a device-driver, it must be placed on a wait-queue before calling the main '`schedule()`' function to schedule another process. This is a scheduling issue, and should not be the (direct) concern of the device driver. In fact, much of the scheduler interaction within the Linux kernel is done to provide simple process synchronisation (most obviously, communicating through a pipe). Calls to device drivers may need to synchronise with hardware events, which again, is usually done using wait-queues (and possibly some unpleasant interrupt handling). Code of this nature simply does not exist in RMoX— at least as far as the operating-system components are concerned. Device drivers, for example, are active processes — if a device driver wishes to perform a delay, or interact with another component (to which it is connected via a channel), it simply does so.

This clear split, between concurrent processes and the mechanisms scheduling them, simplifies operating-system implementation greatly. For scheduling occam processes within RMoX, a modified version of the CCSP [14] run-time kernel is used. The modifications are primarily the disabling of operating-system calls within CCSP itself (such as the mechanisms used to implement blocking system calls and terminal I/O). Within RMoX, any equivalent functionality is provided by active processes, with appropriate low-level support for handling interrupts.

1.3 Paper Overview

This paper is divided into 3 main sections. The first, section 2, presents the overall design of RMoX and its various sub-systems (as networks of occam processes).

Section 3 examines the low-level infrastructure that exists 'underneath' the occam kernel. Much of the work here is related to bootstrapping RMoX, and providing a C (or assembler) interface for accessing the low-level hardware — memory management and interrupts for example.

Given that RMoX is essentially an occam *program* in its own right, running it within the KRoC/Linux environment (as an ordinary Linux process) is possible — with appropriate replacements for the low-level handling and some device-drivers. Section 4 describes UM-RMoX (User-Mode RMoX), that provides this functionality.

RMoX is still in the early stages of development, and cannot nearly be considered complete. Currently, RMoX is sufficiently capable to implement small-scale embedded-systems, although the work to date has been done on Intel Pentium based machines. Much of the framework for future development is in place however. Section 5 discusses the work already done and presents suggestions for the future development of RMoX.

2 Design

The core of RMoX is a client-server network, connected mostly by shared channel-bundles. The 'static' component of the network is comprised of three main server-farms, plus a 'console' process and some auxiliary processes, as shown in figure 1.

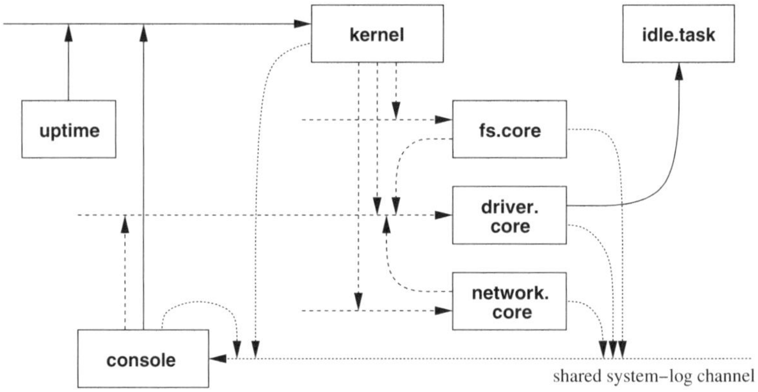

Figure 1: Initial RMoX top-level process network

The '`kernel`' process provides what could be considered the main system interface (system-call interface on Linux). The (client shared) channel-bundle on the left-hand side, initially connected to the '`uptime`' and '`console`' processes, is the main point for system interaction. The channel-type used is defined as:

```
CHAN TYPE CT.KERNEL
  MOBILE RECORD
    CHAN P.KERNEL.IN in?:        -- requests
    CHAN P.KERNEL.OUT out!:      -- responses
:
```

The two variant protocols '`P.KERNEL.IN`' and '`P.KERNEL.OUT`' cover the available range of operating-system 'calls' and responses. In addition to maintaining some local state (for example, the host-name and uptime), the function of the '`kernel`' process is to re-direct requests to the relevant sub-system (the file-system core for example). The kernel process *will* also be responsible for ensuring some security — e.g. that a process attempting to change the host-name is indeed allowed to.

2.1 Device Drivers

At the lowest level of the system are device drivers. The 'driver.core' process itself is a pure server, with the responsibility of maintaining a *farm* of device-driver processes. Figure 2 shows an example of this for an active system. Initially, there are no device driver processes. These are FORKed by the 'driver.core' in response to an appropriate request.

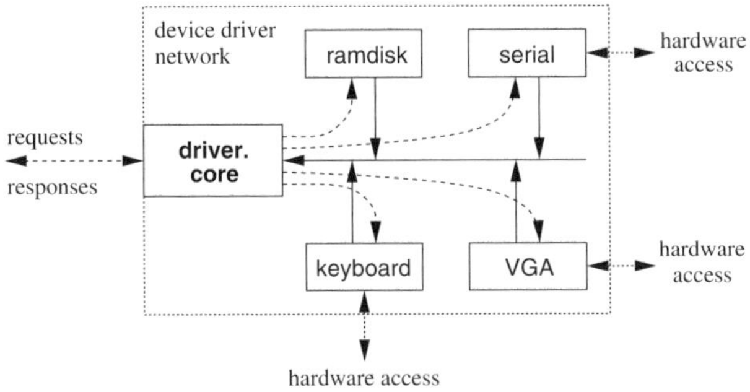

Figure 2: RMoX device driver process network

Requests are made to the 'driver.core' process using the connected client-shared channel-bundle. This is defined as the type 'CT.DRIVER', and in a similar way to the 'CT.KERNEL' channel-type, has two channel fields: one for requests and one for responses:

```
CHAN TYPE CT.DRIVER
  MOBILE RECORD
    CHAN P.DRIVER.IN in?:
    CHAN P.DRIVER.OUT out!:
:
```

Currently, only a small number of operations are supported by 'driver.core', defined in the 'P.DRIVER.IN' protocol:

```
PROTOCOL P.DRIVER.IN
  CASE
    connect.to.driver; MOBILE []BYTE
    start.driver; MOBILE []BYTE
:
```

The range of possible responses is significantly broader however, and reveals much about the way we intend RMoX to work:

```
PROTOCOL P.DRIVER.OUT
  CASE
    driver.started
    no.such.device
    device.busy
    input.device; CT.INPUT!
    shared.input.device; SHARED CT.INPUT!
    output.device; CT.OUTPUT!
    shared.output.device; SHARED CT.OUTPUT!
    block.device; CT.BLOCK!
    shared.block.device; SHARED CT.BLOCK!
:
```

When a 'connect.to.driver' message is sent, and a device driver with the specified name exists, that request is simply forwarded. If the device can be opened, the device driver responds with the client-end of a channel-bundle, of the type 'CT.INPUT', 'CT.OUTPUT' or 'CT.BLOCK', depending on the type of device, that may or may not be shared.

Devices are referred to by name, using a simple hierarchical naming scheme. Many device drivers will be trivial — responsible for only a single piece of hardware. A driver such as 'serial', however, may be responsible for multiple pieces of hardware (each serial port in this case). In this particular case, the 'serial' driver manages a farm of serial-handling processes, as shown in figure 3.

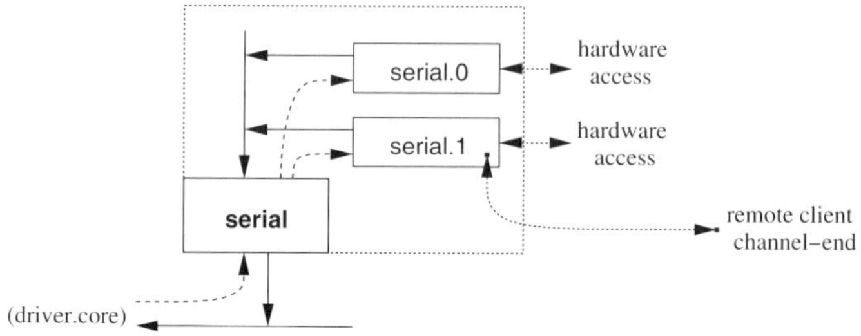

Figure 3: RMoX 'serial' device-driver process network

This example shows the serial network with two sub-processes, and an active connection to the 'serial.1' device. When this device is requested, the 'driver.core' process only examines the name up to the first period ('.') when deciding where to forward the request. The 'serial' device then examines the next component of the name and forwards it appropriately (or returns an error if the device does not exist).

The types 'CT.INPUT' and 'CT.OUTPUT' are used for character devices of the input and output variety respectively. The serial-port drivers within RMoX actually present two devices each — one for input and one for output. These are identified using the suffixes '.input' and '.output' respectively. The 'CT.INPUT' channel-type is defined as:

```
RECURSIVE CHAN TYPE CT.INPUT
  MOBILE RECORD
    CHAN P.INPUT.IN in?:
    CHAN P.INPUT.OUT out!:
    CHAN BYTE data.out!:
    CHAN CT.INPUT! return?:
:
```

The two channels 'in' and 'out' are used primarily for device-specific control, handled using the 'P.INPUT.IN' and 'P.INPUT.OUT' variant protocols. Data generated by a character device is simply sent byte-by-byte down 'data.out'. If a more substantial communication mechanism is required, the 'in' and 'out' channels can be used to provide it.

When a client process has finished using the services of a device driver, it returns its local end on the 'return' channel — using the recursive channel-type feature. This way, a device driver knows when a client process has finished using the device, and that action is synchronised. Currently, there is no return mechanism for devices that are *shared*, although this may be added later on (if the need for it arises).

The 'CT.OUTPUT' type is defined in a similar way, but whose BYTE channel is named 'data.in?' and where the device-specific protocols 'P.OUTPUT.IN' and 'P.OUTPUT.OUT' are used. The 'vga' driver in RMoX is a character output device, initially operating in a very 'dumb' mode. When more of the system starts up (section 2.4), this driver is put into a more complex mode of operation and communication continues on the 'in' and 'out' channels.

Block device drivers use the 'CT.BLOCK' channel-type and are specifically geared towards block-addressable devices (such as disks — including the 'ramdisk'). This channel-type is defined as:

```
RECURSIVE CHAN TYPE CT.BLOCK
  MOBILE RECORD
    CHAN P.BLOCK.IN in?:
    CHAN P.BLOCK.OUT out!:
    CHAN CT.BLOCK! return?:
:
```

The variant protocols 'P.BLOCK.IN' and 'P.BLOCK.OUT' provide mechanisms for reading and writing *blocks* of data, as well as querying the device for its block-size and total number of blocks.

2.2 File Systems

The file-system sub-system is built in a similar way to the device-driver sub-system — i.e. a farm of sub-server processes along with a controlling process ('fs.core').

The interface to the file-system is presented in a UNIX-like manner, supporting the familiar 'mount' and 'umount' operations (which POSIX does not specifically define). When RMoX starts up, there are no mounted file-systems. The first file-system mounted is a ramdisk file-system named 'ramdisk.fs', that is mounted at the root ('/'). Before a file-system can be mounted, the underlying device (if required) must be accessed. The ramdisk file-system uses the 'ramdisk' device-driver, that simply provides read and write access to a number of fixed-size blocks.

In the state where no file-systems are mounted, the majority of requests will return errors (with the exception of mounting a file-system on '/'). The rest of RMoX will continue to operate normally however.

Figure 4 shows the file-system network with two mounted file-systems, 'ramdisk.fs' and 'dev.fs'. 'dev.fs' is a *virtual* file-system that provides access to devices through the file-system interface. The traditional UNIX mechanism for device access is through the use of special (character and block) files, usually found in the '/dev' directory. The use of virtual file-systems for '/dev' within UNIX systems is becoming increasingly popular, however, and it makes good sense to do this — restricting the contents of '/dev' to the devices that are *actually* available, instead of wasting hundreds of directory-entries on devices that are not available.

Internally, the 'ramdisk.fs' process maintains a *block-bitmap* of allocated 'ramdisk' blocks, alongside an array of *inodes*. This is a very simplified version of many traditional UNIX file-systems, except that they retain the block-bitmap and inodes on the device itself (often with a layer of caching to improve performance).

File handling within the ramdisk is performed using FORKed 'file.handler' processes. When a file is opened, a channel-bundle of type 'CT.FILE' is created. The client-end is returned to the process that opened the file and the server-end is given to the newly created 'file.handler' process. In addition to servicing "open file" and similar requests, 'ramdisk.fs' acts as a server for the potentially many 'file.handler' processes. In particular this includes allocating blocks when writing to files, and manipulating inodes.

Figure 5 shows an example process connected to a 'file.handler' within the ramdisk. When such a process has finished with a file, the recursive channel-type feature is used to

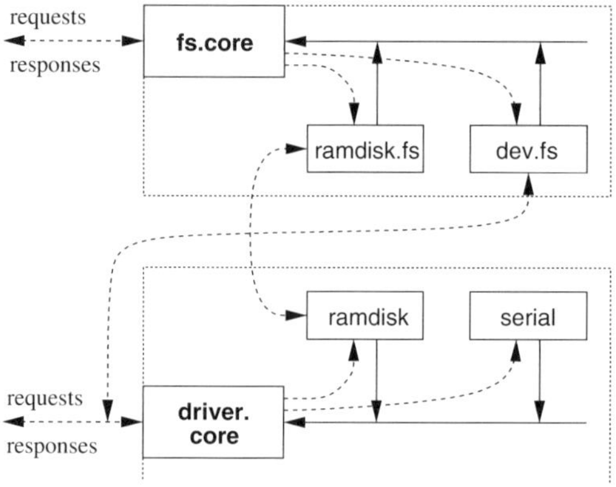

Figure 4: RMoX file-system driver network

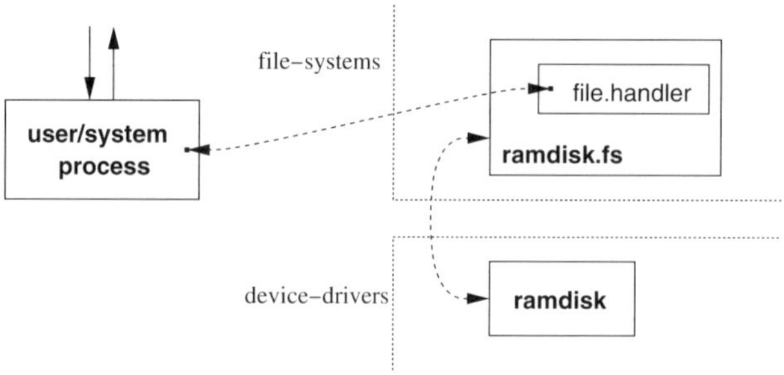

Figure 5: RMoX file handling

return the whole channel to the '`file.handler`' process, that then updates any inode states and shuts-down gracefully.

Directory operations are performed in a similar manner, but use the services of a '`dir.-handler`' process — connected to the client process using a '`CT.DIR`' channel-bundle.

2.3 Networking

The networking infrastructure, specifically the provision of an IP [15] stack, has been largely designed and implemented by Adam Sampson, at student at UKC, as part of occamnet [16].

occam and CSP provide a very natural way of building network infrastructures. It makes good sense to build networking components (such as routers, multiplexors and filters) as communicating processes, that simply communicate 'packets' (plus any additional control information required). The use of `MOBILE` types ensures a minimum copying of data — the communication of a 'packet' will always be a unit-time (pointer-swapping) operation within occam.

Previous work in [17] discusses the building of ATM over DS-Link routing networks using occam and Transputer-based hardware [18, 19]. This is closely related to the work in occamnet, which instead routes IP packets over occam channels. One paper of interest is [20], that describes a method for transforming high-level protocol specifications into workable occam code, ultimately resulting in two communicating state-machines, one for sending and one for receiving. This technique has been used to implement a SLIP [21] protocol driver, that provides support for IP over serial lines (the Ethernet equivalent is ARP [22]).

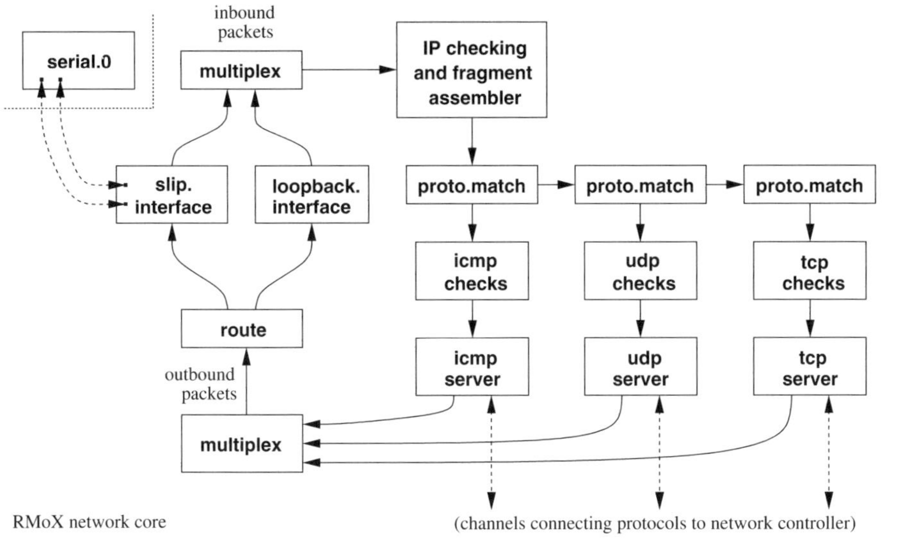

Figure 6: RMoX network core and protocol handling

Figure 6 shows the overall layout of the RMoX/occamnet networking infrastructure, with the SLIP interface at the top-left of the figure. Although it is not done this way currently, the interfaces and protocol stacks will ultimately become dynamic components, giving significantly greater control. At the time of writing, lack of support for some particular types of nested `MOBILE` structures has prevented this dynamic implementation, discussed further in section 5.

Internally, the '`icmp.server`', '`udp.server`' and '`tcp.server`' processes are implemented as process farms, in a style similar to that of RMoX's file-handling (as shown in figure 5).

2.4 The System Console

The '`console`' process in figure 1 provides the traditional "system-console" service. Additionally, it performs the initial system startup. Currently, this is the loading of the '`vga`', '`keyboard`' and '`ramdisk`' device drivers, followed by the mounting of '`ramdisk.fs`' at the file-system root.

Internally, the '`console`' is a network of sub-processes, using the '`vga`' and '`keyboard`' drivers to provide 6 *virtual-terminals*, that each support a basic VT220 functionality. Figure 7 shows this process network.

The first virtual console is used solely for the system log, and is where system messages are displayed (via the shared log channel). When the system first starts, there is only the '`console`' process, with no '`vt.handler`' processes. Until the '`vga`' driver is loaded (and subsequently used for log output), the system log is either simply discarded or written out using a lower-level function (such as any provided by the boot agent, described in section 3).

The '`vga`' driver provides support for six buffers, one of which may be the 'active' (visible) buffer. The switching of buffers is performed by the '`vtkeysw`' and '`vgaplex`' processes, triggered using the keys Alt-F1 through Alt-F6 (in a similar style to Linux).

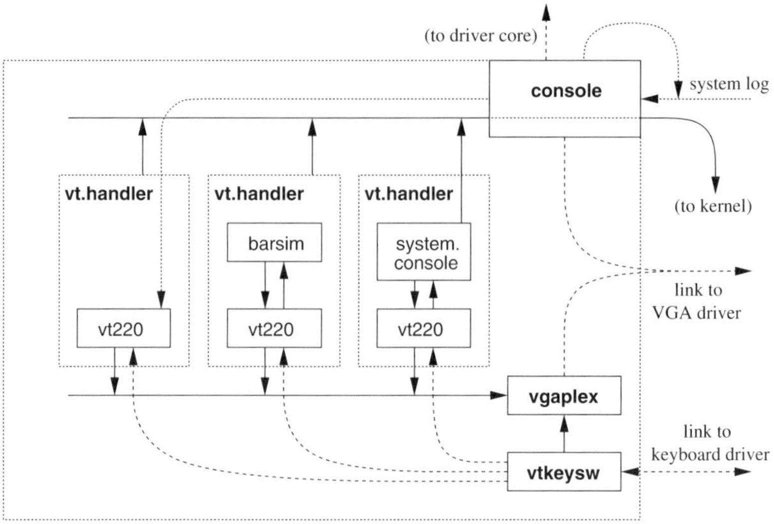

Figure 7: RMoX console process network

The last virtual console is used for the '`system.console`' process. This provides a very basic shell-like functionality, in particular, a set of built-in commands that communicate with the main '`kernel`' process to provide some basic system control (for example, '`mount`' and '`umount`'). Also provided are some more complex built-in commands, such as '`ls`', '`cat`' and '`wfln`' (write text to file). This 'shell' is particularly primitive and does not provide support for features such as pipes and input/output re-direction.

The other virtual terminals are currently either unused or used for demonstrators. Ultimately, these will provide system login points, once those components have been implemented. The processes within the virtual terminals are run at a lower priority than the rest of the operating-system. Although not currently the case (given the applications within RMoX at the time of writing), when the CPU load reaches 100%, we would wish the operating-system components to be scheduled in preference to user applications.

3 Implementation

So far, only the occam components of RMoX have been examined. Much more, however, is required in order to provide operating-system functionality.

The scheduling of occam processes is performed using a modified version of the standard CCSP run-time kernel, that uses lower-level components to provide access to memory regions, I/O regions and interrupts. The use of this extra layer provides a good hardware abstraction for CCSP, that will ultimately aid in the porting of RMoX to other architectures — RMoX is currently restricted to Intel Pentium [23] based architectures.

The current version of RMoX uses the Flux OSKit [24] to provide this functionality. The OSKit can provide much more, however, including schedulers, device drivers, file-systems and network protocol stacks. RMoX currently only uses some of the more basic features of the OSKit — those that provide access to memory, I/O regions and interrupts. The OSKit also provides the boot-loader for RMoX, a fairly complex procedure on Pentium based hardware.

Figure 8 shows the layering of components within RMoX. Hardware is either accessed directly (from RMoX device drivers), or indirectly from CCSP via the OSKit. Accessing hardware directly from occam is in fact relatively trivial. Much of this is related to occam's original application on the Transputer, where there was no need for an operating-system — the hardware performed scheduling and communication directly.

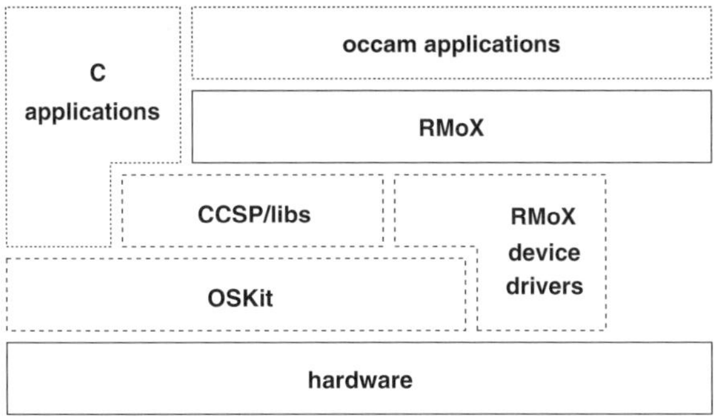

Figure 8: Components of RMoX

Section 3.1 discusses the accessing of (IBM-PC style) hardware from within occam. The provision for supporting concurrent C applications is discussed in section 3.2, although this has not been implemented yet.

3.1 Hardware Access from occam

At the lowest level, RMoX expects to have access to interrupts, memory-mapped I/O regions, and port-mapped I/O regions. The Transputer provided its interrupt mechanism using event-pins [25], mapped directly to communication in occam (using placed channels).

For RMoX, we currently use the OSKit's mechanism for accessing interrupts — which is to call a program-specified function asynchronously when the interrupt occurs. Internally, CCSP maintains an array of interrupt counters and an array of process 'hooks', one for each interrupt[1]. When an interrupt occurs, the appropriate counter value is incremented and a special 'interrupt' sync-flag [26] is set.

When CCSP 'sees' the interrupt flag, it re-schedules any corresponding process waiting for that interrupt, and clears the interrupt counter to zero. If there is no process waiting for an interrupt, nothing happens, and the counter remains unchanged. To synchronise on an interrupt, occam programs call a CCSP built-in that performs the wait (or returns immediately if interrupts have happened), declared with the (external) PROC signature:

```
PROC C.oos.wait.int (VAL INT int, RESULT INT count)
```

This waits for the interrupt specified by 'int', returning the number of interrupts (since the last call) in 'count'. Only one (occam) process may wait for each interrupt. If interrupt sharing is required, it must be implemented in occam (trivial).

The more natural representation of interrupts in occam is that of a channel communication. This can easily be engineered into a process, and adds very little overhead. However, there is no reason to use such an intermediary process, unless an RMoX component needs to ALT against an interrupt.

Accessing I/O memory and ports is somewhat more direct. To enable access to physical memory-ranges or I/O ranges, RMoX provides the following external C functions:

[1]The current implementation provides for 32 hardware interrupts, more than the 16 used by the majority of IBM-PC compatible (uni-processor) hardware.

```
PROC C.oos.mem.map.phys (VAL INT addr, size, VAL BOOL cache,
                         RESULT INT addr, res)
PROC C.oos.io.reserve (VAL INT start, size, RESULT INT res)
PROC C.oos.io.free (VAL INT start, size)
```

Once reserved, memory or I/O ports can be explicitly 'PLACED'. Memory is mapped by the placement of ordinary occam data types — typically arrays or 'PACKED' RECORD types. I/O ports are accessed using 'PLACED PORT's in occam, and are accessed in the same way as channels (using input and output).

The 'keyboard' device driver, for example, uses both interrupts and I/O ports to communicate with the hardware. The main body of this process is simply:

```
INT kyb.ctrl:
INT res:
SEQ
  C.oos.io.reserve (KEYBOARD.IOADDR, KEYBOARD.IOADDR.LEN, res)
  ASSERT (res = 0)

  PLACED [KEYBOARD.IOADDR.LEN]PORT BYTE kyb.control AT KEYBOARD.IOADDR:
  WHILE TRUE
    BYTE kyb.status, kyb.val:
    SEQ
      -- wait for interrupt
      C.oos.wait.int (KEYBOARD.INTR)

      -- read data
      kyb.control[KEYBOARD.STATUS] ? kyb.status
      IF
        (kyb.status /\ KEYBOARD.STATUS.OBF) <> 0
          SEQ
            kyb.control[KEYBOARD.DATA} ? kyb.val
            out ! kyb.val
        TRUE
          SKIP
```

In the actual implementation, a loop is used to retry keyboard reads (up to $10,000$ times) if the 'output buffer full' (OBF) flag is not set.

3.2 Running C Applications in RMoX

RMoX, as viewed by the OSKit, is simply a large monolithic process — CCSP scheduling the occam processes. The OSKit, however, provides support for its own scheduling of applications, using a number of freely-available operating-system schedulers (such as those from Linux and OpenBSD).

Given suitable interfacing, these could be used to provide support for concurrent C applications, allowing a UNIX/POSIX environment to be run on-top of RMoX. Figure 9 shows the proposed method — using a special 'posix.sysif' component that is simultaneously an occam process (within RMoX) and a POSIX system-call interface (within the C process environment).

In this type of setup, the Flux OSKit would provide the main low-level scheduler, scheduling POSIX-interfaced processes alongside RMoX. Unlike C processes, we would not normally allow RMoX to be preemptively scheduled. This ensures that whenever RMoX (or any of its C/occam components) *are not* executing, then RMoX is in a known, safe state — the point at which it invoked the low-level scheduler.

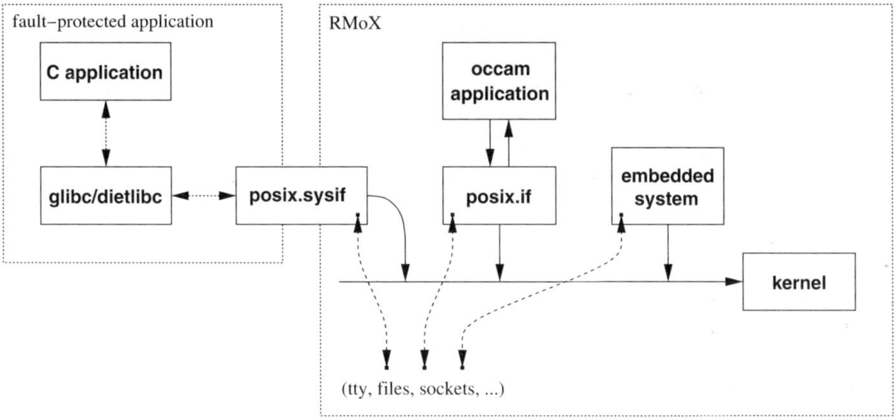

Figure 9: Running C applications alongside RMoX

This way, if a hardware interrupt occurs while a C process is executing, it can be pre-empted immediately, with RMoX resuming at a point where any process waiting for that interrupt will be scheduled first (assuming the absence of higher-priority processes). If a hardware interrupt occurs whilst RMoX is executing, the latency will depend on the frequency at which occam components interact with the scheduler (parallel granularity). Within RMoX, this is typically less than 1 micro-second (on modern hardware, with CPU speeds measured in giga-hertz).

At the time of writing, very little has been implemented with regards to supporting external C processes. Currently, RMoX has no underlying scheduler — it is the only 'process'[2] running. Instead of re-working parts of the Flux OSKit to meet RMoX's requirements (which is still a potential approach), a new low-level scheduler has been developed as part of another student project at UKC. This is discussed further in section 5.1.

4 User-Mode RMoX

User-Mode RMoX is a port of the RMoX kernel and certain drivers to the Linux operating system. In User-Mode RMoX, RMoX itself runs as a user-land process in the host operating system (Linux) and does therefore not have direct access to hardware. Instead, use is made of specific 'user-mode' device-drivers, that emulate any required devices (requiring blocking interaction with Linux in many cases). User-Mode RMoX is largely analogous to the User Mode Linux (UML) port [27].

User-Mode RMoX was initially developed to expedite the RMoX development cycle, which as in any operating system development, is rather cumbersome. Instead of booting from a floppy-disk or network (on raw hardware), which is relatively slow, RMoX is simply run as a Linux program — with almost instantaneous startup.

The reduction in RMoX development time is not the only motivation for User-Mode RMoX. Simply having a user-mode implementation provides easy access for development and general experimentation. Furthermore, User-Mode RMoX makes a good application for extensive testing of the new KRoC/occam features, dynamic channel and process creation in particular.

[2]In this context, a 'process' refers to all code and data contained within a particular *virtual machine*, whose protection is provided for by the hardware.

For RMoX itself, User-Mode RMoX can provide access to devices that are not currently supported within RMoX natively. For example, access to a complete file-system, networking and any (Linux supported) device in general.

This is done by providing file-system, device-driver and networking components that interface with the corresponding Linux/POSIX functionality — essentially glue-code. Interfacing in this way supports the development of components that range from providing the whole functionality, to those that provide only a basic functionality. For example, the RMoX socket interface can either be provided directly (by mapping one-on-one to UNIX sockets), or by using the proposed RMoX networking infrastructure (section 2.3), in combination with a virtualised low-level packet-interface (using the Linux 'tun-tap' driver, for example). In particular, this provides a good environment for developing higher-level RMoX components, that might require functionality not currently available when RMoX is run on raw-hardware.

4.1 User-Mode Alternatives

Other ways of achieving similar (user-mode) functionality do exist, however. For instance, being able to run an unmodified RMoX kernel in user-land is more preferable than running a version that has been modified. Attempts at this, however, have been largely unsatisfactory — for various reasons.

The most straightforward way to run the unmodified RMoX kernel in the host operating system is by the use of a PC *virtualiser*: VMWare, Plex86 and Bochs have all been used with varying degrees of success. In the process of virtualising, however, performance suffers — such systems must provide an entire virtual machine, on top of an existing operating-system. Additionally, there may be compatibility problems. In the case of VMWare, the virtual machine provided is incompatible with RMoX to the extent that attempts to boot RMoX result in VMWare crashing — sometimes taking the host operating-system (Windows or Linux) with it[3]

One advantage of PC virtualisers is that they provide "real" virtual hardware. This can be particularly useful when developing device-drivers whose hardware is particularly sensitive to mis-programming — for example, setting up a DMA (direct memory access) transfer incorrectly, often resulting in severe memory corruption. Device drivers, on the whole, are highly hardware dependant, and the virtual 'hardware' might not accurately model the real hardware.

The Flux OSKit [24], whose functionality RMoX uses, includes its own mechanisms for making a port of kernels built on top of it. This is provided by special 'user-mode' versions of OSKit libraries, plus some additional initialisation code. The modifications within the OSKit to support this are comparable with the modifications in RMoX that make it 'User-Mode RMoX'.

One potential benefit of porting RMoX to the user-mode OSKit is the availability of some OSKit drivers, that are also available in the 'real' OSKit implementation — and could be made available to RMoX. However, RMoX currently does not use any of the device-drivers available in the OSKit, largely as we intend to replace the OSKit with a more dedicated (and simpler) layer in the future (discussed in section 5.1).

4.2 Implementation

The porting of RMoX to a user-mode system is mostly trivial. This can largely be attributed to the fact that RMoX is essentially an occam process network, running on top of a modified CCSP (the run-time system that provides scheduling, communication and dynamic memory

[3]This may very well be due to VMWare's focus on virtualising Windows and Linux.

allocation). Modifying RMoX to run on the standard run-time system is simply a matter of removing or replacing the hardware-specific parts.

Drivers within RMoX access the hardware using memory-mapped I/O, port-based I/O and interrupts, as already discussed in section 3.1. In User-Mode RMoX, the I/O memory-mapper is replaced with a version that simply returns free memory blocks, which the driver can then probe (under the false impression that it is accessing hardware). In most cases, RMoX drivers will simply fail when initialised in the 'virtual' environment, correctly reporting that to the RMoX component(s) attempting to use them.

The only drivers which must be completely replaced for User-Mode RMoX are the 'keyboard' and 'vga' drivers, that provide the basic display and input functionality. In RMoX proper, these drivers use port-mapped I/O and interrupts (for the keyboard driver) to access the hardware directly. The interface to User-Mode RMoX is provided through 'xtextwindow' — a set of X11 routines that provide the display (in a standard X11 window), and the keyboard (through events received on that window).

The user-mode and 'real' RMoX systems are built from the same set of source files — with the exception of replacement device drivers and the user-mode interface. Use is made of the recently added pre-processor capability for this, that supports the conditional compilation of code based on the definition of a pre-processor constant (very much like the C pre-processor's '`#ifdef`' directive). The constant that selects a user-mode version of RMoX, '`USER.MODE.RMOX`', is set in the 'Makefile' (passed via the KRoC command-line to the occam compiler).

The changes required within RMoX (including its supporting code and build system), to support a user-mode implementation, are small — up to the point of providing specialised user-mode drivers. The provision of 'fake' I/O memory-mapping, and dummy handling for port-mapped I/O and interrupts, inside the run-time system (CCSP), also requires few changes.

4.3 Current and Future User-Mode Sub-Systems

The screen and keyboard drivers are perhaps the most important user-mode drivers, as without those it is not possible to interact with a running user-mode RMoX (until a method of accessing RMoX from the network becomes available, that is).

Two versions of these screen and keyboard drivers are currently supported — an X11 based interface ('xtextwindow'), and a text-console based interface. The X11 interface displays a window on the host display, providing keystrokes to the user-mode 'keyboard' driver and displaying data sent to the 'vga' driver. Unfortunately, concurrent interaction in Xlib (the X11 client library for C), is impractical — the mechanisms which Xlib uses internally to protect from race-hazards is non-compatible with the mechanism used to implement blocking system-calls [28] (that are required for the blocking interaction with Xlib).

This problem is solved by allowing device-drivers to request connections to each other, via the main '`driver.core`' process. The user-mode 'vga' driver is responsible for (serial) interaction with the underlying C code (and Xlib). The user-mode 'keyboard' driver requests a connection to the 'vga' driver, that returns a channel-end to which keyboard events are sent. As long as the structuring of inter-driver requests is strictly client-server, as is the case here, there is no possibility for deadlock. There are good reasons for including this functionality, however. For example, a software RAID [29] driver that connects to multiple physical disk drivers.

4.3.1 The Console Interface

The console interface for User-Mode RMoX is somewhat simpler than the X11 interface. RMoX, as a KRoC occam program, does not have the three standard top-level channels (keyboard, screen and error), and passing these channels down through the driver-core to the user-mode 'keyboard' and 'vga' processes would be messy.

Fortunately, KRoC itself implements the handling for the three top-level channels using three *real* occam processes (that access the UNIX keyboard, screen and error streams through standard interfaces — external C calling [30] and blocking system-calls). The user-mode 'keyboard' and 'vga' drivers simply replicate and extend this functionality — in order to support the RMoX interface. This also includes supporting VT220 escape sequences generated by RMoX during the initial startup (before its own VT220 handling processes, within the 'console', are activated). This handling is also required in the X11 drivers, but given the relatively small number of different escape sequences generated, both implementations are reasonably simple.

The console interface is somewhat faster then the X11 interface, but requires a correctly sized terminal window to display output correctly — this particularly applies when running within an 'xterm' or similar virtual terminal.

4.3.2 Host File-System Access

User-Mode RMoX provides a complete file-system in the form of "host-fs" — the host file-system interface. This provides, for RMoX, all the POSIX defined file and directory handling operations (with the exception of memory-mapping), translating requests to Linux file-handling system calls. The 'host-fs' file-system driver is written entirely in occam, and uses the KRoC file-library for file and directory access [31] (that has undergone minor improvements and has been extended slightly, to support all the operations required by RMoX/POSIX).

Some complications arise from differences between RMoX and Linux directory handling. RMoX allows the 'seek' operation to be performed on directories — Linux does not. This is handled by having the 'host-fs' driver cache directory entries, such that 'seek' and 'read' operations can be handled in unit-time. No limits are imposed on the number of directories which can be cached in this way, although they could be, if required.

A virtual 'block' device driver is provided with User-Mode RMoX, that gives seekable block-structured access to a file on the host file-system. Such a file can easily be 'mounted' under Linux, using the 'loop-back' interface — provided, of course, that the file contains a file-system which Linux supports. More usefully, a file-system can be created on the file (from within Linux), and subsequently used for developing RMoX's own native file-system drivers (for the Linux second-extended file-system, for example). Since Linux supports seekable access to block-devices through the file-system, this mechanism can be used to provide User-Mode RMoX with access to physical disk partitions — assuming that the User-Mode RMoX process has sufficient privileges under Linux to open such devices.

Use could be made of this in a dual-boot RMoX and Linux/User-Mode RMoX system, with a partition used exclusively for RMoX. In this way, the partition can be accessed in both environments — either by using the user-mode 'file' block-device, or using a native RMoX block disk device-driver (that, currently, is only partially implemented).

The User-Mode Linux port provides some interesting variations from standard block device-drivers, such as sparse block devices and COW (copy-on-write) block devices [27]. Incorporating these ideas into User-Mode RMoX at some stage would be beneficial — allowing more conservative disk-usage for host 'file' file-systems (the size actually used by files and other information, not the size of the virtual device).

5 Conclusions and Future Work

At the time of writing, RMoX is very much incomplete, but much of the ground-work for future development is in place. Functional, but basic, instances of some components exist — such as the ramdisk file-system and serial device-driver (in addition to the keyboard and VGA drivers).

Looking towards the future, it is hoped that the development effort on RMoX can be stepped-up, now that a functional user-mode implementation exists, under which much development can be done.

At some point, we intend to replace the OSKit layer with a more specialised implementation — with some consideration to portability. The OSKit currently only supports Intel i386 based platforms. We aim to be broader than that, particularly considering the range and scale of hardware and applications for which RMoX, and CSP-based technologies in general, are suitable.

The long-term possibilities for RMoX are vast, particularly if our claims about security, scalability and reliability hold true in real-life situations. The first (significant) demonstrator produced will be RMoX combined with a modified version of the occam web-server [32], which we hope will perform relatively well in comparison with other web-servers, given the comparative immaturity of RMoX.

5.1 A Dedicated Low-Level Scheduler

One of the main motivations behind providing our own low-level scheduler, besides portability, is the added level of simplicity that it offers. The Flux OSKit is a very general-purpose tool-kit — as it was designed to be — and RMoX only uses a small proportion of its functionality.

The reason for wanting a low-level scheduling functionality (that involves hardware support for memory protection), is to support the concurrent execution of UNIX/POSIX processes — typically those written in C — as discussed in section 3.2. Such separate processes must still interact with RMoX however, since it provides the kernel (POSIX) functionality. In particular, we want RMoX to retain control at all times — i.e. it must never be preempted.

An experimental scheduler and memory-manager has been developed by Ramsay Taylor as a final-year project at UKC [33]. Currently, however, it is not sufficiently complete for use with RMoX, although a basic version is not far away.

5.2 Distribution and Replication

Figure 8 contains a gap. In the future, we intend to fill this space with functionality that provides distribution and replication for networked devices, as dynamically and transparently as possible. At one level, this could be applied to networks of workstations to provide the combined resources of all the nodes — a general-purpose distributed operating-system.

At another level, for example (and very hypothetically), such functionality could be used to inter-link the embedded control-systems of Earth orbiting satellites, to a variety of ends — from ensuring that they avoid each other (where possible/useful) to providing transparent replication of services in cases of failure.

Techniques for distributed operating-system control are already well-investigated, particularly in the networks and distributed systems field of research. It is highly likely that much of this previous work will be relevant when building a distributed RMoX.

5.3 User-Mode RMoX

Work is currently underway to provide a network interface for User-Mode RMoX, at either the Ethernet or IP level, using the 'tun-tap' driver in Linux[4]. This provides what is essentially a virtual 'network-card', allowing development and testing of the RMoX networking infrastructure (protocol stacks in particular) within the user-mode environment.

Although RMoX currently has little use for pointer interfaces (typically mice and pen-based devices), the provision of a virtual 'mouse' driver from the X11 interface should be trivial, using a similar mechanism to the user-mode 'keyboard' driver, that collects X11 key events from the 'vga' driver. Providing a similar driver for the console user-mode interface is trickier, but can be done by collecting data from 'gpm', a program that provides 'cut and paste' functionality in the Linux virtual consoles. Collecting similar events from X11 virtual terminals (such as an 'xterm') is hard, and can only be done by connecting to the xterm's X11 window and collecting keyboard events manually (through Xlib). The use of a mouse in the RMoX console environment is fairly limited, however, especially when we intend to provide graphical interfaces in the future (both for the raw-metal and user-mode versions of RMoX).

Work is currently underway to provide a native (occam) port of the MINIX [34] file-system, that is small, portable and widely implemented. As there is currently no support in RMoX for a real physical block device (such as a hard-disk or floppy-disk), the MINIX file-system driver is being developed in User-Mode RMoX using a virtual 'file' block device to emulate the missing physical device support.

5.4 Future Uses for User-Mode RMoX

The current use of User-Mode RMoX is largely to improve the 'test-debug' cycle time of RMoX itself. However, as RMoX matures, it is conceivable that the role of User-Mode RMoX will shift away from a development-only tool to more general uses. One perceived use is that of student investigations into concurrent operating-system design and implementation, within a CSP/occam framework. Both the X11 and console interfaces can be run remotely, making such investigation easily accessible — the only requirements are a Linux system and remote terminal or X11 display clients.

One potential future use, in conjunction with the user-space KRoC.net infrastructure [35], is to provide an interim mechanism for future experiments with distribution and replication of RMoX— with one intention being the creation of a clustered operating-system environment, as described in section 5.2. On a single system, multiple instances of User-Mode RMoX could be run to provide a 'virtual cluster'. Such setups would also be useful for testing distributed applications, which we later intend to run on real hardware clusters. Also, for demonstrating and experimenting with cluster computing in CSP/occam, where a real cluster might not available.

Acknowledgements

We would like to thank Peter Welch, Dyke Stiles, Adam Sampson and Ramsay Taylor from UKC, for their various input, ideas and code. We also wish to thank the WoTUG/CPA community, who have provided much discussion and enthusiasm for this work. In particular, Alex Brodsky, Matt Pedersen and Peter Welch, who came up with the name "RMoX".

---

[4]Similar functionality also exists in other operating-systems, as it provides a convenient way of allowing applications to access raw IP/Ethernet interfaces — typically with a different IP than that of the host-system, which simply forwards packets to/from the 'tun-tap' interface(s).

References

[1] C.A.R. Hoare. Communicating Sequential Processes. *Communications of the ACM*, 21(8):666–677, August 1978.

[2] C.A.R. Hoare. *Communicating Sequential Processes*. Prentice-Hall, 1985. ISBN: 0-13-153271-5.

[3] A.W. Roscoe. *The Theory and Practice of Concurrency*. Prentice Hall, 1997. ISBN: 0-13-674409-5.

[4] Inmos Limited. occam 2.1 Reference Manual. Technical report, Inmos Limited, May 1995. Available at: `http://www.wotug.org/occam/`.

[5] Brian W. Kernighan and Dennis M. Ritchie. *The C Progamming Language*. Prentice Hall, Englewood Cliffs, 1 edition, 1978. ISBN: 0-13-110163-3.

[6] T. Jim, G. Morrisett, D. Grossman, M. Hicks, J. Cheney, and Y. Wang. Cyclone: A safe dialect of C. In *Proceedings of the 2002 USENIX Annual Technical Conference*, Monterey, California, USA, June 2002. Available at: `http://www.usenix.org/events/usenix02/jim.html`.

[7] Daniel P. Bovet and Marco Cesati. *Understanding the Linux kernel.* O'Reilly UK, 2nd edition, January 2003. ISBN: 0-596-00213-0.

[8] P.H. Welch, J. Moores, F.R.M. Barnes, and D.C. Wood. The KRoC Home Page, 2000. Available at: `http://www.cs.ukc.ac.uk/projects/ofa/kroc/`.

[9] F.R.M. Barnes and P.H. Welch. Mobile Data, Dynamic Allocation and Zero Aliasing: an occam Experiment. In Alan Chalmers, Majid Mirmehdi, and Henk Muller, editors, *Communicating Process Architectures 2001*, volume 59 of *Concurrent Systems Engineering*, pages 243–264, Amsterdam, The Netherlands, September 2001. WoTUG, IOS Press. ISBN: 1-58603-202-X.

[10] F.R.M. Barnes and P.H. Welch. Prioritised Dynamic Communicating Processes: Part I. In James Pascoe, Peter Welch, Roger Loader, and Vaidy Sunderam, editors, *Communicating Process Architectures 2002*, WoTUG-25, Concurrent Systems Engineering, pages 331–361, IOS Press, Amsterdam, The Netherlands, September 2002. ISBN: 1-58603-268-2.

[11] F.R.M. Barnes and P.H. Welch. Prioritised Dynamic Communicating Processes: Part II. In James Pascoe, Peter Welch, Roger Loader, and Vaidy Sunderam, editors, *Communicating Process Architectures 2002*, WoTUG-25, Concurrent Systems Engineering, pages 363–380, IOS Press, Amsterdam, The Netherlands, September 2002. ISBN: 1-58603-268-2.

[12] F.R.M. Barnes and P.H. Welch. Prioritised dynamic communicating and mobile processes. *IEE Proceedings – Software*, 150(2), April 2003.

[13] International Standards Organization, IEEE. Information Technology – Portable Operating System Interface (POSIX) – Part 1: System Application Program Interface (API) [C Language], 1996. ISO/IEC 9945-1:1996 (E) IEEE Std. 1003.1-1996 (Incorporating ANSI/IEEE Stds. 1003.1-1990, 1003.1b-1993, 1003.1c-1995, and 1003.1i-1995).

[14] J. Moores. CCSP – a Portable CSP-based Run-time System Supporting C and occam. In B.M. Cook, editor, *Architectures, Languages and Techniques for Concurrent Systems*, volume 57 of *Concurrent Systems Engineering series*, pages 147–168, Amsterdam, the Netherlands, April 1999. WoTUG, IOS Press. ISBN: 90-5199-480-X.

[15] J. B. Postel. Internet protocol. RFC 791, Internet Engineering Task Force, September 1981.

[16] Adam T. Sampson. The design and development of occamnet, 2003. Project report, Computing Laboratory, University of Kent.

[17] M.D. May, P.W. Thompson, and P.H. Welch. *Networks, Routers and Transputers*, volume 32 of *Transputer and occam Engineering Series*. IOS Press, 1993.

[18] M. Homewood, D. May, D. Shepherd, and R. Shepherd. The IMS T800 Transputer. *IEEE Micro*, pages 10–26, October 1987.

[19] Inmos Limited. *The T9000 Transputer Instruction Set Manual.* SGS-Thompson Microelectronics, 1993. Document number: 72 TRN 240 01.

[20] Nathalie Supornpaibul. Protocol implementation in occam. *Transputer Applications*, 1989.

[21] J. L. Romkey. Nonstandard for transmission of IP datagrams over serial lines: SLIP. RFC 1055, Internet Engineering Task Force, June 1988.

[22] J. B. Postel and J. F. Reynolds. Standard for the transmission of IP datagrams over IEEE 802 networks. RFC 1042, Internet Engineering Task Force, February 1988.

[23] Intel Corporation. *Intel Architecture Software Developer's Manual, Volume 2: Instruction Set Reference*, 1999. Available at: `http://developer.intel.com/design/PentiumIII/manuals/`.

[24] Bryan Ford, Godmar Back, Greg Benson, Jay Lepreau, Albert Lin, and Olin Shivers. The flux OSKit: A substrate for kernel and language research. In *Symposium on Operating Systems Principles*, pages 38–51, 1997. Software available from: `http://www.cs.utah.edu/flux/oskit/`.

[25] Inmos Limited. *The T9000 Transputer Hardware Reference Manual.* SGS-Thompson Microelectronics, 1993.

[26] P.H. Welch and D.C. Wood. The Kent Retargetable occam Compiler. In Brian O'Neill, editor, *Parallel Processing Developments, Proceedings of WoTUG 19*, volume 47 of *Concurrent Systems Engineering*, pages 143–166. World occam and Transputer User Group, IOS Press, Netherlands, March 1996. ISBN: 90-5199-261-0.

[27] Jeff Dike. User-mode Linux. In *Proceedings of the 5th Annual Linux Showcase & Conference*, Oakland, CA, USA., November 2001. Available from `http://www.linuxshowcase.org/dike.html`.

[28] F.R.M. Barnes. occwserv – an occam Web-Server, September 2001. CPA-2001 SIG presentation, University of Bristol, UK. Available at: `http://frmb.org/publications.html`.

[29] Peter M. Chen, Edward K. Lee, Garth A. Gibson, Randy H. Katz, and David A. Patterson. RAID: High-performance, reliable secondary storage. *ACM Computing Surveys*, 26(2):145–185, 1994.

[30] David C. Wood. KRoC – Calling C Functions from occam. Technical report, Computing Laboratory, University of Kent at Canterbury, August 1998.

[31] Fred Barnes. *Socket, File and Process Libraries for occam.* Computing Laboratory, University of Kent at Canterbury, June 2000. Available at: `http://www.cs.ukc.ac.uk/people/rpg/frmb2/documents/`.

[32] F.R.M. Barnes. occwserv: an occam web-server. In *Communicating Process Architectures 2003*, WoTUG-26, Concurrent Systems Engineering, IOS Press, Amsterdam, The Netherlands, September 2003.

[33] Ramsay G. Taylor. Implementing Memory Management for Highly Concurrent Systems, 2003. Project report, Computing Laboratory, University of Kent.

[34] Andrew S. Tanenbaum. MINIX: A UNIX clone with source code for the IBM PC. *login: the USENIX Association newsletter*, 12(2):3–9, March/April 1987. ISSN: 1044-6397.

[35] M. Schweigler, F.R.M. Barnes, and P.H. Welch. Flexible, Transparent and Dynamic occam Networking with KRoC.net. In *Communicating Process Architectures 2003*, WoTUG-26, Concurrent Systems Engineering, IOS Press, Amsterdam, The Netherlands, September 2003.

Communicating Process Architectures – 2003
Jan F. Broenink and Gerald H. Hilderink (Eds.)
IOS Press, 2003

Scheduling for ILP in the 'Processor-as-a-Network'

D. K. Arvind and S. Sotelo-Salazar
ICSA, School of Informatics, University of Edinburgh, Edinburgh EH9 3JZ, Scotland

Abstract. This paper explores the idea of the processor as an asynchronous network, called the micronet, of functional units which compute concurrently and communicate asynchronously. A micronet-based asynchronous processor exposes spatial as well as temporal concurrency. We analyse the performance of the 'processor-as-a-network' by comparing three scheduling algorithms for exploiting Instruction Level Parallelism (ILP). Schedulers for synchronous architectures have relied on deterministic instruction execution times. In contrast, ILP scheduling in micronet-based architectures is a challenge as it is less certain in advance when instructions start execution and when results become available. Performance results comparing the three schedulers are presented for SPEC95 benchmarks executing on a cycle-accurate model of the micronet architecture.

1 Introduction

The instruction execution times in a clocked or synchronous processor is fixed at the design stage. They are expressed in terms of numbers of clock cycles, which is both precise and deterministic (except, perhaps, in the case of memory instructions with caches), which forms the basis of compiler optimisations for Instruction Level Parallelism (ILP). In contrast, in a clock-free, micronet-based, asynchronous architecture [1], the operations proceed at their own speed which is dependent on the nature of the data, local delays and the availability of architectural resources. As a result the execution times of instructions would vary, albeit within a range, in a non-deterministic manner. This poses an interesting problem for an ILP compiler, which can no longer assume exact and deterministic instruction execution times for the purposes of scheduling and optimisation.

Instruction scheduling for architectures with different types of resources is known to be NP-hard, and a large body of work exists for clocked processors [2] [3] [4] [5]. This paper analyses the performance of the 'processor as a network', and, in particular, strategies for ILP scheduling in a micronet-based asynchronous processor (MAP). The PTD scheduler [6] had previously been proposed for scheduling instructions for architectures with uncertain latencies. The performance of the PTD scheduler is evaluated against two well-known list schedulers - the Gibbons-Muchnik (GM) [3] and the Balanced [4] schedulers. In the rest of this paper: Section 2 describes the architecture of the micronet-based processor; Section 3 presents scheduling strategies for ILP in micronet-based architectures, and the algorithmic complexity of the PTD scheduler is derived; Section 4 describes the Evaluation Framework which produced the results presented in Section 5.

2 Micronet-based Asynchronous Processor (MAP) Architectures

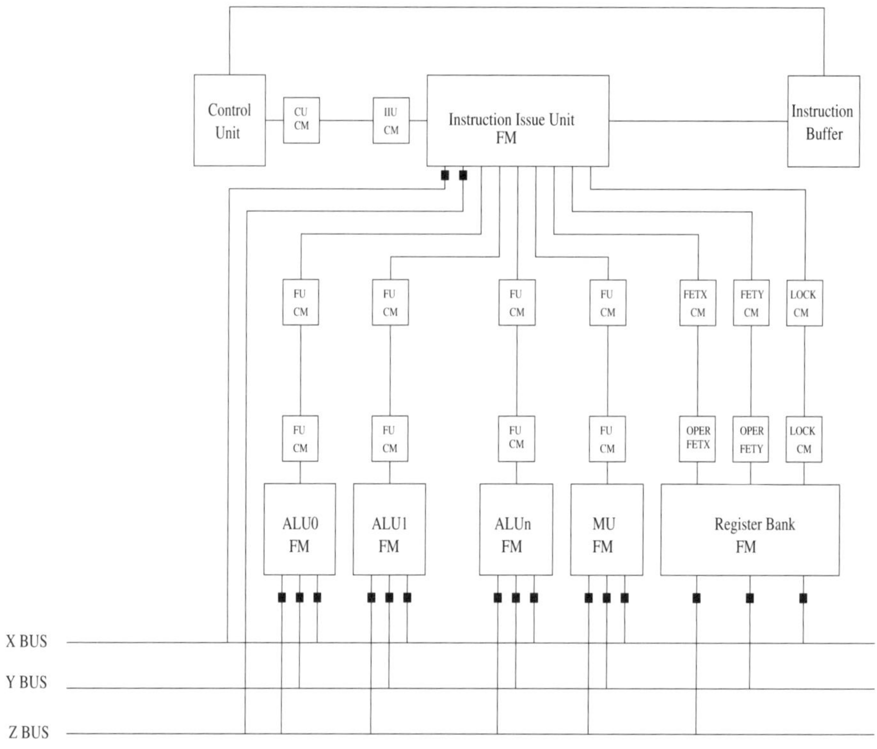

Figure 1: A MAP datapath

The micronet is a network of entities which compute concurrently and communicate asynchronously, with minimal centralised control. MAP is a family of processors based on the micronet operational model. The MAP datapath, as shown in Figure 1, is a *network* of execution units, such as the Instruction Issue Unit, Control Unit, Register Bank, and functional units such as the Arithmetic Unit(s) (AU), Logic Unit (LU), Floating Point Unit (FPU) and the Memory Unit (MU). The 32-bit datapath executes a subset of the MIPS instruction set.

Each execution unit is composed of a *Functional Microagent (FM)* which executes a specific micro-operation, and communicates with other units via *Communicating Microagents (CM)*, by employing a four-phase handshaking protocol. The Instruction Buffer caches instructions for the Instruction Issue Unit, and the Control Unit mediates their operations. On the issue of an instruction, the appropriate control signals are asserted and the destination register is locked, and the instruction is considered issued once the signals have been acknowledged. The CMs between the register bank and the X and Y buses communicate to place the register values, as specified in the instruction, on the X and Y buses, respectively. The ALU will operate on the values and the results are placed on the Z-bus and written back to the destination register, which is then unlocked. In its simplest form, the MAP architecture does not assume queues for the operands and the results in the execution units.

The issue unit issues one instruction at a time in an in-order manner. It is responsible for issuing instructions as soon as the operands become ready, *i.e.*, when they are written

Component Type		Latency time		Cycle time	
		Minimum	Maximum	Minimum	Maximum
Issue unit	(IU)	1.00	2.00	0.50	1.00
Read buses	(RF)	2.00	4.00	0.50	1.00
Write buses	(RF)	2.00	4.00	0.50	1.00
Arithmetic unit	(AU)	4.00	8.50	0.50	1.00
Logical unit	(LU)	2.00	7.00	0.50	1.00
Floating point unit	(FPU)	6.00	8.00	0.50	1.00
Memory unit	(MU)	10.00	20.00	0.50	1.00

Table 1: Latency and Cycles times (in ns) for the architectural components

into the register file. When the operands are ready, the issue unit inspects if both *read* buses and a functional unit of the appropriate type are all available. If this is the case, then the instruction is issued; otherwise, it is stalled, and the outcome depends on which resources were unavailable, as described next.

If a functional unit of the appropriate type is unavailable, but the *read* buses and operands are, then the instruction will be issued and the operand fetch will proceed normally. It will stall once the operands are read, and will remain so until the functional unit becomes available to execute the operation.

The second scenario occurs when any of the source operands are not ready (data requirement), or the *read* buses are unavailable (resource requirement). In either case, both the issue unit and the instruction will remain stalled until the data and resource requirements are satisfied.

Once a functional unit completes its execution, the result is sent to the register as soon as the Z bus becomes available. Otherwise, the functional unit stalls and remains in a "busy" state until it can deliver its result to the register.

A register *locking* scheme is employed to ensure that data coherency is maintained during the execution of the instruction. Such a scheme guarantees correctness of the data in the presence of asynchronous accesses to the register file. In MAP, each register has a lock bit which when set disables any read or write accesses to that particular register. Since there is no guarantee as to when a register will be unlocked, the issue unit cannot issue instructions which are dependent on a locked register, *i.e.*, if there is a RAW or WAW dependency, and will therefore remain stalled.

Conversely, when the issue unit proceeds to issue an instruction, its destination register is locked until the functional unit commits the result of the instruction to the register. If a subsequent instruction is stalled waiting on that data, then it will change status and issue if the other conditions are satisfied and will in turn lock the destination register of that instruction.

The delay values for the architectural components, as listed in Table 1, capture the range of latency values for the functional units, together with their cycle times, which is defined as the delay for components to become ready after completion of an operation. The distribution of latency times over the interval is determined by the type of functional unit, *e.g.*, the memory unit has a bimodal distribution which models the cache behaviour - the extreme values represent cache hits and misses; the distribution of latencies for the arithmetic units is based on the graph in Figure 4 in [7]; and the distribution is assumed to be uniform in the case of the Logic Unit.

2.1 Analysis of the MAP architecture

The MAP architecture, as shown in Figure 1, is a scalar architecture that features a single, in-order, instruction issue unit, with the results being committed fully out-of-order. Once instructions are issued, they proceed at their own speed, which is determined by the availability of data and resources. More than one instruction in flight implies that instructions may overtake.

The MAP architecture is a hybrid of certain features of VLIW and superscalar classes of architectures. Like VLIW architectures, the issue unit in the MAP architecture relies on the compiler to statically determine the schedule of independent instructions for execution. Similar to superscalar architectures, the MAP architecture is required to prevent run-time hazards. It differs from VLIW architectures in that it does not have to issue a number of independent instructions or no-operations in lock-step. Instead, the MAP architecture issues instructions as soon as the resources and data dependencies allow it to, and relies on the compiler to identify a schedule which minimises the instruction issue stalls and deals instructions at a rapid enough rate to keep all the functional units busy.

The compiler takes the strain in identifying a schedule which minimises the program execution time by maximising the instruction level parallelism and minimising stalls either in the issue unit or the functional units. Unlike a superscalar architecture, dyanamic data-forwarding is not implemented as this would require the operand fetch stage to be synchronised with the write-back stage. This synchronisation will inevitably slow down the faster stage, *i.e.* the fetch stage, when two dependent instructions are fetched and issued in succession. In such cases this synchronisation will take place whether or not data forwarding is implemented. However, if instructions are not scheduled one after the other, or if there is more than one instruction awaiting the result of the first instruction, then the write-back stage of the first instruction will be held up unnecessarily.

Once an instruction is issued in a micronet model, *i.e.* data and resource requirements are fulfilled, then it runs to completion without any synchronisation. Unlike VLIW architectures, there is no need to include "bubbles" in the schedule which is the case when no-ops are issued if the compiler cannot identify independent instructions for those time slots, *i.e.* those functional units will remain idle for an entire instruction cycle. Micronets have no need to issue no-ops which do not contribute to the execution of the program; instead, the compiler for the MAP architecture is required to schedule independent instructions such that the issue unit does not stall, or stalls minimally if it does.

The following section introduces the scheduling problem for a MAP architecture and the proposed solutions.

3 Instruction Scheduling for MAP Architectures

For a general, p-functional unit processor, the complexity of the scheduling problem using a non-preemptive approach is NP-complete (p is assumed to be greater than 2; in the case of $p = 2$, the complexity is polynomial, if the functional units are identical and the latencies are all the same). The problem is NP-hard in the case of a processor with functional units of different types and differing latencies. Scheduling hueristics are therefore employed which exploit domain knowledge to prune this exponentially increasing search space. The *list* scheduler uses a priority list of *ready* instructions. The hueristic will decide the best candidate instruction to be scheduled next, and when the instruction is removed then all its successors become available for consideration in the next cycle.

Gibbons and Muchnick [3] is an example of a list scheduler in which the ready instructions are prioritised on the basis that the candidate instruction will not cause an interlock with

Types of dependencies	Consecutive instructions	Separated by one inst.
True dependency with a load instruction	3	1
True dependency with a branch instruction	2	0
Any other true dependency	1	0
Resource dependency from a memory instruction	1	0

Table 2: Degree of the penalties depending of the type of dependency

the previous one, and given a choice, the candidate instruction is more likely to interlock with instructions after it.

The *Balanced* scheduler [4] was originally devised to take account of unpredictable memory access latencies. It is based on the idea of computing weights for load instructions based on the number of available independent instructions. These are scheduled, as in a traditional list scheduler, with independent instructions being distributed behind loads to buffer against unpredictable memory accesses. This idea is generalised to a MAP architecture in which all the instructions have unpredictable latencies. The priority for ready instructions is based on a weighted sum of values derived from heuristics tailored to the micronet model. These include the following: whether the instruction uses the same resources as the previous scheduled one; the number of immediate successors of the instruction; the length of the longest path from the instruction to the leaves of the graph; and, the number of source registers which are freed should the instruction be scheduled, which effectively takes account of register pressure.

The Penalise True Dependence (PTD) scheduler [6] had been proposed as a method for scheduling instructions for MAP architectures. The essence of the scheduler is to identify true data and resource dependencies in an instruction schedule and re-order the instructions where possible, so as to reduce the stalls in the instruction issue unit due to them. The PTD scheduler calculates a *penalty measure* which reflects the degree of resource contentions and stalls due to data dependencies. The scheduler moves instructions which would lower the penalty measure (Table 2 lists the degree of the penalties). The scheduler traverses the schedule to evaluate optimisations on every instruction that is penalised. In order to reduce the penalty measure the scheduler must find independent and unrelated instructions to place in between the dependent ones.

The instruction schedule may be visualised as a "horizontal" sequence of instructions which are executed in order, from left to right. When a penalised instruction is encountered, then an independent, unrelated instruction is searched on both sides of the offending instruction, first starting on the left, and if unsuccessful, switching to the right of the instruction. There are two necessary conditions for an instruction to be considered a suitable candidate for movement ahead of the penalised instruction. Firstly, the instruction in question has to be independent of the penalised instruction, and secondly, the instruction has to be independent of all instructions scheduled in between the candidate and the penalised instructions. These conditions are necessary to preserve the semantics of the code and are known as *valid* conditions. They allow only valid movement of instructions in which the order of execution is preserved, but the performance of the outcome of such movements has to be analysed further. The *safety* conditions guarantee that a valid movement is chosen which results in the penalty measure being reduced.

Benchmark	Total lines	GM. Time	Bal. Time	PTD Time	Percen.
intmm	151	0.0836	0.0839	0.0680	22.94 %
livermore	1915	3.7947	3.8388	2.7817	36.42 %
fract	5336	3.5940	3.5588	2.1920	62.35 %
li	16832	8.7857	8.1169	4.4349	83.02 %
puzzle	936	0.7781	0.6369	0.4711	35.19 %
compress	1236	0.9757	0.7300	0.5494	32.87 %
go	83838	65.2829	60.1321	44.0446	36.53 %
m88k	34089	22.0481	20.6159	12.6411	63.09 %

Table 3: Average compilation times (in seconds) of the benchmarks

The main reason for searching candidate instructions starting from the left-hand side of the penalty is because there is a greater likelihood of finding one faster. The exit of the basic block can be seen as a synchronisation point, and may therefore offer fewer options, and conversely, the entry of the basic block may offer a greater scope for suitable candidates. The scheduler aims to first reduce penalties with the higher delay-costs to the issue unit, *i.e.*,reduce the higher penalty values according to Table 2, which corresponds to data dependencies due to loads. The second pass attempts to reduce penalties due to branch instructions in the basic block, if it ends with such an instruction. And, in the final pass, the scheduler tries to reduce the remaining penalties with a value one.

The complexity of the PTD scheduler is derived to be $\theta(et^2 + n - e)$ (see Section 3.1 for details), where e is the number of penalties and n is the number of instructions in a basic block, and t is the distances (in terms of number of instructions) between the penalised and candidate instructions. The PTD scheduler compares favourably with the other two, as its complexity is governed by the number of penalties in a basic block (rather than the number of instructions), and which reduces as the algorithm progresses. In comparison, the number of iterations is constant in the case of the other two traditional techniques, and is proportional to the number of instructions in the basic block.

Table 3 gives a comparison of the average compilation times for the three schedulers. It gives the average of five compilation times (in seconds) of the scheduling sections in the three schedulers (this was obtained using the gethrtime function from the time.h standard library). The last column represents the percentage improvement of the PTD scheduler against the faster of the other two schedulers. The compilation times are on average 39% faster, with notable improvements of more than 60% for the fract and m88k benchmarks, and reaching a peak of 83% in the case of the li benchmark. The standard deviations revealed a narrow range of variations in the compilation times.

3.1 Algorithmic Complexity of the PTD Scheduler

The structure of the PTD scheduler is different from the GM and Balanced schedulers as the algorithm is driven by the penalties in the code. The complexity of the PTD scheduler is derived next.

The while sections in the functions *PTD_resource_phase* (Algorithm 1 in Appendix A), *PTD_consecutive_phase* (Algorithm 2) and *PTD_nonconsecutive_phase* (Algorithm 3) traverse the basic block stopping at every penalty. These sections have a complexity of $\theta\,(\,et^2\,+\,(n\,-\,e)\,)$, where n is the number of instructions in the basic block, and e is the number of penalties. The term, et^2, corresponds to the time spent searching for a candidate instruction and checking its dependencies when the functions *PTD_arrange_left* and *PTD_arrange_right* are called. The term, $(n - e)$, covers the instructions that are not penalised.

The `repeat` loops (lines `3-16`, `17-30` and `31-44` in Algorithms 1, 2 and 3, respectively) ensure that at least one scheduling pass is performed. The `repeat` sections continue until there are no reductions in the penalty measure. The number of times these sections are repeated is denoted by c. Thus, the above term becomes $\theta\,(\,cet^2\, +\, c\,(n - e)\,)$.

However, the parameters c, e and t are not general. Basic blocks have different number of penalties (e) of a given type and particular number of retries (c) for each basic block. Similarly, penalties have different instruction distances (t) when searching for a candidate.

The `repeat` block is replicated eight times in the three scheduling phases. Therefore, the complexity of the PTD scheduler is

$$\theta\,(\,8\,cet^2 + 8\,c\,(n - e) + 8\,nc + 3\,n\,) \tag{1}$$

The terms, $8\,nc$ and $3\,n$, represent the computation of the penalty measure throughout the three scheduling phases, both inside and outside the `repeat` sections, respectively.

Equation 1 has a few important simplifications. Observations of the scheduling process show that the number of times the `repeat` sections are looped is not greater than three or four. Therefore, c can be considered to be a small constant ($c = 2, 3, 4$). Since the search for candidate instructions start from the instruction neighbouring the penalised one, it is expected that the search (on either side) does not reach $n/2$. Furthermore, if the candidate is found on the left-hand side of the penalised instruction, the search on the right-hand side is not necessary. Thus, the factor t can be considered to be: $t \ll n$. As for the term e, it is often the case that there are not as many penalties of the same type as there are instructions, which results in $e < n$.

The upper and lower bounds of Equation 1 are defined by two opposite scenarios. The upper bound is represented by pure sequential code. It takes place when there are as many penalties as instructions ($e = n$), and there are no candidates found for all of those penalties ($t = n - 1$) and there can only be two retries ($c = 2$). The upper bound of Equation 1 is therefore

$$\theta\,(\,16\,n\,(n - 1)^2 + 19\,n\,) \tag{2}$$

However, since the same type of penalty cannot affect more than one `repeat` section, only one term is governed by $cn\,(n - 1)^2$, while the other seven share the term $c\,n$. The former term dominates, so the equation has an upper bound of n^3. In practice, the upper bound becomes:

$$\theta\,(\,2\,n\,(n - 1)^2 + 33\,n\,) \tag{3}$$

The lower bound of Equation 1 is represented by a purely independent code. There are no penalties in this case ($e = 0,\ \Rightarrow t = 0$) and no retries ($c = 0$). The complexity is therefore

$$\theta\,(\,3\,n\,) \tag{4}$$

The `if` conditions before the `repeat` sections avoid any scheduling attempt if there are no penalties. Only the initial PTD measures take part in the complexity. The lower bound of the PTD scheduler is therefore of the order of n.

If all the constants are removed from Equation 1, then the complexity of the PTD scheduler becomes

Benchmark	Functions	Basic Blocks	Instructions	Initialisation phase
intmm	6	11	196,074	12.000 %
livermore	16	48	1,601,672	84.504 %
fract	70	505	3,392,216	2.421 %
li	386	2,228	15,207,508	2.007 %
puzzle	20	137	16,055,217	0.509 %
compress	18	128	16,269,122	20.847 %
go	396	8,880	18,242,718	7.906 %
m88k	259	3,841	34,323,842	7.504 %

Table 4: Characteristics of the benchmarks

Benchmark	Arithmetic	Logic	Memory	Floating	Branch
intmm	70.11 %	5.13 %	20.04 %	0.00 %	4.72 %
livermore	73.75 %	6.60 %	13.35 %	0.00 %	6.30 %
fract	26.40 %	8.22 %	19.64 %	42.85 %	2.89 %
li	21.50 %	23.30 %	41.21 %	0.05 %	13.94 %
puzzle	52.08 %	22.12 %	11.43 %	0.00 %	14.37 %
compress	38.52 %	26.64 %	24.56 %	3.10 %	7.18 %
go	46.29 %	20.76 %	18.87 %	0.00 %	14.08 %
m88k	36.07 %	24.62 %	27.86 %	0.00 %	11.45 %
Average	45.59 %	17.17 %	22.12 %	5.75 %	9.37 %

Table 5: Breakdown of the benchmarks in terms of instruction types

$$\theta\,(\,et^2 + n - e\,) \tag{5}$$

In normal conditions, however, the parameters e and t have particular values with respect to n. As the algorithm progresses, the number of penalties is reduced, so e becomes: $e \ll n$. Similarly, t is much smaller than n ($t \ll n$), since, in general, the candidate is meant to be found from a close neighbour. Therefore, as the algorithm progresses in normal conditions, Equation 1 is found to be of the order of n.

4 The Evaluation Framework

The compilation environment is based on the Stanford University Intermediate Format (SUIF) compiler [8]. The benchmarks are characterised in Tables 4 and 5, and were derived mainly from the SPEC95 suite of benchmarks. The MAP architecture was modelled at the instruction level and simulated is an event-driven stochastic simulator that read and executed assembly language instructions generated by the SUIF compiler. Each instruction is associated with a number of events that emulate the stages in the micronet datapath for its execution. The events are created dynamically and their latency depend on the type of instruction and resource contentions at run-time.

The C benchmark programs are compiled in SUIF. Next, a loader program converts the resulting assembly code so that global memory references and labels fit into a global referencing scheme, and the resulting output is fed into the instruction-level simulator of the micronet architecture for evaluation. The output of this path is considered to be the *base* case since the code is not scheduled after code generation. Next the base case is fed into a scheduling phase using one of the three schedulers and the output is fed to the simulator as before. Each result reported in the next section was the average of five runs.

5 Results

Figures 2 - 5 compare the reduction in the instruction issue stalls for the three schedulers for processor configurations consisting of one or more AUs. They represent the percentage improvement with respect to the base cases, *i.e.* unscheduled code executing on 1 AU, 2 AU, 3 AU and 4 AU, respectively. The causes of the issue stalls are broken down as due to general data dependencies (`Data`), data dependencies due to branch instruction (`Branch`), and those due to resource contention for the read buses (`Bus`), or for a functional unit (`Rsc`). The figures demonstrate that the three schedulers are successful in reducing the issue stalls due to data dependencies (`Data` and `Branch`) - the bars always lie along the positive y-axis. As the code is optimised, the causes of issue unit stalls shift from data dependencies to resource contentions (`Bus` and `Rsc`), which is to be expected as the functional units and their buses get busier. As the architecture scales the stalls due to bus contention become less important; in fact, these stalls are practically negligible in the 4 AU configuration (Figure 5).

The stalls due to contentions for functional units (`Rsc`) are also reduced as the architecture scales; however, their reduction is not as clear-cut as in the previous case as the scaling is confined only to arithmetic units. Benchmarks such as `li`, which have more memory instructions, cannot take advantage of the larger amount of concurrency in the AU units in architecture. The `fract` benchmark has a similar limitation since it contains a large proportion of floating point instructions (see Table 5). When the `fract` benchmark was simulated with four floating point units (FPU), the average reduction in the issue stalls due to resource contention went down from -17.91% (1 FPU) to -9.80% (4 FPU), in the case of the PTD scheduler.

The `livermore` benchmark is dominated by arithmetic instructions (see Table 5). The large negative percentages for this benchmark in the case of the 4 AU configuration (see Figure 5) presents scope for further improvement. This was confirmed for the PTD scheduler when the benchmark was simulated with a configuration containing 5 AU, and the issue stall was reduced from -11.22% (5 AU) to -10.18% (4 AU) (the remaining stalls are due to memory operations only). The `go` benchmark presents an interesting scaling pattern: for greater than 2 AU, the benchmark suffers a degradation in the `data` stalls which is mainly due to the limited parallelism in the benchmark. In contrast, the `bus` and `rsc` stalls improve as the architecture scales. `Puzzle` is recursive in nature which explains the high number of branch instructions in Table 5. In the absence of sufficient parallelism in the benchmark the issue stalls does not improve with an increase in the number of AUs. In contrast, the loop-oriented benchmarks such as `intmm` and `livermore` show significant improvement as the number of AUs are scaled.

A characteristic of the PTD scheduler is that it is more effective in reducing issue stalls due to resource dependencies than those due to data. This is clearly in evidence in Figure 3 with the `compress` benchmark. Both the GM and the Balanced schedulers have a significant impact on the issue stalls due to data dependencies (`data`), reducing it by almost 50%. The PTD scheduler does not exhibit the same increase (almost 40%), but it reduces the stalls due to resources to around -1%, whereas the other two schedulers manage a less respectable -10% reduction. In the case of the PTD scheduler, the improvements in the instruction issue stall due to data dependencies is restrained by the overlapping penalties. In contrast, the PTD scheduler tackles resource dependencies by applying penalties to consecutive instructions of the same type when there are not enough functional units of that type. The net result is that the overall improvement in the issue stall due to PTD compares well with the other two

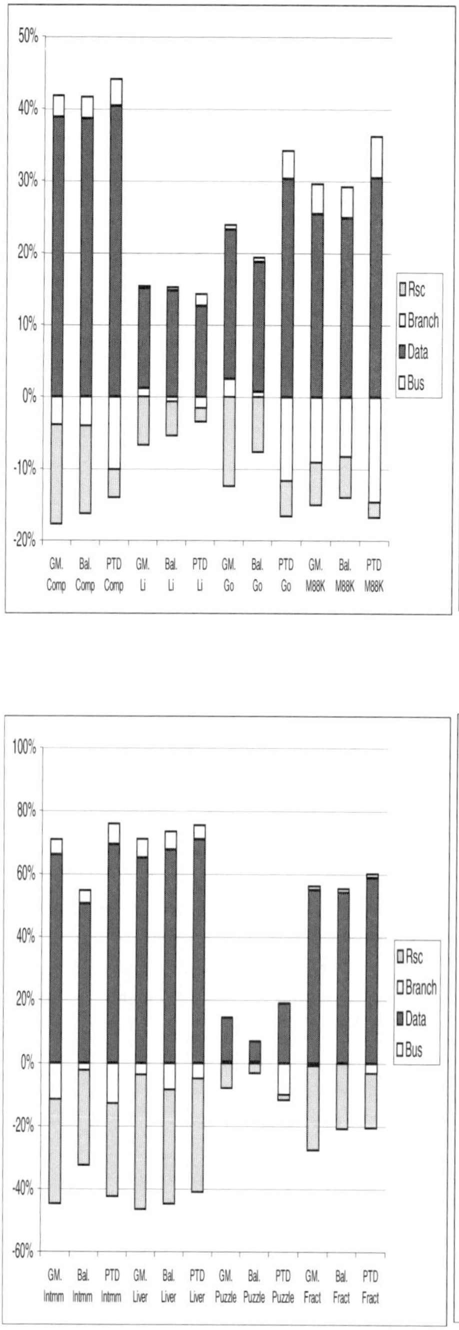

Figure 2: Percentage improvement in the instruction issue stalls (1 AU)

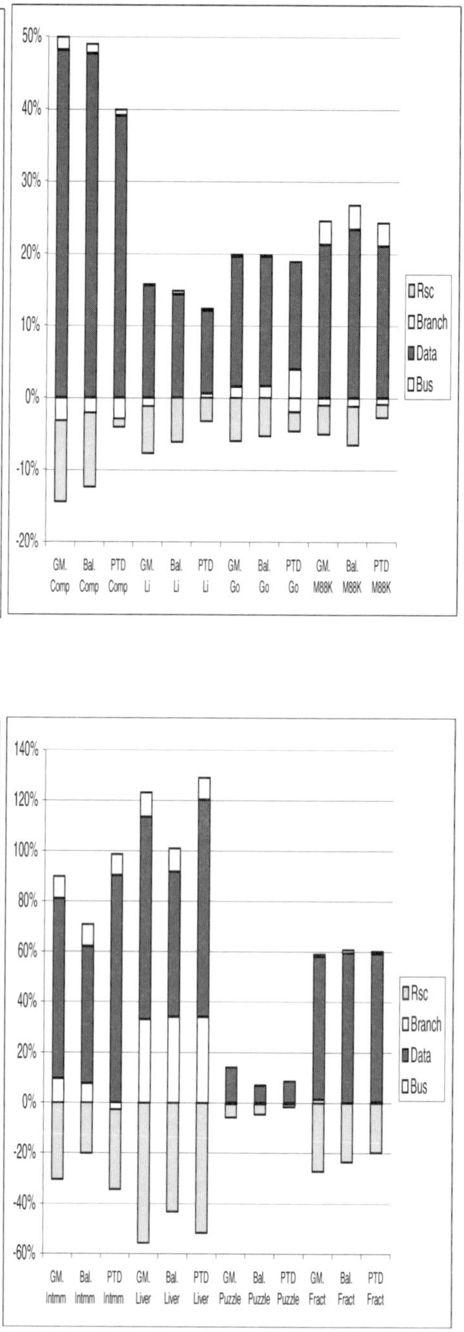

Figure 3: Percentage improvement in the instruction issue stalls (2 AU)

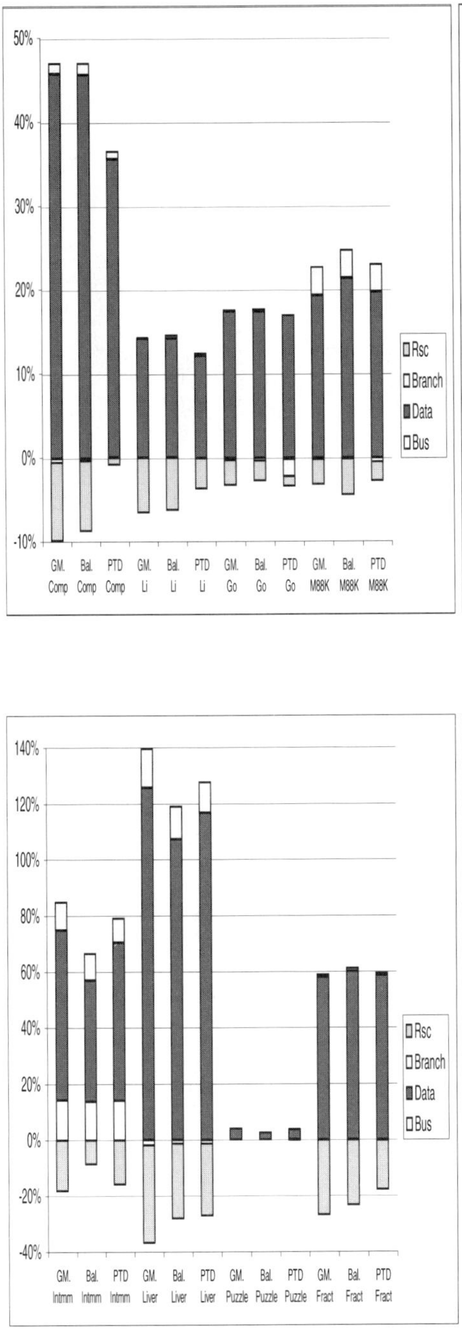

Figure 4: Percentage improvement in the instruction issue stalls (3 AU)

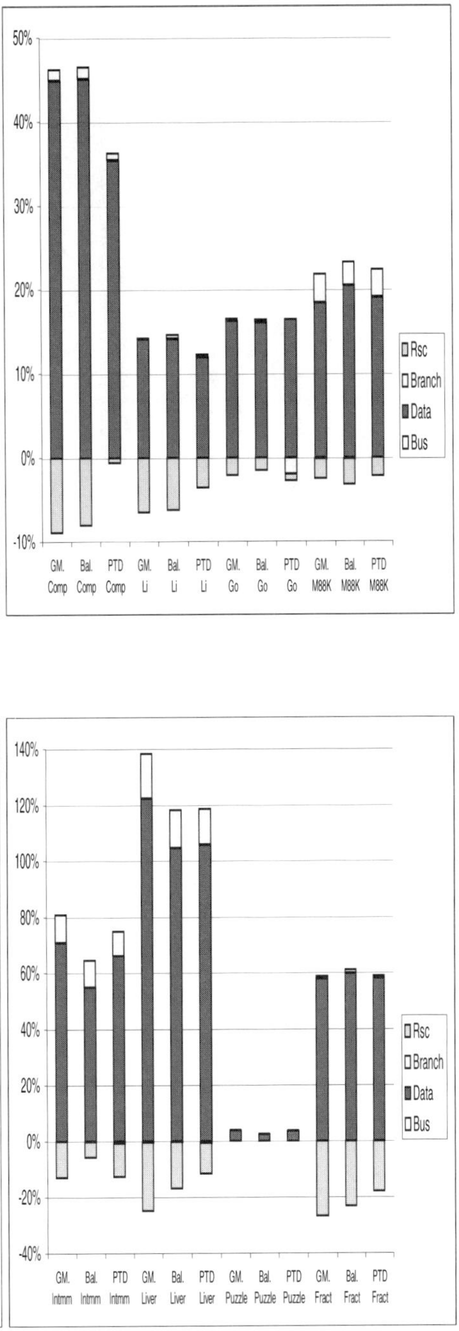

Figure 5: Percentage improvement in the instruction issue stalls (4 AU)

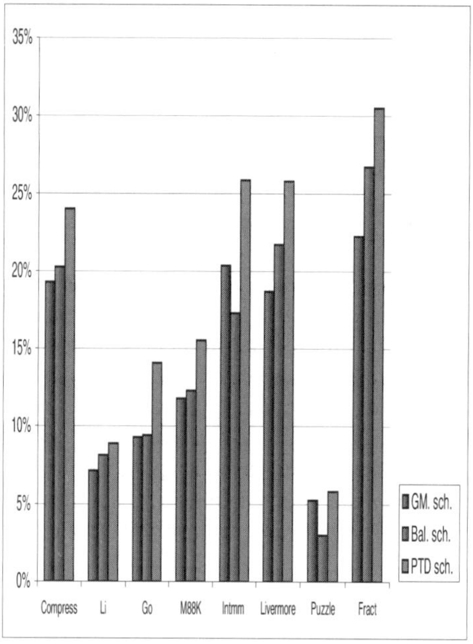

Figure 6: Percentage improvement in execution times due to scheduling for ILP (1 AU)

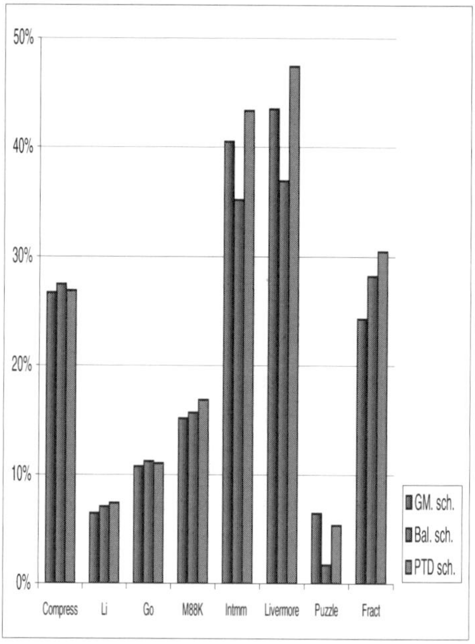

Figure 7: Percentage improvement in execution times due to scheduling for ILP (2 AU)

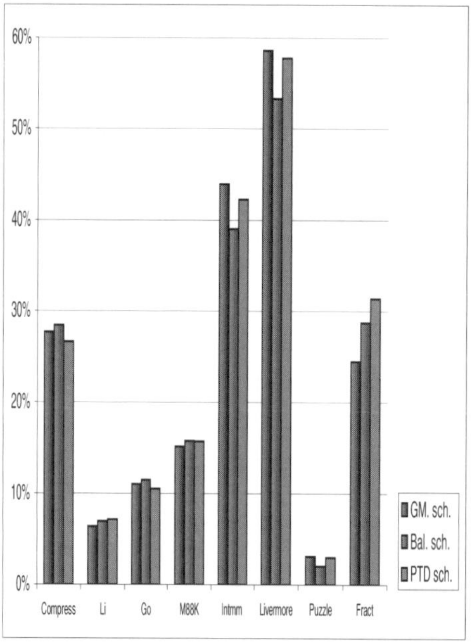

Figure 8: Percentage improvement in execution times due to scheduling for ILP (3 AU)

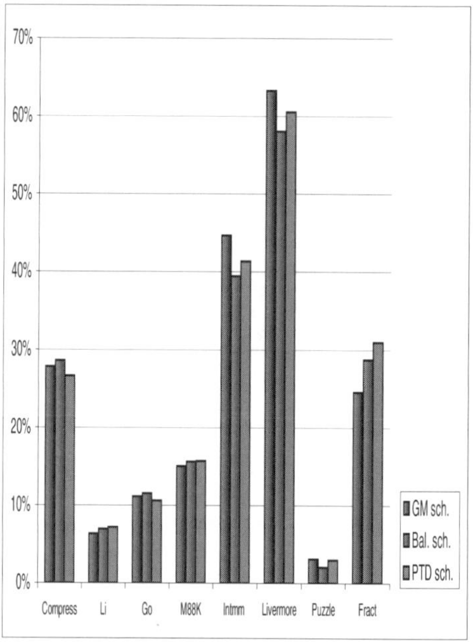

Figure 9: Percentage improvement in execution times due to scheduling for ILP (4 AU)

schedulers.

Figures 6 - 9 illustrate the percentage improvement in the program execution times due to instruction level parallelism for configurations ranging from 1 to 4 AUs. In each configuration, the improvement is compared against the unscheduled code executing on the same number of AUs. All the three schedulers benefited from memory disambiguation. The PTD scheduler consistently outperforms the other two by 4% on average in the 1 AU configuration. This falls to within 2% as the architecture scales. This effect is due to the overlapping penalties produced by the PTD scheduler as it optimises the code. They confirm the correlation between the reduction in issue unit stalls and the improvement in overall execution times, as predicted.

6 Conclusions

This paper has analysed the performance of three ILP schedulers for the 'processor-as-a-network'. Instruction scheduling for such an architecture is important in order to exploit fine-grain temporal and spatial concurrency. The PTD scheduler has addressed the problem of statically generating efficient instruction schedules for a micronet target with uncertain instruction latencies. The complexity of the PTD algorithm improves on other well-known list-based schedulers and performs better than them when resources are constrained and at least as well when the resources are scaled. In this paper the application of the PTD scheduler has been confined to a basic block. It has been demonstrated recently that PTD can be extended as a global scheduler in conjunction with techniques such as code motion beyond basic block boundaries to produce better schedules for micronet-based asynchronous architectures. This is both a feasible and attractive proposition thanks to the efficient compilation times of the PTD scheduler.

Acknowledgements

Salvador Sotelo-Salazar was supported by a postgraduate studentship from the Science and Technology National Council of Mexico (CONACYT).

References

[1] Arvind, D. K. and Rebello, V. 1994. Instruction Level Parallelism in Asynchronous Processor Architectures. In *Proc. of the 3rd International Workshop on Algorithms and Parallel VLSI Architectures*, pp 80-91, Eds. M, Moonen and F. Cathoor, Leuven, Belgium, Aug. 1994, Elsevier.

[2] V. Bala and N. Rubin. Efficient scheduling using finite state automota. In *28th Annual International Symposium on Microarchitecture (MICRO-28)*, pp 46-56, Nov. 1995.

[3] P. B. Gibbons and S. S. Muchnik. Efficient instruction scheduling for a pipelined architecture. In *SIGPLAN '86 Symposium on Compiler Construction*, 21(7):11-16, July 1986.

[4] D. R. Kerns and S. J. Eggers. Balanced scheduling: Instruction scheduling when memory latency is uncertain. In *ACM SIGPLAN'93, Language on Programming Language Design and Implementation*, pp 278-89, June 1993.

[5] K. V. Palem and B. B. Simons. Scheduling time-critical instructions on RISC machines. In *ACM Transactions on Programming Languages and Systems*, 15(4):632-58, Sep. 1993.

[6] D. K. Arvind and S. Sotelo-Salazar. Scheduling instructions with uncertain latencies in asynchronous architectures. In *3rd International Euro-Par Conference*, pp 771-78, Springer-Verlag, Passau, Germany, Aug. 1997.

[7] D. J. Kinniment. An evaluation of asynchronous addition. In *IEEE Trans. on Very Large Scale Integration (VLSI) Systems*, 4(1):137-40, March 1996.

[8] M. W. Hall *et al.* Maximizing multiprocessor performance with the SUIF compiler. In *IEEE Computer*, Dec. 1996.

Appendix A

Algorithm 1: PTD_resource_phase_root Algorithm

```
measure = PTD_measure(root, resource_phase)
IF measure > 0 THEN
     REPEAT
       node = root
       last_measure = measure
       WHILE node ~= NULL DO
         IF penalty_resource(node) = 3 THEN {PENALTY: MEMORY INST.}
           PTD_arrange_left_resource(node)
         ENDIF
         IF penalty_resource(node) = 3 THEN {PENALTY: MEMORY INST.}
           PTD_arrange_right_resource(node)
         ENDIF
         node = next(node)
       ENDWHILE
       measure = PTD_measure(root, resource_phase)
     UNTIL measure = last_measure

     REPEAT
       node = root
       last_measure = measure
       WHILE node ~= NULL DO
         IF penalty_resource(node) = 1 THEN {PENALTY: OTHER TYPES}
            PTD_arrange_left_resource(node)
         ENDIF
         IF penalty_resource(node) = 1 THEN {PENALTY: OTHER TYPES}
            PTD_arrange_right_resource(node)
         ENDIF
         node = next(node)
       ENDWHILE
       measure = PTD_measure(root, resource_phase)}
     UNTIL measure = last_measure
ENDIF
```

Algorithm 2: PTD_consecutive_phase(root) Algorithm

```
measure = PTD_measure(root, first_phase)
IF measure > 0
   REPEAT
      node = root
      last_measure = measure
      WHILE node ~= NULL DO
        IF penalty_consecutive(node) = 3 THEN {PENALTY: LOAD INST.}
           PTD_arrange_left_data(node)
        ENDIF
        IF penalty_consecutive(node) = 3 THEN {PENALTY: LOAD INST.}
            PTD_arrange_right_data(node)
        ENDIF
```

```
          node = next(node)
        ENDWHILE
        measure = PTD_measure(root, first_phase)
    UNTIL measure = last_measure

    REPEAT
        node = root
        last_measure = measure
        WHILE node ~= NULL DO
          IF penalty_consecutive(node) = 2 THEN {PENALTY: BRANCH INST.}
             PTD_arrange_left_data(node)
          ENDIF
          IF penalty_consecutive(node) = 2 THEN {PENALTY: BRANCH INST.}
             PTD_arrange_right_data(node)
          ENDIF
          node = next(node)
        ENDWHILE
        measure = PTD_measure(root, first_phase)
    UNTIL measure = last_measure

    REPEAT
        node = root
        last_measure = measure

        WHILE node ~= NULL DO
          IF penalty_consecutive(node) = 1 THEN {PENALTY: OTHER INST.}
             PTD_arrange_left_data(node)
          ENDIF
          IF penalty_consecutive(node) = 1 THEN {PENALTY: OTHER INST.}
             PTD_arrange_right_data(node)
          ENDIF
          node = next(node)
        ENDWHILE
        measure = PTD_measure(root, first_phase)
    UNTIL measure = last_measure
ENDIF
```

Algorithm 3: PTD_nonconsecutive_phase(root) Algorithm

```
measure = PTD_measure(root, second_phase)
IF measure > 0 THEN
 REPEAT
     node = root
     last_measure = measure

     WHILE node ~= NULL DO
      IF penalty_nonconsecutive(node) = 3 THEN {DISTANCE: 1 INST.}
         PTD_arrange_left_data(node)
      ENDIF
      IF penalty_nonconsecutive(node) = 3 THEN {DISTANCE: 1 INST.}
         PTD_arrange_right_data(node)
      ENDIF
      node = next(node)
     ENDWHILE
    measure = PTD_measure(root, second_phase)
 UNTIL measure = last_measure

 REPEAT
```

```
      node = root
      last_measure = measure

      WHILE node ~= NULL DO
        IF penalty_nonconsecutive(node) = 2 THEN {DISTANCE: 2 INST.}
           PTD_arrange_left_data(node)
        ENDIF
        IF penalty_nonconsecutive(node) = 2 THEN {DISTANCE: 2 INST.}
           PTD_arrange_right_data(node)
        ENDIF
        node = next(node)
      ENDWHILE
      measure = PTD_measure(root, second_phase)
   UNTIL measure = last_measure

   REPEAT
      node = root
      last_measure = measure

      WHILE node ~= NULL DO
        IF penalty_nonconsecutive(node) = 1 THEN {DISTANCE: 3 INST.}
           PTD_arrange_left_data(node)
        ENDIF
        IF penalty_nonconsecutive(node) = 1 THEN {DISTANCE: 3 INST.}
           PTD_arrange_right_data(node)
        ENDIF
        node = next(node)
      ENDWHILE
      measure = PTD_measure(root, second_phase)
   UNTIL measure = last_measure
ENDIF
```

Communicating Process Architectures 2003
Jan F. Broenink and Gerald H. Hilderink (Eds.)
IOS Press, 2003

Accurate Calculation of Deme Sizes for a Parallel Genetic Scheduling Algorithm

Michelle Moore
Department of Computing and Mathematical Sciences
Texas A&M University-Corpus Christi
mmoore@sci.tamucc.edu

Abstract. The accuracies of three equations to determine the size of populations for serial and parallel genetic algorithms are evaluated when applied to a parallel genetic algorithm that schedules tasks on a cluster of computers connected via shared bus. This NP-complete problem is representative of a variety of optimisation problems for which genetic algorithms (GAs) have been shown to effectively approximate the optimal solution. However, empirical determination of parameters needed by both serial and parallel GAs is time-consuming, often impractically so in production environments. The ability to predetermine parameter values mathematically eliminates this difficulty. The parameter that exerts the most influence over the solution quality of a parallel genetic algorithm is the population size of the demes. Comparisons here show that the most accurate equation for the scheduling application is Cantú-Paz' serial population sizing calculation based on the gambler's ruin model [1]. The study presented below is part of an ongoing analysis of the effectiveness of parallel genetic algorithm parameter value computations based on schema theory. The study demonstrates that the correct deme size can be predetermined quantitatively for the scheduling problem presented here, and suggests that this may also be true for similar optimisation problems. *This work is supported by NASA Grant NAG9-140.*

1 Introduction

Many cluster computing applications require efficient scheduling of tasks. The scheduling model described here is being developed for use inside a computational parallelisation and optimisation tool that runs on a Beowulf-type cluster of PCs. The cluster is used for a variety of applications in computational science. Several of these applications require "production" runs, where information about execution time and communication time is known. The overall execution time for these projects tends to be large, and computing resources are limited. Time expended to obtain a good schedule for these tasks is well spent.

This research builds on work that examined task scheduling with sequential GAs [11]. In [12], high quality schedules were reported when a distributed GA was implemented by dividing the population of the GA into multiple "demes" with each deme executing on a separate processor. In [13], results obtained by Goldberg [6], Goldberg, Deb, and Clark [7] and recent breakthroughs by Cantú-Paz [1] were applied for the first time to the scheduling problem, with encouraging results. This paper presents an in-depth examination of the sizing equations developed in [7] and [1] in terms of their applicability to the scheduling problem.

The remainder of this paper is organised as follow. Section 2 describes the scheduling problem model and its representation within the genetic metaphor. Section 3 describes the genetic algorithm operators and parameters that were held constant to provide a stable environment for the experiments. Section 4 presents the three sizing equations and defines the variables of the equations. Section 5 describes the design of the experiments, and the problem-specific variable values used in the sizing calculations. Section 6 discusses the results of the experiments. Section 7 presents conclusions and directions for further study.

2 The Task Scheduling Problem

2.1 The Computational Model

The cluster consists of *m* processors $(p_1,...,p_m)$ connected by a shared Ethernet bus. Tasks contain an execution time estimate and a communication time estimate (e,c). The *n* tasks can be represented as $(t_1,...,t_n)$ or as $((e_1,c_1),...,(e_n,c_n))$.

Tasks are created and scheduled by an initial processor designated p_1. Tasks assigned to p_1 do not require the scheduling of communication time. The time required to send the task assignments to the processors is assumed to be constant and is therefore not considered. All tasks executions must complete before the corresponding result communication can be placed on the bus. Any final computations that must occur after all results are communicated back to p_1 are unaffected by the particular schedule, and are not included in the schedule quality evaluations.

The quality or "fitness" of the schedule is the total time required for all tasks to complete execution. An exhaustive search of all possible schedules (even using branch-and-bound) will require time bounded by $O(m^n)$ where *n* is the number of tasks to schedule, and *m* is the number of processors to schedule. The addition of a communication bus that also must be "scheduled" increases the complexity of the problem. Since the problem is NP-complete, approximation algorithms for scheduling attempt to create schedules as close to the optimal as possible.

Processors are identical. Tasks are independent. Only one message may be on the bus at a time. The result message is sent as soon as possible after task execution completion. The system is non-pre-emptive. A processor will not remain idle if a task is available. Only one task will execute on a processor at a time.

2.2 The Genetic Metaphor

A schedule is represented as a "chromosome" consisting of *n* genes, where *n* is the number of tasks to be scheduled. Each position of the chromosome corresponds to a task, and the values at each position correspond to a processor number. A schedule for 6 tasks on 4 processors might appear as 3 4 2 2 1 2, indicating that t_1 is executed on p_3, t_2 is executed on p_4, and so on. Chromosomes are also referred to as strings. The cardinality of the string alphabet is the number of digit values that may appear on the string. In the example above, the cardinality of the alphabet is 4. Many genetic algorithms use binary alphabets, but in general, we may refer to a χ-ary alphabet where the cardinality is χ.

3 Genetic Operators and Parameters

Genetic Algorithms employ metaphors from biology and genetics to "evolve" an initial population of solutions (schedules in this case) into high quality solutions. In the original serial version of this algorithm, the initial population of schedule chromosome strings is randomly generated. The fitness, or quality of each schedule is evaluated. The "fitter" schedules are saved and used as a "mating pool". These schedules are paired randomly to create new schedules until there are enough "offspring" to return the population to its original size.

The fitness of an individual chromosome string is evaluated as *fitness = -1 * Schedule-Execution-Time*. In each generation, individuals with fitness greater than or equal to the average fitness are allowed to survive to the next generation. This is a rank-based selection method that combines truncation selection and $(\gamma + \lambda)$ selection. The top $1/\nu$ of the population is selected. Then the $\gamma = 1/\nu$ parents mate to create λ offspring. The union of γ and λ constitute the next generation. In this manner, very fit individuals may survive to mate repeatedly.

Mating is accomplished using uniform crossover with a .5 crossover probability. In other words, when creating an offspring from a pair of schedules, the offspring's processor assignment for each task is equally likely to match either parent. The mutation rate is .02. This means that after crossover, the GA allows for a .02 probability for each task of being reassigned to a randomly chosen processor. The process of selection, mating, crossover, and mutation iterates for a set number of generations, predetermined by the user of the scheduler.

To allow schedules to be generated more quickly, the serial scheduling algorithm was rewritten in LAM MPI to execute on multiple cluster processors. In the distributed algorithm, the initial population of schedules is distributed among the available cluster nodes. Each of these subpopulations, or demes, is allowed to evolve in isolation for a set number of generations called an "epoch". After each epoch, the more fit individuals of each deme are copied and distributed among the other demes, replacing the least fit individuals. After each migration, the local iterations of selection, mating, crossover, and mutation resume. After the execution of the preset number of generations in an epoch elapse, a new migration occurs. Again, the amount of time this continues is predetermined by the user of the scheduler.

Migration occurs after selection and before mating every three generations. The migration rate is set to the maximum that can occur for the deme with the most survivors after selection. Migration follows what is called a "best-worst policy", since migrants take the place of the least fit individuals in each deme.

4 The Sizing Equations

Three sizing equations were applied, and the resulting schedules were analysed. Detailed explanation of the derivation of the first equation can be found in [7]. The second and third equations are described extensively in [1]. Equation (1) and Equation (2) were designed for serial GAs, but the population size for each deme, n_d can be found by dividing the serial population size n by the number of demes r. Equation (3) was designed specifically for parallel genetic algorithms. Consideration of collateral noise (the noise of other partitions when deciding between best and second best building block) is built into the equations. External noise may exist if fitness cannot be directly calculated. The scheduling application did not require adjustments for external noise.

In the past, the determination of population and deme sizes for practical applications of GAs required a great deal of what was frequently termed "empiricism", i.e., trial and error. The three equations below represent state-of-the-art sizing estimations based on statistical analysis of the behaviour of parallel genetic algorithms for simple theoretical applications. Goldberg, Deb, Clark, Cantú-Paz, and others have based their in-depth analysis on the somewhat controversial schema theorem. The focus of this study of the equations was not to confirm or dispute the underlying validity of the theorem, but to determine if any of the

equations could be adapted to provide useful a priori sizing values for the cluster scheduler.
$n_d = n / r$

Equation (1) $n = 2\, z^2 \chi^k m' \; (\sigma^2_{rms} / d^2)$
Equation (2) $n = -2\chi^{k-1} \ln(\alpha)\, (\sigma_{bb}\, sqrt(\pi\, m') / d)$
Equation (3) $n_d = \chi^k (\ln(1 - P_{bb})) / ((sqrt(\delta)\, \tau' + 1) \ln(q/p))$

These equations and their application to the scheduling problem are discussed individually below.

4.1 Equation (1)

$$n = 2\, z^2 \chi^k m' \; (\sigma^2_{rms} / d^2)$$

The symbols used in Equation (1) are defined below.

(1.1) α ~ probability of making the wrong choice between two competing schema

(1.2) $z(\alpha)$ ~ ordinate of a unit one-sided, normal deviate at α

(1.3) z^2 ~ square of a one-sided, normal deviate at α

(1.4) χ ~ cardinality of the chromosome string alphabet

(1.5) k ~ building block size, the number of fixed positions in the schema

(1.6) lval ~ length of the chromosome (schema)

(1.7) m ~ number of partitions in the string, number of building blocks, lval / k

(1.8) m' ~ (m - 1)

(1.9) f_{max} ~ maximum fitness function value possible

(1.10) f_{min} ~ minimum fitness function value possible

(1.11) σ^2_f ~ overall variance of all schemas for each fixed position, $= \sum_{i=1}^{m} \sigma^2_{fi}$
$\approx (f_{max} - f_{min})^2 / 12$. [7]

(1.12) σ^2_{rms} ~ root mean squared of the sub function variances, $= \sigma^2_f / m$

(1.13) d ~ signal difference for peak to peak (mean to mean) function difference between the best and the second best competing individuals, this affects the likelihood of choosing the better individual

4.2 Equation (2)

$$n = -2\chi^{k-1} \ln(\alpha)\, (\sigma_{bb}\, sqrt(\pi\, m') / d)$$

The symbol π represents the actual constant value 3.14159...

The additional symbol used in Equation (2) is defined below.

(2.1) σ_{bb} ~ sqrt (σ^2_f) the average building block standard deviation

4.3 Equation (3)

$$n_d = \chi^k(ln(1- P_{bb})) / ((sqrt(\delta)\tau'+1) \ ln(q/p))$$

The additional symbols used in Equation (3) are defined below.

(3.1) δ ~ degree of a deme, the number of neighbours it has

(3.2) τ ~ number of epochs, the generations between migrations constitute an epoch

(3.3) τ' ~ (τ - 1)

(3.4) q ~ probability of losing a copy of the best building block between generations

(3.5) p ~ (1 - q), $\approx$ 1/2 + ψ / (sqrt(2π)) , where ψ = d / (σ_{bb} sqrt(2m'))

(3.6) Q ~ required problem solution quality, = (1 - α) m

(3.7) Pbb ~ probability that one partition in one deme is correct, probability of success per deme, $\approx$ Q / m - sqrt (ln (r)) / sqrt (2m)

5 Experiment Design

In order to evaluate the quality of the schedules that were produced, actual optimal schedules were found for the data sets. Because of the exponentially increasing amount of time required to find the comparison optimal schedules, the number of tasks in the tests sets was kept small. Data sets consisted of 20 normally distributed, randomly generated (e,c) task pairs. The mean of the execution time was 25 with a standard deviation of 2%. Three communication means were used, 10, 25, and 40, (std = .02) to vary the relative effects of communication overhead. The sizing program generated population and deme sizes for each of the equations using probabilities of success of .7, .75, .80, .85, .90, .95, .99. or corresponding z^2 of 0.2809, 0.4489, 0.7056, 1.0816, 1.6641, 2.7225, 5.4289. Five runs were averaged for each communication mean at each confidence level. This was done in order to smooth out stochastic variance. Size values were generated at each confidence level, with each of the three communication means.

The following variable values were set.

r	=	4
χ	=	3
k	=	1
lval	=	20
m	=	20
m'	=	19
d	$\approx$	(1 - μ_e) / 2
f_{max}	=	if μ_c = 10, lval $\mu_e + \mu_c$
		if μ_c = 25, lval $\mu_e + \mu_c$
		if μ_c = 10, lval $\mu_c + \mu_e$

$$f_{min} = \begin{cases} \text{if } \mu_c = 10, & 1/3 \text{ lval } \mu_e + 2\mu_c \\ \text{if } \mu_c = 25, & 1/2 \text{ lval } \mu_e + \mu_c \\ \text{if } \mu_c = 40, & 1/3 \text{ lval } \mu_c + \mu \end{cases}$$

τ = 3
δ = 3
p = .508743

6 Results

The population sizes were tested in the parallel genetic scheduling algorithm. Figures 1-3 compare the expected solution quality as predicted by the equations to the average measured solution quality at each deme size. An expected quality of 70% meant that using the deme size calculated at the 70% confidence level was expected to produce schedules able to execute in an amount of time exceeding the optimal by 30%. The quality achieved with deme sizes from Equation 2 follows the expected values the most closely. Figure 4 shows the deme sizes calculated by each of the equations. Both Equations 2 and 3 are very low compared to those calculated by Equation 1. Since larger demes take longer to process, all other things being equal, Equations 2 and 3 would both appear better. Figure 1 indicates that Equation 1 overestimates the required population size at the lower quality levels. However, Equation 3 underestimates the deme size at the 99% quality level, and overestimates at lower levels (Fig. 3). In terms of deme size and closeness to quality estimates, Equation 2 appears superior.

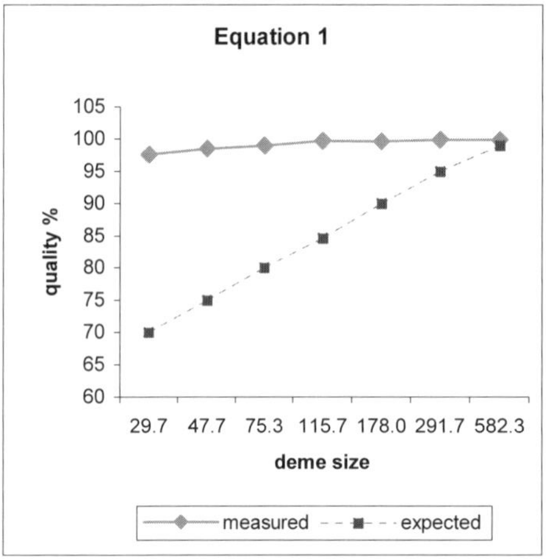

Figure 1: Performance of Equation 1

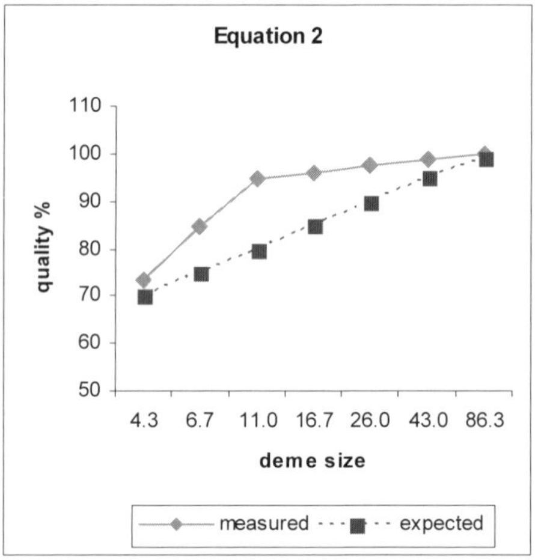

Figure 2: Performance of Equation 2

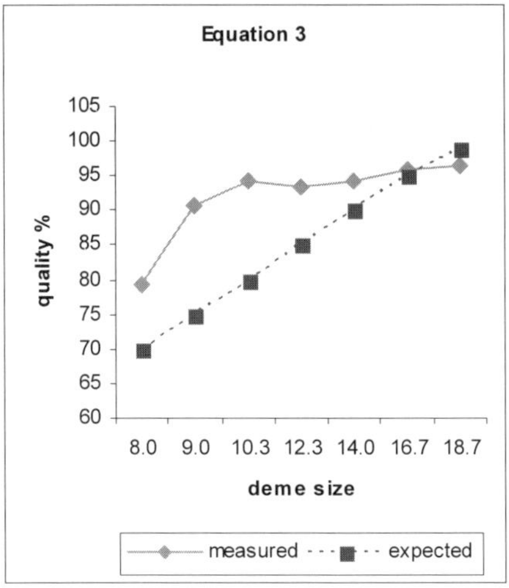

Figure 3: Performance of Equation 3:

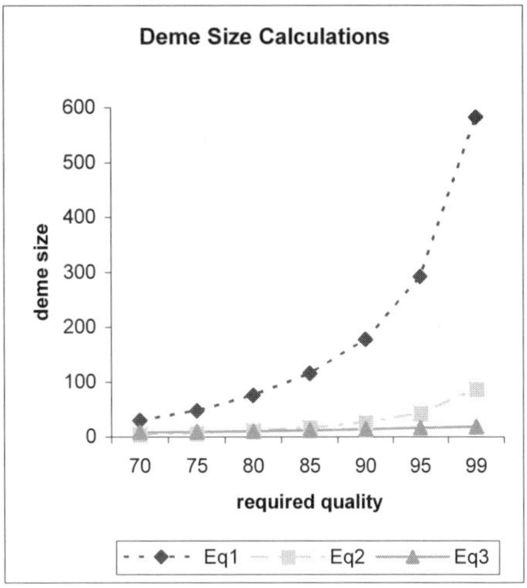

Figure 4: Comparison of Deme Size

Viewing Figure 5, it is important to note that all of the equations lead to deme sizes that produce good schedules. The farthest miss of required accuracy was Equation 3 which reached 95% accuracy when 99% was needed. The other two equations specify deme sizes that always allow the scheduler to meet the 99% accuracy requirement, with the scheduler finding the optimal schedule frequently, and average schedules better than 99.5% in all other cases.

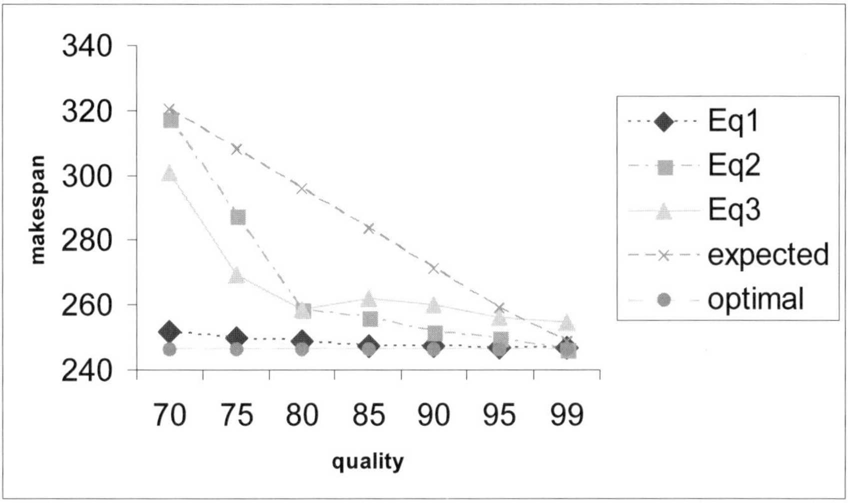

Figure 5: Schedule Execution Times Compared to Optimal and Expected

7 Conclusions and Future Work

Equation 2, which combined analysis using the gambler's ruin model and schema theory, provided the best deme sizes when the population sizes it generated were divided by the number of demes. The accuracy of the scheduling algorithm was as good as or better than that expected in all cases. In addition, the measured quality was closer to the expected solution quality than that of the other two equations. Equation 2 is a more accurate model and is successful at attaining desired quality levels without overestimating population size. This means that schedules using the sizing results will take only the time needed to provide the desired result quality. This can be especially important in time-critical applications. In addition, the ability to determine population sizing values mathematically rather than by trial and error will reduce the time required to determine schedules for new applications.

Other than the communication time mean and the execution time mean for the data, all values required by the equations can be calculated using theoretical analysis or knowledge of the problem domain. This suggests that the equations can be applied to problems where very little is known about the specific problem data. If the mean is known, function max and function min can be estimated. Where max and min are known, the mean can be estimated.

There are many directions for future work. There are various equations for deciding in advance the optimal values of many of the other parallel GA parameters. These will also be applied to the parallel scheduling problem. In addition, it will be interesting to see if calculated size values are equally successful with multi-objective, multi-constrained applications.

Acknowledgements

Research Assistants Brian McCord, Tzintzuni Garcia, William Jackson, Simon San Miguel, and Jason Picarazzi have contributed and continue to contribute to this project. *This work is supported by NASA Grant NAG9-1401.*

References

[1] Cantú-Paz, E., *Efficient and Accurate Parallel Genetic Algorithms,* Dordrecht, The Netherlands: Kluwer Academic Publishers Group, 2000.

[2] Coffman, E., Introduction to Deterministic Scheduling Theory, *Computer and Job-Shop Scheduling Theory,* New York, NY: John Wiley and Sons, 1976.

[3] De Jong, K. and Spears, W., Using Genetic Algorithms to Solve NP-complete Problems, *Proceedings of the Third International Conference on Genetic Algorithms*, 1989, 124-132.

[4] Eshelman, L., Caruana, R. and Schaffer, J., Biases in the Crossover Landscape, *Proceedings of the Third International Conference on Genetic Algorithms,* 1989, l0-19.

[5] Goldberg, D., *Genetic Algorithms Search, Optimization, and Machine Learning,* Reading, MA: Addison-Wesley, 1989a.

[6] Goldberg, D., Sizing Populations for Serial and Parallel Genetic Algorithms, *Proceedings of the Third International Conference on Genetic Algorithms,* 1989b, 70-79.

[7] Goldberg, D., Deb, K., and Clark, J., Genetic Algorithms, Noise, and the Sizing of Populations, *Complex Systems, 6,* 332-362.

[8] Holland, J., *Adaptation in Natural and Artificial Systems,* Ann Arbor, MI: University of Michigan Press, 1975.

[9] Horowitz, E. and Sahni, S., Exact and Approximate Algorithms for Scheduling Nonidentical Processors, *Journal of the ACM,* vol. 23, no. 2, 1976, 317-327.

[10] Hou, E., Hong, R. and Ansari, N., Efficient Multiprocessor Scheduling Based on Genetic Algorithms, *Proceedings of the 16th Annual Conference of the IEEE Industrial Electronics Society,* 1990, 1239-1243.

[11] Kidwell (Moore), M., Using Genetic Algorithms to Schedule Distributed Tasks on a Bus-Based System, In Forrest, S. (Ed.), *Genetic Algorithms: Proceedings of the Fifth International Conference*, San Mateo, CA: Morgan Kaufmann, 1993.

[12] Moore, M., Parallel Genetic Algorithms to Find Near Optimal Schedules for Tasks on Multiprocessor Architectures, In Chalmers, A., Mirmehdi, M., and Muller, H. (Eds.) *Communicating Process Architectures 2001*, Concurrent Systems Engineering Series, IOS Press, vol. 59, 2001, pgs, 27-36.

[13] Moore, M., An Accurate and Efficient Parallel Genetic Algorithm to Schedule Tasks on a Cluster, *Proceedings of the International Parallel and Distributing Processing Symposium,* 2003.

[14] Salleh, S. and Zomaya, A., *Scheduling in Parallel Computing Systems,* Dordrecht, The Netherlands: Kluwer Academic Publishers Group, 2000.

[15] Syswerda, G., Uniform Crossover in Genetic Algorithms, *Proceedings of the Third International Conference on Genetic Algorithms,* 1989, 2-9.

Communicating Process Architectures 2003
Jan F. Broenink and Gerald H. Hilderink (Eds.)
IOS Press, 2003

A multimodal robotic control law modelled and implemented with the CSP/CT framework[†]

Gerald H. HILDERINK, Dusko S. JOVANOVIC, Jan F. BROENINK
Twente Embedded Systems Initiative,
Drebbel Institute for Mechatronics and Control Engineering,
Faculty of EE-Math-CS, University of Twente,
P.O.Box 217, 7500 AE, Enschede, the Netherlands
g.h.hilderink@utwente.nl

Abstract. We use several formal methodologies for developing control applications at our Control Engineering research group. An important methodology we use for designing and implementing control software architecture is based on CSP concepts. These concepts allow us to glue multidisciplinary activities together and allow for formal stepwise refinement from design down to its implementation. This paper illustrates a trajectory and shows the usefulness of CSP diagrams for a simple mechatronic system. The simulation tool 20-SIM is used for creating the control laws and our CTC++ package is used for coding in C++.

1. Introduction

1.1 The need for a new approach

At the Control Engineering research group of the University of Twente, concurrent architectures are used in control (robotic, mechatronic) applications for more than two decades. Good experiences with applying transputers [1] programmed with occam [2] gave rise to the investigation of building control software with CSP concepts; transputers and occam are founded on CSP. CSP stands for Communicating Sequential Processes, which is a theory in the form of a process-algebra for specifying and analysing concurrent systems [3]. In order to exploit occam/CSP-like concurrent programming (developing concurrent software) in sequential programming languages like C/C++ and Java, the *Communicating Threads* (CT) libraries were developed [4]. CT allows for concurrent programming where multithreading in execution is encapsulated in CSP-like abstractions of processes and channels. Relying on these libraries, a software developer is freed from explicit dealing with low-level synchronization issues in a concurrent software architecture [5]. CT for C is called CTC, for C++ is CTC++ and for Java is CTJ.

Engineers engaged with designing control laws for a control application use block diagrams. Block diagrams provide an abstract graphical notation for modelling control laws offering a data-flow oriented and a plug-and-play design concept.

There is a big gap between a block diagram and its implementation; there are

[†] This research is supported by PROGRESS, the embedded system research program of the Dutch organization for Scientific Research, NWO, the Dutch Ministry of Economic Affairs and the Technology Foundation STW.

discontinuities between data-flow and control-flow. Furthermore, the engineer has little influence in the flow of control of the execution framework. In order to bridge the gap so that the engineer can specify any executable behaviour of the control software we use an intermediate graphical language, namely CSP diagrams. A CSP diagram specifies data-flow (continues on the block diagram notation) and control-flow in an application and eliminates the discontinuities between the two. This language allows the engineer to refine a design towards the demands of the system. This way the engineer gets a better understanding of concurrency that affects the real world in relation to the real-time software.

This paper reports a stepwise refinement trajectory from a control design to its implementation with the help of CSP diagrams. We use the simulation tool 20-SIM to design block diagrams of the system dynamics and the control laws. 20-SIM allows translating the equations to C-code. Of course we only translate the control laws since these have to be computed by the computer system. The paper mainly focuses on the way CSP diagrams are used for systematically designing control software. We use a simple mechatronic setup, named JIWY, which is a controlled motion device with two joints that rotate according to the movement of a joystick.

The needs for modelling concurrency are most commonly:

- managing complexity with multiple control loops,
- dealing with multiple frequencies,
- priorities and pre-emptive behaviours,
- event-driven control or reactiveness,
- guaranteeing real-time behaviour,
- dealing with distributive and parallel architectures.

The CSP diagrams deal with these issues at a high level of abstraction. We will show that this abstraction directly maps on software using CTC++. The path of code-generation is under control of the engineer *and* under control of the code-generation tool in a formal way.

1.2 The tools used in the refinement trajectory

We use a few languages and tools in the refinement trajectory from control laws to their implementation. As previously mentioned we use block diagrams to specify the control laws, the CSP diagram language is used to specify the concurrent executable framework, and C++ with CTC++ is used to implement the control software. For modelling block diagrams and C-code generation of the control laws we use the tool 20-SIM, see Section 1.2.1. Designing CSP diagrams is done by a drawing tool because there is no a specific design tool for CSP diagrams available. We are working on a specific design tool designing and analysing CSP diagrams. For coding we use a workbench that usually comes with the computer platform or a simple text editor tool for typing the C++ code. We have developed C++ template files that is used by 20-SIM, which takes care of most of the coding. Finally, a C++ compiler-linker and a program loader are used.

We will not elaborate on using these tools in this paper. We are primarily concerned with the design trajectory. This trajectory relies on identifying processes (i.e. physical processes, controller loop processes, control mode processes, and support processes) and identifying communication between these processes (i.e. signals, synchronization, interrupt handling, AD-DA conversion, etc.). The design forms the solution. In order to deal with concurrent software architectures we use the CSP concepts. Concurrency that boils from above affects every stage in the design to the implementation. Using CSP diagrams and CT is our solution to implementing concurrent software architectures.

1.2.1 20-SIM

20-SIM is a modelling and simulation tool developed by ControlLab Products B.V. [6], a spin-off company of the Control Engineering group of the University of Twente. It is a standard MS Windows application consisting of several integrated modules that supports modelling and design of mechatronics products in many aspects.

Modelling of a system can be performed by means of one or more 20-SIM modelling languages: bond graphs [7], block diagrams, iconic diagrams, equations [8] and importing Matlab models is supported as well. The tool complies well with the demand of offering a time-efficient and elaborate feedback to the user on the modelling/design decisions, by means of simulation plots, animated graphs and 3-D animations.

A recent functionality of 20-SIM is to generate C-code from internal equation model derived from a selected submodel. The C-code is sequential and needs to be inserted in a larger framework written in C or C++. This framework is specified in template files. Template files contain target-specific code, like device drivers, and special keywords that are replaced by generated C-code from 20-SIM submodel.

A C++ template is created which extends the 20-SIM code-generation. This C++ template allows us to generate individual submodel objects. A submodel object contains a data structure and methods that operate on this data structure, including the control law equations. The resulting C++ files can be compiled, linked and loaded on the target computer system. In Section 3.2 we discuss aspects of the C++ template for generating control processes, i.e. 20-Processes.

1.2.2 CSP diagrams

Since the introduction of the graphical modelling language based on CSP by Hilderink [9] the language has slightly evolved to become applicable for designing control software in the form of CSP diagrams. CSP diagrams show processes and their interrelationships by a graphical notation. A CSP diagram expresses an execution model of the resulting network of processes. The designer can specify the desired nature of concurrency of the real-time control software.

In this research we investigated the continuity between control models in 20-SIM, CSP diagrams, and the resulting code. The use of CSP diagrams is intended to bridge the gap between control models and the final code. CSP diagrams allow the designer to specify the connectivity between control processes and their execution order. CSP diagrams are based on data-flow concepts and control-flow concepts which can be used as an addition to the block diagrams used in control engineering.

One of many advantages that come with using CSP diagrams is that the graphical CSP notation offers straightforward code generation. Since our effort to create tools for drawing and analyzing CSP diagrams is in a begin stage, we do not have developed automated code-generation yet. Only 20-SIM can generate code which we will extensively use in this research. Coding of the execution framework is mainly done by handcraft and the results of this research will be used for the development of an automated code-generation tool of CSP diagrams. The coding by hand is not a difficult task since the mapping between a CSP diagram and object-oriented code with our CT library is almost one-on-one. See Section 3.

1.2.3 CT – Communicating Threads package

The Communicating Threads (CT) package consists of an object-oriented framework providing high-level design patterns based on CSP. These design patterns encapsulate multithreading from the user by means of processes, channels, and compositional constructs. It basically adds process-oriented concepts to the object-oriented paradigm which are useful for reasoning about concurrency and real-time behaviors. Real-time and embedded computer issues have been carefully designed in CT. For example, it integrates real-time scheduling into the application rather than using it from a real-time operating system. In fact, CT does not require a real-time operating system and it can run on bare microprocessors, microcontrollers, or digital signal processors (DSPs). The CT framework has been ported to Java (CTJ), to C (CTC) and to C++ (CTC++). For CTC and CTC++ the memory footprint is low and the scheduling overheads are efficient and fast. We use CTJ for educational purposes and CTC/CTC++ for real-time laboratory setups. The examples presented in this paper are based on CTC++.

1.3 Overview

The refinement trajectory using block diagrams in 20-SIM and CSP diagrams for JIWY is described in Section 2. Section 3 elaborates on the implementation of CSP diagrams and generated 20-SIM C-code using C++ and CTC++. The mapping between CSP diagrams and the resulting code is discussed, while detail on the 20-SIM code-generation process is not discussed. Conclusions are found in Section 4

2. Designing concurrent controllers

2.1 Case study JIWY

As a case study to illustrate the stepwise refinement trajectory, a classical example of a multimodal controller has been chosen, which is a two-degree-of-freedom robotic end-effector, called JIWY, see Figure 1. The construction contains two revolute joints that allow mounted device to rotate on a horizontal axis and a vertical axis. The joints are equipped with DC motors and incremental encoders. We use a PC under DOS, an analog joystick, a amplifier/circuit box to drive the motors and to convert sensor signals. Details on JIWY can be found in [10].

JIWY uses incremental encoders and therefore the centre position for each axis has to be calculated. It is required that the outer boundaries of each axis are measured first in order to determine the centre position of the axis. A joint will first rotate to the left and when it reaches the end stop this process terminates. It remembers the maximum value. The next process rotates the joint with constant velocity to the right until it reaches the end stop. Again this process remembers the maximum value. Then the main servo position motion controller for controlling JIWY takes over. It uses the two maximum values from the alignment processes to determine the exact centre position of the setup. The main controller can be stopped by the user by pressing a specific button on the joystick. After the main controller is stopped by the user it is required that the joints return back to their centre position as a safe position. This process is called homing. After homing the motors will be disabled.

This functional description gives rise to four control modes or sub control laws per joint. One velocity control law for aligning left, one velocity controller for aligning right, a motion controller reacting on the joystick input, and a position controller for homing to the safe centre position.

Figure 1 Photo of the JIWY end-effector

2.2 Process design and control law design

The design trajectory is a continuous interaction between process design and control law design. See Figure 2. Each design discipline can be carried out separately, but the whole results in an executable control software framework. The process design embraces the context (or overview of controllers), integration of control processes, and target depending issues. For this, CSP diagrams are used. The control law design focuses on the system dynamics and the control laws. The control laws are exclusively designed in 20-SIM using block diagrams. In the following these two design disciplines are interleaved.

2.2.1 Modes of control

The identified control modes are designed as four different control processes. Figure 3 shows these processes and their interrelationships in a CSP diagram. Figure 3a shows the communication diagram representing the data-flow between these four processes; the arrows render the communication relationships or channels between processes. Figure 3b shows the compositional diagram of the control-flow between these processes. Each joint operates in parallel and so there is a parallel relationship between the horizontal and vertical processes. The little circle at the end of the parallel relationship denotes a parenthesis which determines a group of control processes for the horizontal joint. The parenthesis symbol

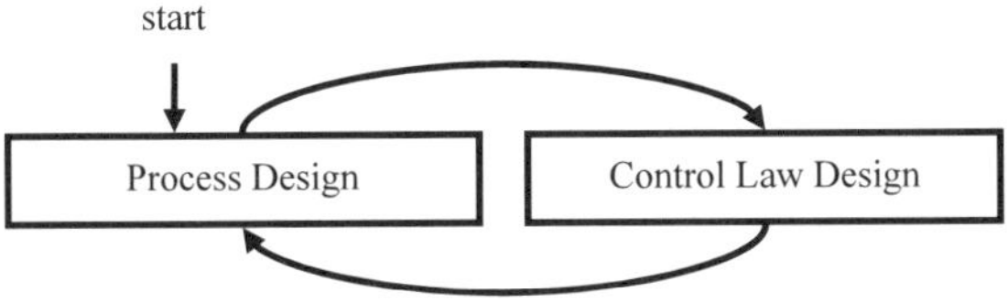

Figure 2 Proces design and control law design

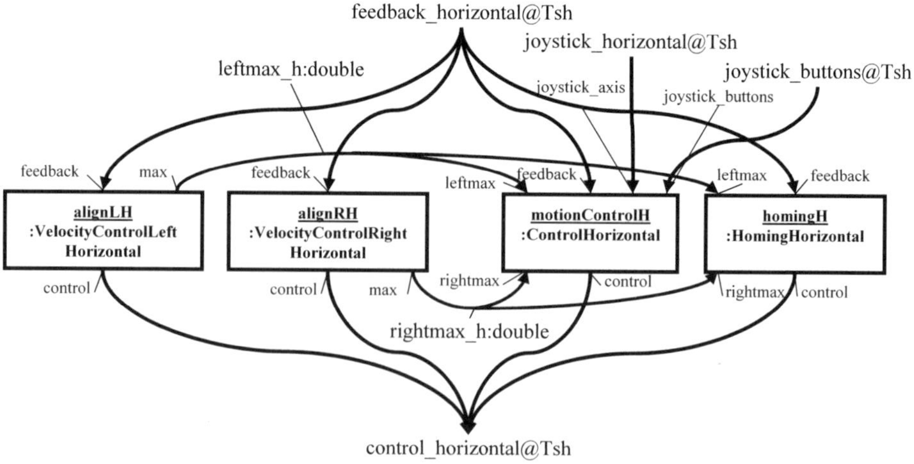

(a) Four sub-controllers and their communication relationships for the horizontal joint

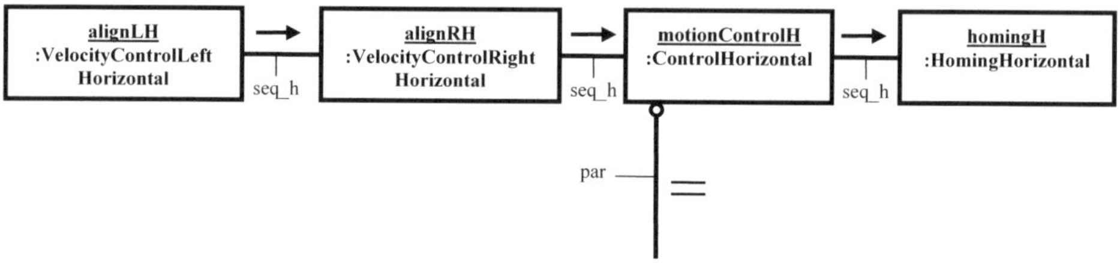

(b) Composition relationships between the four sub-controllers for the horizontal joint

Figure 3 CSP diagram of the horizontal controller

forms one parent anonymous process. This is similar for the vertical joint, which is not shown in the figure. In order to describe this system we will mainly restrict ourselves to the horizontal joint. The vertical joint is identical but with slightly different parameters than the horizontal joint.

Here, suffix "@Tsh" specifies that we deal with a sampled data system, where the samples of measured values arrive at equidistant moments in time. Clearly, each sample must be processed before the next arrival. This suffix means that the external channels are triggered on sampling period T_{sh} for the horizontal control loop. T_{sv} is the sampling period for the vertical control loop, which is not shown here. This information is used to set up timing, see Section 3.7.

A process can be in a different scope than another process when:

- a process resides at a different level in hierarchy than the other process,
- one process is in a different context than the other process, e.g. process in software and process in hardware.

A process that is out of scope is replaced by a *port name* at the end of the arrow. See the names `feedback_horizontal`, `control_horizontal`, `joystick_horizontal`, and `joystick_buttons` in Figure 3a. These names identify process-interface elements or ports of the parent process to which the CSP diagram belongs. These names are used inside the parent process and are connected to communication relationships outside the process. In Figure 3a these external processes reside in hardware. A port of the parent process acts also as a local label of the communication relationship.

In order to show the ports of a process to which communication relationships are connected, the name of a port can be shown next to the arrow head or arrow tail, e.g. `feedback`, `max`, `leftmax`, etc. These names are used locally by the process at the end of the communication relationship. Each port has properties such as its name, direction and communication type, e.g. {`feedback_horizontal, channelin, double`} and {`control_horizontal, channelout, double`}. The communication types of ports at each end of a communication relationship should be of the same type otherwise they are incompatible and cannot be connected. For example, port {`max, channelout, double`} on `alignLH` is compatible to port {`leftmax, channelin, double`} on `motionControlH`. The label `leftmax:double` identifies the actual communication object, i.e. a variable of type double, between the ports `max` and `leftmax`. These labels can be hidden individually in order to hide detail.

A communication diagram is responsible for the declarations of the processes, the internal communication relationships and the ports to external communication relationships. A composition diagram is also responsible for the declarations of processes, the compositional relationships, conditions, and guards. In Section 3 we will see them appearing in the code with the use of CTC++. For this each element requires an identifier.

2.2.2 Motion controller

The `motionControlH` process in Figure 3 is the main sub-controller that receives the set points from the joystick that is used by the user. This process only terminates when the stop button on the joystick is pressed. This process contains a 20-SIM process `servoHorizontal` which computes one sample in the control law. See Figure 4. The looping is specified by the μ-process.

In Figure 4a communication relationships between the external ports and the ports of the 20-SIM process `servoHorizontal` are specified. A 20-SIM process can be recognized by the name `20Process` at the end of the process class name, like

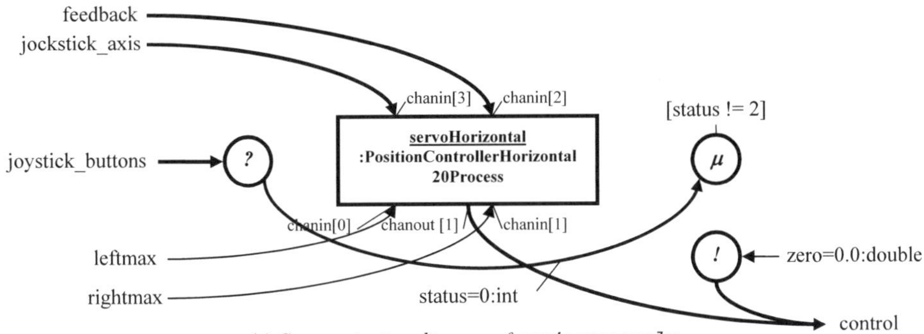

(a) Communication diagram of `motionControlH`

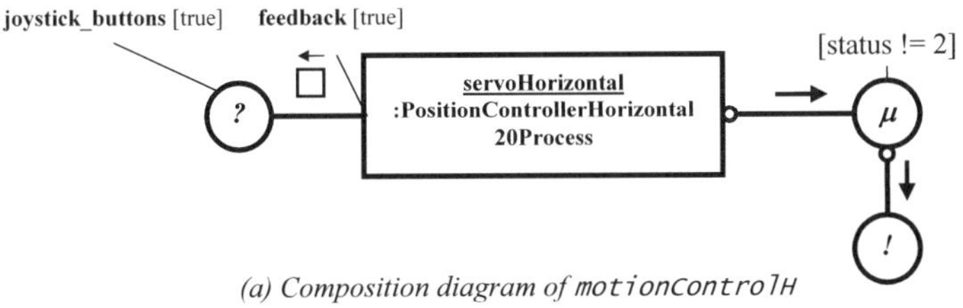

(a) Composition diagram of `motionControlH`

Figure 4 CSP diagram of process `motionControlH:ControlHorizontal`

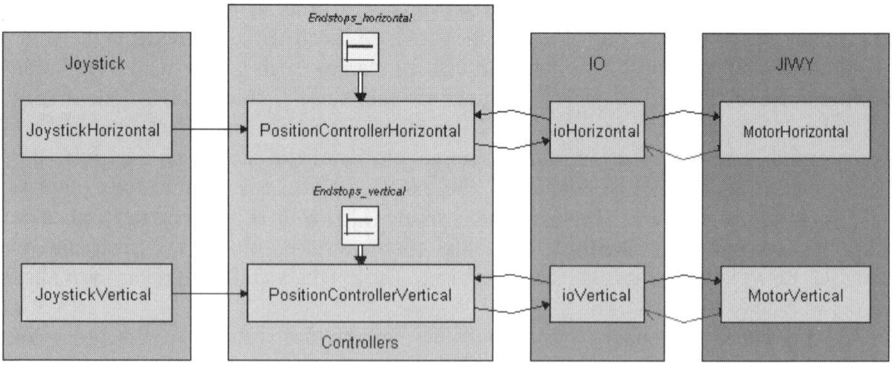

Figure 5 20-SIM model of the position motion controller

`PositionControllerHorizontal20Process`. A 20-SIM process or simply 20-Process is a CSP process with special treatment, namely:

- it is a single-shot process performing a 20-SIM submodel for one sample;
- the process-interface (ports) consists of a channel-input and a channel-output array;
- it is generated by 20-SIM.

The 20-SIM process ports are rendered by indexed channel array names. These channel array names rise from the 20-SIM submodel; chanin[i] → u[i] and y[j] → chanout[j], where u[] is the model input vector $\bar{u}$ and y[] the model output vector $\bar{y}$.

Two communication relationships have been added, namely `status` and `zero`. Here `status` holds the status of the joystick buttons when it is pressed. The looping process μ will continue repeating to perform the alternative relationship '$\overleftarrow{\square}$' until `status` is 2. `zero` is a constant value 0 in the communication diagram, see Figure 4a. Therefore these variables are declared and initialized with the default value in the code.

The alternative construct will wait until the channels `feedback` or `joystick_buttons` become ready. Then it makes a (prioritized) choice between reading the joystick button and computing one sample in the controller. The construct will select the controller when feedback is ready to communicate and when the joystick button is not pressed. If the joystick was pressed and feedback is not yet ready to communicate then the joystick buttons will be read. If both the joystick button was pressed and the feedback is ready then reading the buttons from the joystick button gets a higher priority. In this case the loop terminates right before the entire process terminates, it will send a zero to the actuator in order to release any steering of the joint.

Separate from CSP diagrams, a simulate-able model in 20-SIM has to be designed. A 20-SIM model embraces all relevant dynamics of the system and the controller itself. The model can be simulated and the feedback can be used to understand and improve the dynamics of the system. The identified processes can be found in the 20-SIM model in the form of mathematical blocks.

The position motion controller is the main operational mode of this servo system. The model renders the context of close-loop system where the two axes are servo-controlled by a joystick as a position set-point generator. Figure 5 presents a 20-SIM model of this control mode (for both axes). The joystick input is simulated by some function as described in submodel `JoystickHorizontal` and `JoystickVertical`. The `ioHorizontal` and `ioVertical` submodels model and simulate the hardware input/output interfacing between the physical system and the control software to make simulation more realistic compared to

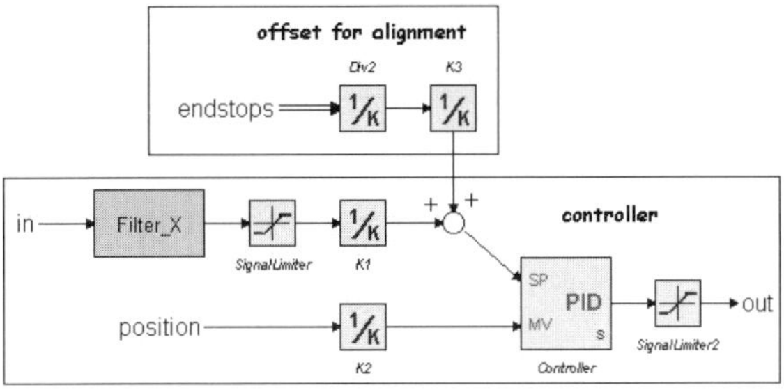

Figure 6 Block diagram of a position controller

the real setup. The physical system is modelled by the `MotorHorizontal` and the `MotorVertical` submodels using bond-graphs.

The submodels `PositionContollerHorizontal` and `PositionControllerVertical` are the software controllers that will be code generated by 20-SIM as 20-SIM processes. The block diagrams of these submodels slightly differ with different parameter values. One block diagram of the control law is depicted in Figure 6.

2.2.3 Alignment controller

The alignment modes are based on velocity controllers. For each rotation and joint the same velocity controller model is used. Again these controllers use slightly different parameter values.

Process `alignLH` has been worked out in Figure 7. This process contains a 20-SIM process `vleftHorizontal` and a loop construct which repeats itself until `stop` is true. The variable `stop` becomes true when the end-stop has been reached. This is specified by the 20-SIM submodel. Process `alignRH` is identical to `alignLH` except that the motor rotates right.

We separated the classes `VelocityControlLeftHorizontal` and `VelocityControlRightHorizontal` because the submodels are also separated in 20-SIM. This allows changing parameters independently of each other as a result of parameter fine-tuning. For example, the resistance at rotating left could be different than when rotating right. The 20-SIM model for left alignment and for right alignment is depicted in Figure 8.

The submodels `VelocityContolLeftHorizontal`, `VelocityContolLeftVertical`, `VelocityContolRightHorizontal`, `VelocityContolRightVertical` are the controllers that will be code generated by 20-SIM, called 20-SIM processes or 20-Processes. One of the block diagrams is depicted in Figure 9.

2.2.4 Homing controller

The homing controller uses the same position controller process as in the motion controller, see Figure 4. This process requires a set-point that is set to the centre position (i.e. zero). Figure 10 shows the CSP diagram. The 20-SIM model is given in Figure 5.

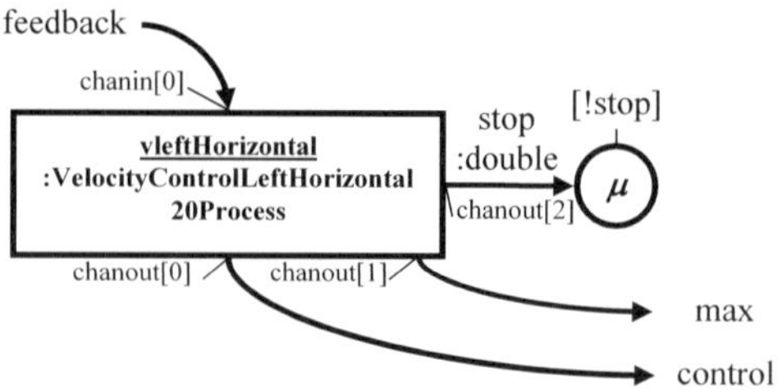

(a) Communication diagram of `alignLH`

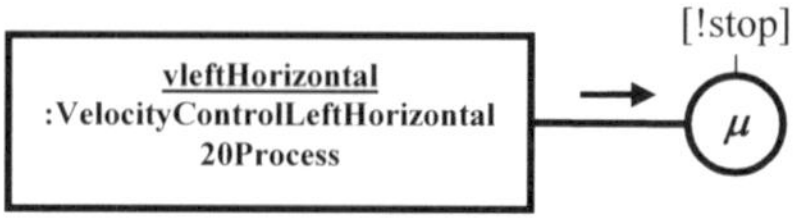

(b) Composition diagram of `alignLH`

Figure 7 CSP diagram of `alignLH:VelocityLeftHorizontal`

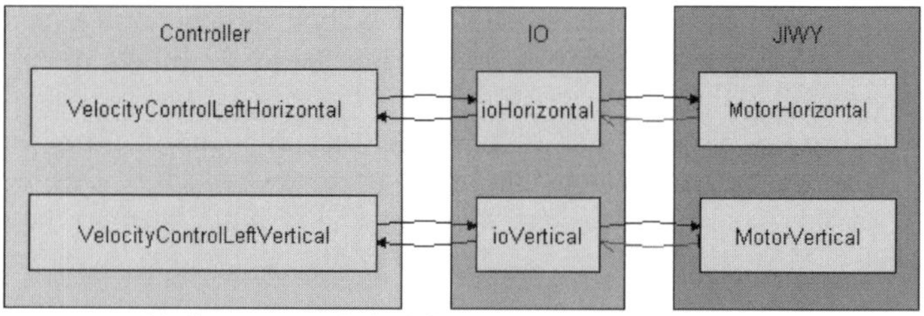

(a) Alignment mode for left rotation

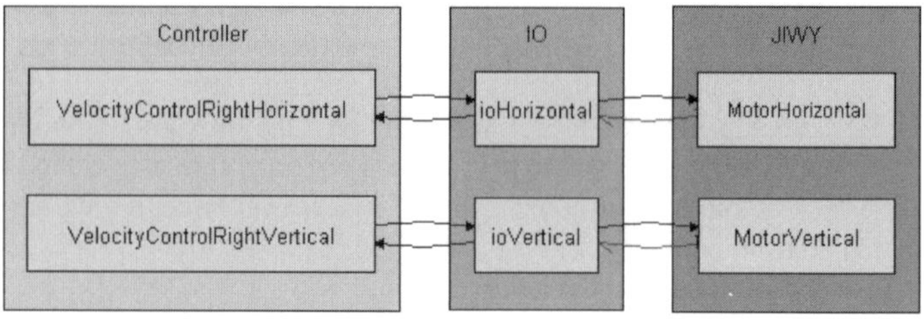

(b) Alignment mode for right rotation

Figure 8 Alignment modes for left and right rotations

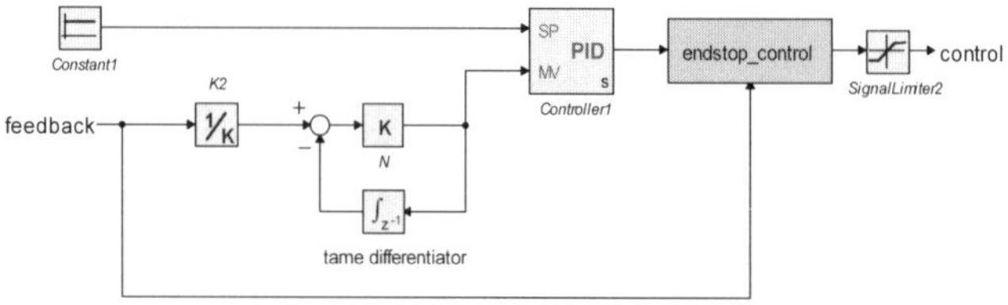

Figure 9 Block diagram of a velocity controller

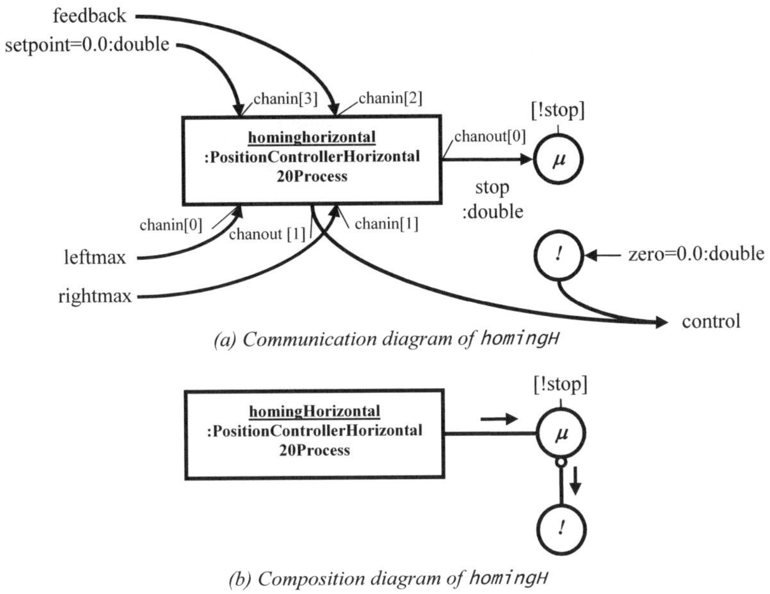

Figure 10 CSP diagram of `homingH:HomingHorizontal`

3. Implementation

3.1 20-SIM submodels

A system is usually designed by separating concerns in submodels in 20-SIM. The submodels which are essential to the control software must be code-generated in C-code. In our case, a submodel is code generated with 20-SIM using the C++ template which results in a C++ class: `MySubmodel` class.

A 20-SIM submodel requires a vector of input signals $\bar{u}$ and a vector of output signals $\bar{y}$. These signals are mapped on array of variables `u[]` for input-only and `y[]` for output-only.

```
double u[n], y[m];
```

The constructor requires no arguments and the constructor sets up the data structure.

```
submodel = new MySubmodel;
```

After construction of the object the begin state of the model needs to be initialized by

```
submodel->Initialize(u, y, 0);
```

Once the model has been initialized the control law equations can be performed by invoking the `Calculate()` method. This method requires the input vector and the output vector of signals (i.e. arrays of variables). This method must be called for each sampling periode.

```
submodel->Calculate (u, y);
```

The 20-SIM submodel is a black box which behavior can be observed and studied by simulating the submodel in 20-SIM. No immediate knowledge of the generated code is required since modifications to control laws are done in block diagrams and not directly in C-code. Therefore we omit the implementation of `MySubmodel`.

3.2 20-SIM process

A 20-SIM process is a CSP process that is generated by 20-SIM. A 20-SIM process has the task to invoke the equations of the submodel object when input data is available and in a particular order of execution. A 20-SIM process is a single-shot process, see Section 2.2.2. This means that the output values are calculated once according to the control laws and the process immediately terminates afterwards. A 20-SIM process can be executed successively according to its compositional relationships with other processes. A 20-SIM process can be used in any kind of compositional relationship; in parallel, in sequence or by choice. In case the 20-SIM process is executed in parallel it must pass data and synchronize on CSP channels (or blocking channels) in order to prevent any race-hazards between process input and output. These channels provide reactiveness to internal and external events. In case a 20-SIM process is in sequence to another process then unblocking channels (`ChannelVar`) should be used to pass data between these processes because blocking channels will cause deadlock.

A 20-SIM process of submodel `MySubmodel` has the following default constructor:

```
MySubmodel20Process::MySubmodel20Process(ChannelIn<double> **chanin,
ChannelOut<double> **chanout)
{
  this->chanin  = chanin;
  this->chanout = chanout;
  this->u       = (double *) malloc (n * sizeof (double));
  this->y       = (double *) malloc (m * sizeof (double));

  //--- Execute Initialize submodel
  submodel = new MySubmodel;
  submodel->Initialize(this->u, this->y, 0);
}
```

The constructor requires two channel arrays (or vectors); an array of channel-inputs and an array of channel-outputs. The arrays and each array element must be declared outside the process. The constant number n is the number of input-channels and m is the number of output-channels. These constants are determined and filled-in by the 20-SIM code-generator.

The 20-SIM process performs a standard procedure:

- read from channels to $\bar{u}$
- calculate submodel with inputs $\bar{u}$ and output $\bar{y}$
- write to channels from $\bar{y}$
- terminate

The code of this procedure is given in the `run()` method of the process below:

```
void MySubmodel20Process::run(void)
{
  //--- input
  for(int i=0; i<1; i++)
    if (chanin[i] != NULL)
      chanin[i]->read(&(u[i]));

  //--- perform calculation
  submodel->Calculate (u, y);

  //--- output
  for(int i=0; i<1; i++)
    if (chanout[i] != NULL)
      chanout[i]->write(&(y[i]));

  //--- done and now terminate
}
```

The arrays `u[]` and `y[]` are used internally by the submodel and the 20-SIM process and their content is not influenced other than by channel inputs. Therefore, the constructor declares these arrays exclusively for the 20-SIM process.

3.3 Top-level construct

As mentioned in Section 2 the CSP diagram declares all entities. The declarations of the top-level CSP diagram in Figure 3 and the vertical part can be found in the main source file of JIWY. Since we do *not* have a code-generator this is done by hand. The class names and identifiers specified in the CSP diagram are found in the source code.

The communication relationships, processes, and constructs are declared as follows:

```
//--- channel pointer declarations
ChannelIn<double>   *feedback_horizontal = NULL;
ChannelIn<double>   *feedback_vertical   = NULL;
ChannelOut<double> *control_horizontal  = NULL;
ChannelOut<double> *control_vertical    = NULL;
ChannelIn<double> *joystick_horizontal  = NULL;
ChannelIn<double> *joystick_vertical    = NULL;
ChannelOut<int>    *joystick_buttons     = NULL;

//--- internal channel declarations ('chan' prefix to variable names)
ChannelVar<double> chanleftmax_h  = new ChannelVar<double>;
ChannelVar<double> chanrightmax_h = new ChannelVar<double>;
ChannelVar<double> chanleftmax_v  = new ChannelVar<double>;
ChannelVar<double> chanrightmax_v = new ChannelVar<double>;

//--- Analog joystick
AnalogJoystick *joystick = new AnalogJoystick();
joystick_horizontal = new AnalogJoystickX(joystick);
joystick_vertical   = new AnalogJoystickY(joystick);
```

```
joystick_buttons     = new AnalogJoystickButtons(joystick);

//--- DAQSTC, National Instruments 6024E IO Board
DAQSTC *daqstc = new DAQSTC();
daqstc->Initialise();

//--- Analog Output
daqstc->SetAOTM(AOTM::Primary, AOTM::CPUDriven);
control_horizontal = daqstc->GetDAC(DAC::DAC0);
control_vertical   = daqstc->GetDAC(DAC::DAC1);

//--- Two counters for sensors
feedback_horizontal = daqstc->GetCounter(GPC::Counter0);
feedback_vertical   = daqstc->GetCounter(GPC::Counter1);
```

Firstly, the pointers to the external channels and the internal channels are declared. Secondly, the external channels are declared and are assigned to the external channel pointers, see italic code. The italic code cannot be derived from the CSP diagram and is added apart from diagram-to-code translation. These external channels are based on hardware-specific drivers with channel-interfaces and with CSP-valid channel semantics. Therefore, this code is considered to be target dependent, where as the control processes remain hardware-indepedent. The communication relationships `leftmax_h`, `rightmax_h`, `leftmax_v`, and `rightmax_v` are prefixed by `chan`. We will see later that this is required to distinguish a channel from a variable used in conditional expression.

Usually CSP channel will cause deadlock when they are used between processes in sequential composition. Therefore to avoid deadlock we use unblocking channels only for communication between processes in sequential composition. These communication relationships are `leftmax_h`, `rightmax_h`, `leftmax_v`, and `rightmax_v`. These become `ChannelVar` channels which are the unblocking versions of `Channel`. `ChannelVar` is a subclass of `Channel` and therefore a `ChannelVar` can replace a `Channel` in code; they can be intertwined because they have the same channel-interface. See listing above. These unblocking channels can be analysed in CSP as an one-place override buffer within the communication relationship.

The processes are declared in the main source file as follows:

```
//--- declare all processes
ControlHorizontal *motionControlH =
   new ControlHorizontal (joystick_horizontal, joystick_buttons,
feedback_horizontal, control_horizontal, chanleftmax_h, chanrightmax_h);
ControlVertical   *motionControlV =
   new ControlVertical (joystick_vertical, joystick_buttons, feedback_vertical,
control_vertical, chanleftmax_v, chanrightmax_v);
VelocityControlLeftHorizontalProcess  *alignLH =
   new VelocityControlLeftHorizontalProcess(feedback_horizontal,
control_horizontal, chanleftmax_h);
VelocityControlRightHorizontalProcess *alignRH =
   new VelocityControlRightHorizontalProcess(feedback_horizontal,
control_horizontal, chanrightmax_h);
VelocityControlLeftVerticalProcess    *alignLV =
   new VelocityControlLeftVerticalProcess(feedback_vertical, control_vertical,
```

```
chanleftmax_v);
VelocityControlRightVerticalProcess    *alignRV =
   new VelocityControlRightVerticalProcess(feedback_vertical, control_vertical,
chanrightmax_v);
HomingHorizontal *homingH =
   new HomingHorizontal(feedback_horizontal, control_horizontal, chanleftmax_h,
chanrightmax_h);
HomingVertical    *homingV =
   new HomingVertical(feedback_vertical, control_vertical, chanleftmax_v,
chanrightmax_v);
```

The arguments correspond to the port names of the process-interface. These names can be found in the interface description of the process classes. See for example the process-interface specified by the constructor-interface in Section 3.4. Here the port names are substituted by the local names of the channel pointers.

The compositional construct of the Figure 4 with CTC++ is:

```
//--- create CSP relationships
Sequential *seq_h = new Sequential(alignLH,alignRH,motionControlH,homingH,NULL);
Sequential *seq_v = new Sequential(alignLV,alignRV,motionControlV,homingV,NULL);
Parallel   *par   = new Parallel(seq_h, seq_v, NULL);

//--- timer code, see Section 3.7
```

The entire process will be executed by invoking `run()` on the top-level construct:

```
par->run();
```

Thus before the `run()` method of the top-level construct is invoked, all processes, channels, constructs, timing, and objects have been created. Once the `run()` method is invoked the real-time run bodies perform according to the communication relationships *and* the compositional relationships in the CSP diagrams. After the top `run()` method terminates the declared entities can be deleted and the program can gracefully terminate. This top-level process is called a *network building process*. In our approach this top-level process is the only hardware dependent process in the software since it is the only one that sets up hardware dependent objects. All other processes are hardware independent since they solely use channels and external channels access hardware devices. Whether channels are internal or external is invisible to the processes since channels have a common and most simplified fundamental interface for channel communication. Of course, processes may depend on the data that flows through channels.

3.4 Motion controller process

The process-interface of `motionControlH` is specified by the constructor below. This constructor assigns its ports to the ports of the child processes. The internal channels for `setpoint`, `stop`, and `zero` are declared. Of course the 20-SIM process `servoHorizontal` is declared.

```
ControlHorizontal::ControlHorizontal(ChannelIn<double> *joystick_axis,
ChannelIn<double> *joystick_buttons, ChannelIn<double> *feedback,
ChannelOut<double> *control, ChannelIn<double> *leftmax, ChannelIn<double>
*rightmax)
{
  //--- create channel-input array and a channel-output array
  this->chanin  = new ChannelIn<double> * [4];
  this->chanout = new ChannelOut<double> * [2];

  chanstatus = new ChannelVar<int>(0); //(!)

  chansetpoint = new ChannelVar<double>(0.0);
  chanstop = new ChannelVar<double>;
  chanzero = new ChannelVar<double>(0.0);

  chanin[0]  = leftmax;
  chanin[1]  = rightmax;
  chanin[2]  = feedback;
  chanin[3]  = joystick_axis;
  chanout[0] = NULL;
  chanout[1] = control;

  //--- create the 20-SIM process
  servoHorizontal = new PositionControllerHorizontal20Process(chanin, chanout);

  //--- set up the alternative construct process
  alt = new Alternative();
  alt->add(new Guard(feedback));
  alt->add(new Guard(joystick_buttons));
}
```

The run() method performs the CSP diagram of Figure 4. Here we see that status and chanstatus are used. We need to distinguish between a channel and a variable. This is similar for zero and chanzero.

```
void ControlHorizontal::run(void)
{
  int status = 0;
  double zero = 0;

  do
  {
    switch(alt->select())
    {
      case 0: //--- controller process
          servoHorizontal->run();
        break;
      case 1: //--- stop button process
        joystick_buttons->read(&status);
        chanstatus->write(&status); // (!)
```

```
      break;
    }
    chanstatus->read(&status); // (!)
  } while (status != 2);
  //--- release output by setting to zero
  chanzero->read(&zero); // (!)
  control->write(&zero);
}
```

The `switch(alt->select()) {}` clause performs the alternative construct. The lines marked with "*// (!)*" can be eliminated as part of optimization. This optimization increases the performance of the application by using variables instead of `ChannelVar` channels. This optimization requires further research and is not discussed in this paper. Anyway, CSP diagrams contain enough information to allow these kinds of optimizations.

3.5 Alignment controller process

The process-interface of `alignLH` is specified by the constructor below

```
VelocityControlLeftHorizontal::VelocityControlLeftHorizontal(ChannelIn<double>
*feedback, ChannelOut<double> *control, ChannelIn<double> *max)
{
  this->chanin  = new ChannelIn<double> * [1];
  this->chanout = new ChannelOut<double> * [3];

  chanstop = new ChannelVar<double>;

  chanin[0]  = feedback;
  chanout[0] = control;
  chanout[1] = max;
  chanout[2] = chanstop;

  vleftHorizontal = new VelocityControlLeftHorizontal20Process(chanin, chanout);
}
```

Figure 7b and some information of Figure 7a translates to the following process body.

```
void VelocityControlLeftHorizontal::run(void)
{
  do
  {
    vleftHorizontal->run();
    chanstop->read(&stop);
  } while (!stop);
}
```

3.6 Homing controller process

The CSP diagram in Figure 10 is translated to:

```
HomingHorizontal::HomingHorizontal(ChannelIn<double> *feedback,
ChannelOut<double> *control, ChannelIn<double> *leftmax, ChannelIn<double>
*rightmax)
{
  this->chanin  = new ChannelIn<double> * [4];
  this->chanout = new ChannelOut<double> * [2];

  chansetpoint = new ChannelVar<double>(0.0);
  chanstop = new ChannelVar<double>;
  chanzero = new ChannelVar<double>(0.0);

  chanin[0]  = leftmax;
  chanin[1]  = rightmax;
  chanin[2]  = feedback;
  chanin[3]  = chansetpoint;
  chanout[0] = chanstop;
  chanout[1] = control;

  homingHorizontal = new PositionControllerHorizontal20Process(chanin, chanout);
}

void HomingHorizontal::run(void)
{
  do
  {
    homingHorizontal->run();
    chanstop->read(&stop);
  } while (!stop);
  //--- release output by setting to zero
  chanzero->read(&zero);   // (!)
  control->write(&zero);
}
```

3.7 Timing and sampling

The CSP diagrams are based on untimed CSP. We have created an environmental process which takes part accepting every communication event in the software. We can command the environmental process to accept events at certain moment in time with periodical interval. This will be based on external channels. The environmental process will control a timer and on interrupts of the timer the drivers inside the external channels will be executed in sequence and at highest priority (atomic).

```
Environment::at(feedback_horizontal, starttime, Tsh);
Environment::at(joystick_horizontal, starttime, Tsh);
Environment::at(control_horizontal, starttime, Tsh);
Environment::at(feedback_vertical, starttime, Tsv);
Environment::at(joystick_vertical, starttime, Tsv);
Environment::at(control_vertical, starttime, Tsv);
```

The `starttime` is set to a value that is long enough for all initializations to be finished and until it waits for the first events to occur. Usually we set the `starttime` to `10000` (=10000 μsec, or 100 ms). The sampling time for the horizontal control loop `Tsh` and for the vertical control loop `Tsv` are set for the timed external channels. See @Tsh for the external channels `feedback_horizontal`, `joystick_horizontal`, and `control_horizontal` in Figure 3. When processes do not arrive on the timed channel before the environmental process want to accept the communication event, then a `TimeoutException` will be thrown by the channels and the processes will be released and terminate unsuccessfully. In this paper we omit exception handling. JIWY is hard real-time and timeout-exceptions do not occur so far. In further research we will investigate the exception framework under various kinds of errors in the system.

4. Conclusions

Guidance for transformation of the control engineering block diagram language to the control software processes and composition with other vital software components starts coming in sight. The CSP diagrams provide links for refinements of the substantial control system objectives towards complete runnable computer code and language for composing the computer code as a concurrent ensemble. The CT libraries catch the necessary constructs for building the compositional network structure, timing aspects and a CSP-consistent hardware access framework. The control code is generated from CSP diagrams and 20-SIM submodels.

This way, following an intuitive reasoning path, the interdisciplinary communication in development phases of a mechatronic project, is freed from serious design flow discontinuities.

References

[1] INMOS. Inmos Web Site, www.inmos.com,

[2] INMOS. occam 2 Reference Manual. C. A. R. Hoare *International Series in Computer Science*, Prentice Hall. 1988. ISBN 0-13-629312-3.

[3] A.W. Roscoe. The Theory and Practice of Concurrency. R. Bird *Series in Computer Sciences*, Prentice-Hall. 1998. 0-13-674409-5.

[4] G.H. Hilderink. Communicating Threads for Java (CTJ) home page, http://www.rt.el.utwente.nl/javapp, 2002.

[5] G.H. Hilderink, A.W.P. Bakkers and J.F. Broenink. A Distributed Real-Time Java System Based on CSP. In *The Third IEEE International Symposium on Object-Oriented Real-Time Distributed Computing (ISORC-2000)*, IEEE Computer Society. Newport Beach, California: pp. 400-407, 2000.

[6] 20-SIM. Control Labs Products, www.20sim.com, 2003.

[7] P.C. Breedveld. Multibond graph elements in physical systems theory. In *J. Franklin Inst.* **319:** pp. 1-36, 1985.

[8] J.F. Broenink. 20-Sim software for hierarchical bond-graph/block-diagram models. In *Simulation Practice and Theory*. **7:** pp. 481-492, 1999.

[9] G.H. Hilderink. A Graphical Modelling Language for Specifying Concurrency based on CSP. In *IEE*

Proceedings Software, IEE. **150:** 108-120, 2002. ISSN 1462-5970.

[10] D.S. Jovanovic, G.H. Hilderink and J.F. Broenink. A Communicating Threads (CT) Case Study: JIWY. In V. S. Sunderam *Communicating Process Architecture 2002*, IOS Press. University of Reading, UK. **60:** pp. 311-320, 2002. ISSN 1383-7575.

Author Index

Arantsson, B.	37	Jacobsen C.	269
Arvind, D.K.	289	Jovanovic, D.	319
Avila, J.	157	Lawrence, A.E.	97
Barnes, F.R.M.	199, 251, 269	Marshall, S.W.	1
Boosten, M	185	Moore, M.	305
Broenink, J. F.	111, 235, 319	Moore, P.W.	83
Brown, N.C.C.	139	Nicole, D.	167
Burke, K. W.	91	O'Neill, B.C.	83
Bustacara, C.	157	Orlic, B.	235
Campbell, J.	173	Panesar, G.	125
Cho, Y.-F.	1	Petitpierre, C.	17
Clark, S.	83	Raju, V.	63
Cronie, H.S.	225	Rong, L.	63
Duller, A.W.G.	125	Schaller, N.C.	1
East, I.R.	33	Schweigler, M.	199
Ellis, S.	167	Slump, C.H.	225
Gonzalez, E.	157	Smith, M.L.	91
Hancock, S.	167	Sotelo-Salazar, S.	289
Happe, H.H.	49	Stiles, G.S.	63, 173
Hilderink, G.H.	111, 319	Towner, D.	125
Hoeksma, F.W.	225	Vinter, B.	37, 49, 269
Hughes, C.E.	91	Welch, P.H.	139, 199